Diode Lasers

Series in Optics and Optoelectronics

Series Editors: **R G W Brown**, Queens University Belfast, UK
E R Pike, Kings College, London, UK

Other titles in the series

Applications of Silicon-Germanium Heterostructure Devices
C K Maiti and G A Armstrong

Optical Fibre Devices
J-P Goure and I Verrier

Laser-Induced Damage of Optical Materials
R M Wood

Optical Applications of Liquid Crystals
L Vicari (Ed.)

Stimulated Brillouin Scattering
M Damzen, V I Vlad, V Babin and A Mocofanescu

Handbook of Moiré Measurement
C A Walker (Ed.)

Handbook of Electroluminescent Materials
D R Vij (Ed.)

Diffractional Optics of Millimetre Waves
I V Minin and O V Minin

Forthcoming titles in the series

Fast Light, Slow Light and Left-Handed Light
P W Milonni

High Speed Photonic Devices
N Dagli (Ed.)

Transparent Conductive Coatings
C I Bright

Photonic Crystals
M Charlton and G Parker (Eds.)

Other titles of interest

Thin-Film Optical Filters (Third Edition)
H Angus Macleod

Diode Lasers

D Sands
Department of Physics
The University of Hull, UK

Institute of Physics Publishing
Bristol and Philadelphia

© IOP Publishing Ltd 2005

All rights reserved. No part of this publication may be reproduced, stored in a retrieval system or transmitted in any form or by any means, electronic, mechanical, photocopying, recording or otherwise, without the prior permission of the publisher. Multiple copying is permitted in accordance with the terms of licences issued by the Copyright Licensing Agency under the terms of its agreement with Universities UK (UUK).

British Library Cataloguing-in-Publication Data

A catalogue record for this book is available from the British Library.

ISBN 0 7503 0726 9

Library of Congress Cataloging-in-Publication Data are available

Commissioning Editor: Tom Spicer
Production Editor: Simon Laurenson
Production Control: Sarah Plenty and Leah Fielding
Cover Design: Victoria Le Billon
Marketing: Nicola Newey, Louise Higham and Ben Thomas

Published by Institute of Physics Publishing, wholly owned by The Institute of Physics, London

Institute of Physics Publishing, Dirac House, Temple Back, Bristol BS1 6BE, UK

US Office: Institute of Physics Publishing, The Public Ledger Building, Suite 929, 150 South Independence Mall West, Philadelphia, PA 19106, USA

Typeset by Apek Digital Imaging Ltd, Nailsea, Bristol
Printed in the UK by MPG Books Ltd, Bodmin, Cornwall

To Tomomi, without whose support I would not have been able to spend so much time at my computer, and to Taro and Chibi, without whose attention I would have had no excuse to leave my computer.

Contents

Preface		xi
1	**Introduction**	**1**
	References	5
2	**Essential semiconductor physics**	**6**
	2.1 Free electrons in semiconductors	6
	2.2 Formation of bands in semiconductors	7
	2.3 Band theory and conduction	9
	2.4 Electron and hole statistics	11
	2.5 Doping	14
	2.6 Heavy doping	15
	2.7 Recombination and generation	19
	2.8 Energy bands in real semiconductors	20
	2.9 Minority carrier lifetime	23
	2.10 Minority carrier diffusion	24
	2.11 Current continuity	25
	2.12 Non-equilibrium carrier statistics	26
	2.13 Summary	26
	2.14 References	26
3	**Laser fundamentals**	**28**
	3.1 Stimulated emission	31
	3.2 Population inversion in semiconductors	34
	3.3 The p–n homojunction laser	35
	3.4 The active region and threshold current	40
	3.5 Optical properties of the junction	43
	3.6 Output characteristics of the homojunction laser	47
	3.7 Summary	51
	3.8 References	52

4	**Optical properties of semiconductor materials**	54
4.1	A model of the refractive index	54
4.2	The refractive index of a semiconductor laser cavity	61
4.3	Gain in semiconductors	66
	4.3.1 The vector potential and the interaction Hamiltonian	70
	4.3.2 Fermi's golden rule	71
	4.3.2 The matrix element and densities of states	75
4.4	Summary	76
4.5	References	77

5	**The double heterostructure laser**	79
5.1	Introduction	79
5.2	Materials and epitaxy	81
	5.2.1 Molecular beam epitaxy	88
	5.2.1.1 MBE of aluminium gallium arsenide	90
	5.2.1.2 MBE of indium gallium arsenide phosphide	92
	5.2.2 Chemical vapour phase epitaxy	92
	5.2.2.1 Hydride chemical vapour deposition	93
	5.2.2.2 The trichloride process	94
	5.2.2.3 MOCVD	94
5.3	Electronic properties of heterojunctions	96
	5.3.1 Band bending at heterojunctions	99
5.4	The double heterostructure under forward bias	102
	5.4.1 Recombination at interfaces	106
5.5	Optical properties of heterojunctions; transverse mode control and optical confinement	108
5.6	Materials and lasers	115
	5.6.1 InP systems: InGaAs, InGaAsP, AlGaInP	116
	5.6.2 InAs-InSb lasers	125
5.7	Lateral mode control	126
5.8	Summary	128
5.9	References	128

6	**Quantum well lasers**	132
6.1	Classical and quantum potential wells	132
6.2	Semiconductor quantum wells	139
6.3	Quantised states in finite wells	147
6.4	The density of states in two-dimensional systems	150
6.5	Optical transitions in semiconductor quantum wells	153
	6.5.1 Gain in quantum wells	157
6.6	Strained quantum wells	163
6.7	Optical and electrical confinement	167
6.8	Optimised laser structures	177
6.9	Summary	180
6.10	References	181

7	**The vertical cavity surface emitting laser**		**185**
	7.1	Fabry–Perot and waveguide modes	199
	7.2	Practical VCSEL cavity confinement	202
	7.3	Oxide confined devices	207
	7.4	Long wavelength VCSELs	213
	7.5	Visible VCSELs	216
	7.6	Summary	218
	7.7	References	219
8	**Diode laser modelling**		**222**
	8.1	Rate equations; the idealised DH laser	222
	8.2	Gain compression	224
	8.3	Small signal rate equations	228
	8.4	Modelling real laser diodes	232
		8.4.1 InGaAsP/InP quantum well lasers	232
		8.4.2 Separate confinement heterostructure quantum well laser	235
		8.4.3 Three level rate equation models for quantum well SCH lasers	245
	8.5	Electrical modelling	249
	8.6	Circuit level modelling	253
	8.7	Summary	258
	8.8	References	258
9	**Lightwave technology and fibre communications**		**260**
	9.1	An overview of fibre communications and its history	260
	9.2	Materials and laser structures	268
	9.3	Laser performance	272
		9.3.1 Mode selectivity	272
		9.3.2 Modulation response	273
		9.3.3 Gain switching	275
		9.3.4 Linewidth	277
	9.4	Single wavelength sources	281
		9.4.1 DBR lasers	285
		9.4.2 DFB lasers	287
	9.5	High bandwidth sources	288
	9.6	Summary	290
	9.7	References	291
10	**High power diode lasers**		**293**
	10.1	Geometry of high power diode lasers	294
	10.2	Single emitter broad area diode lasers	297
	10.3	Lateral modes in broad area lasers	298
	10.4	Controlling filamentation	305

		10.4.1 Mode filtering	305
		10.4.2 Materials engineering	309
	10.5	Catastrophic optical damage	311
	10.6	Very high power operation	322
	10.7	Visible lasers	327
	10.8	Near infra-red lasers	331
	10.9	Mid infra-red diode lasers	332
	10.10	Diode pumped solid state lasers	339
	10.11	Summary of materials and trends	342
	10.12	References	344
11	**Blue lasers and quantum dots**		**349**
	11.1	Nitride growth	357
	11.2	Optical and electronic properties of (Al,Ga,In)N	363
	11.3	Laser diodes	370
	11.4	Quantum dot lasers	374
	11.5	Summary	376
	11.6	References	376
12	**Quantum cascade lasers**		**381**
	12.1	Quantum cascade structures	387
	12.2	Minibands in superlattices	390
	12.3	Intersubband transitions	396
	12.4	Intersubband linewidth	398
	12.5	Miniband cascade lasers	400
	12.6	Terahertz emitters	403
	12.7	Waveguides in quantum cascade structures	405
	12.8	Summary	408
	12.9	References	409

Appendix I	**412**
Appendix II	**413**
Appendix III	**417**
Appendix IV	**433**
Appendix V	**440**
Solutions	**442**
Index	**447**

Preface

For many years the laser was commonly perceived as a solution looking for a problem. Quite when this changed is difficult to pin-point, but it certainly is not the case now. The laser has made a tremendous impact on our daily lives. Sometimes it is immediate and obvious, such as its use in life-saving medical treatments. At others it may be less so, but its use in manufacturing, scientific research, telecommunications, and home entertainment all affect our daily lives. The diode laser is, of course, just one type of laser but it is fully a part of this photonic revolution. It will always be associated with communications and information technology, which are its first and biggest applications, but it is also finding increasing use in these other fields, and in some cases competing with other types of lasers.

The diode laser could easily be seen as the poor relation. It has a relatively poor beam quality, and it is not capable of delivering the sort of short, high power pulses that characterise excimer or CO_2 lasers, for example. On the other hand, it is a true continuous wave device. It can offer an unparalleled range of wavelengths from the blue to the far infra-red, and it is part of the electronics revolution that began in the last half of the twentieth century and continues today. CD and DVD players would not be possible without these compact devices, and because of them the diode laser can now be found in virtually every household in the world.

The study of diode lasers is thus only partly a study in laser physics and technology and principally a study in semiconductor physics and technology. The diode laser's astonishing diversity of types and characteristics is due entirely to its being part of the electronics revolution. The rapid development of the nitride material system, for example, could not have occurred without the extensive knowledge of materials science and semiconductor physics, as well as the processing equipment and technology, developed over decades of research into other materials and devices. The same is true of the quantum cascade laser, which is a triumph of materials engineering. This book is therefore as much a book about semiconductor physics as it is a book about lasers.

The contents are organised according to laser type. The essential semiconductor and laser physics given at the outset is just enough to understand the double heterostructure laser. Specialist knowledge, for example, quantised states in 2-D systems relevant to quantum well lasers, is left to the particular chapters dealing with the specific laser type. Each chapter is thus susceptible to being read in isolation. Some of the chapters are quite detailed. but this is deliberate. It is always a question of balance as to how much detail should be included in an undergraduate text. The choice is simple; to gloss over some of the detail and simply present complicated ideas as facts, or to try to convey some understanding of the ideas and concepts. The latter requires detail, but many books written for undergraduates do not contain sufficient for graduates to make the transition to postgraduate research, whilst books written for the research community are often inaccessible to those starting out in their careers because too much prior knowledge is required. The detail in this book is there to bridge this gap.

The book is intended to be comprehensive without being too detailed. 1 have tried as far as possible to cover the whole field, and if there are omissions it is simply because I cannot include everything. I have attempted also to summarise the research in the various fields and to point the interested reader towards original sources. I am indebted to all those whose papers I have read and cited. Without this effort by so many people books like this would not be possible. Unusually for a book of this size I have not called upon too many people for support in its writing. I would like to acknowledge the support of my close friends and family, who have given of their time and patience over the years, and also Tom Spicer, commissioning editor for IOPP, whose support over the last three years has been very important.

David Sands

Hull, 2004

Chapter 1

Introduction

The diode laser was invented in 1962. This presents both students and teachers with a challenge, because in the intervening years laser design has changed so much that today's state-of-the-art devices bear very little resemblance to those earliest devices. In fact, by 1968 the next generation of devices had already rendered some concepts obsolete, but the basic physics of the diode laser is still best explained with reference to these very simple devices. As you can imagine, this does not make it easy for students to make the connection between these ideas and the structures found today. It would be possible, of course, to dispense with the historical details and attempt to explain the ideas in relation to more complicated, but technologically relevant devices, but I believe there are difficulties with this approach. Most notably, the diode laser is taken out of the context in which it was developed. Research is a slow process in which the collective efforts of the many are often just as important as the brilliance of an individual, and unless one appreciates the context in which the research was conducted the nuances of laser design, such as why some structures are favoured over others, can be lost. The historical approach runs through this book.

Robert H Rediker has described his personal perspective on the development of the GaAs diode laser [1,2]. Rediker was involved in research into Ge narrow base diodes but switched to GaAs research in 1958. The first task of the group, bearing in mind the experience already developed on Ge diodes, was to develop the technology for the fabrication of p–n junction diodes, which was achieved by diffusing Zn into n-type GaAs. The first problem that this material posed was immediately apparent. The first diffusion was done in vacuum and Rediker was surprised to see the appearance of metallic droplets formed on the GaAs. These droplets were not Zn but Ga, formed because As had evaporated from the surface leaving the Ga behind. This problem was solved by diffusing in an As ambient.

Semiconductor research is made up of such problems. It is truly a multi-disciplinary subject requiring a grasp of solid state physics, quantum mechanics, materials science, thermodynamics, electronics, optics, and vacuum science, to name a few. This will become apparent as you progress through this book. Incidentally if you are not familiar with the *p–n* junction, simply accept for the present that a layer of one type of material can be created adjacent to another by diffusing in impurity atoms. At the junction of these two layers electrons and their positively charged counterparts, holes, interact to produce light. These ideas will be desribed in greater detail in chapter 2 along with the energy levels that particles occupy. Laser fundamentals, including a description of the homo-junction laser, are described in chapter 3. To return to Rediker, diffused diodes were eventually fabricated but their electrical characteristics were not ideal. In an attempt to understand why, Rediker proposed to look at the luminescence emitted from the forward biased diode at 77 K. The light intensity recorded from these diodes saturated the detectors, and drastic measures to reduce both the gain of the measurement and the amount of light reaching the detector were needed. This was the first demonstration of high efficiency luminescence from GaAs and led directly to the concept of the laser. These results were reported at a conference in July 1962 and by September of that year the laser had been invented. There were four groups reporting laser action in November and December of that year. Accounts by principal members of all four groups of the events of that time appeared in a special issue of the IEEE Journal of Quantum Electronics, Vol 23 (6) 1987 [2,3,4]. It makes fascinating reading, not least because the sense of excitement and challenge of the times comes clearly through the writing. These were people engaged in epoch changing research, though at the time they did not realise it. They were just doing what they did best; sometimes from a position of knowledge, at other times not.

Robert Hall, who worked at the General Electric Research and Development Center, New York, recalls that he had been asked several times in the summer of 1962 whether he thought it possible to fabricate a semiconductor laser and always responded negatively [3]. This was at a time when lasers were big news. Both the HeNe laser and the ruby laser had just been invented in 1960 and these devices required long optical paths and highly reflecting mirrors. It didn't seem sensible to try to achieve something similar with semiconductors. Nonetheless, Maurice Bernard was an annual visitor to the laboratory and helped to express the ideas of population inversion in semiconductors in terms well known to the semiconductor community. Bernard's paper with Duraffourg [5] presented what is commonly known as the Bernard-Duraffourg condition (see appendix I). What changed matters in Hall's mind was the announcement by Rediker's group of high efficiency luminescence in GaAs. This was a major obstacle that had been overcome and set Hall on the task of looking at other factors which might be overcome.

In fact, as Hall reports, he could see several reasons why a laser might not work. *"Foremost was the knowledge that the active region, where degenerate conditions of electrons and holes would have to coexist, would be very thin. Nevertheless, it would have to provide enough [optical] gain to maintain a wave that would fringe out a considerable distance into the lossy* n-*type and* p-*type regions on either side. We did not know how lossy these regions would be, how deeply the light wave would extend into them, or how thick the active layer would be. We did not even know how much gain to expect in this layer since there was no way of guessing how many electrons and holes we might be able to inject into it or how much their effective temperatures might be increased by the injection process itself"*. The uncertainties did not stop here. It was known that heavy doping affected the density of electron and hole states at the band edges, but the consequences for a laser were unknown. It was also known that optical properties of *n*-type and *p*-type GaAs were differerent, again, with unknown consequences.

The approach taken was essentially empirical; fabricate some diodes and see how they worked. One of the big uncertainties was knowing what to look for as evidence of laser action. The team under Hall built a pulse generator capable of delivering 50 A pulses and mounted the diodes in an unsilvered glass dewar so the team could see the operation. Eventually, by monitoring the light output as a function of current the team found that above a certain current the light output began to rise much more rapidly than the current input, and very shortly afterwards the team obtained some far-field radiation patterns that "*were unquestionably the result of coherent light emission*". Hall reports that his results appeared in November, but at the same time reports appeared that a group at IBM had reported stimulated emission.

The author of the IBM paper was Marshall I Nathan [4]. Nathan had the advantage of working in the same organization as Peter Sorokin, who in 1960 and 1961 had reported with Stephenson [6,7] four level lasers using trivalent uranium ions and divalent samarium ions in CaF_2. Following hot on the heels of the HeNe laser and the ruby laser, these developments demonstrated that the laser was not just a novelty but a serious development that was here to stay. Nonetheless, Nathan himself did not take too seriously the efforts at IBM directed towards semiconductor lasers. In his own words, "*I was never asked to work on the contract [from Fort Mammoth, US Army]. However, I would have refused to do so because at the time I regarded my scientific freedom as very important. To be constrained to work on a contract that someone else had written would have been intolerable*". Strong words, but the intolerable was soon tolerated. So much so in fact that he became the advocate of laser research within IBM.

Along the way, Nathan describes the difficulties he faced as his interest in GaAs luminescence developed. Having discovered some GaAs samples in which the luminescence linewidth was about 100 times narrower than other samples which had been reported as exhibiting highly efficient luminescence, he decided that these might be good candidates for laser action. The narrow linewidth is generally indicative of the quality of the material, the narrower the linewidth the better the quality. Nathan discussed the results with Sorokin, who understood very little about semiconductors. Nathan himself understood very little about lasers and the conversation is described as "*difficult but interesting*". Nathan's ideas on producing a semiconductor laser were quite vague and he describes an attempt to produce photo-pumped stimulated emission which in retrospect seemed hopeless from the start.

Neither Nathan nor anyone from his group was at the 1962 conference when Quist and Keyes from Rediker's group reported the high efficiency luminescence. It was reported in the New York Times, though, and word reached the group through this medium. Eventually the group began to see signs of stimulated emission, most obviously the narrowing of the linewidth as the current was increased. In many cases the emission seemed to narrow to a limit of ~30 Å, but in at least one sample the linewidth narrowed to a point below the resolution of the spectrometer, ~2 Å. These results were submitted to Applied Physics Letters. The excitement of the time is described by Nathan. New results were reported "almost daily" and the method of cleaving was dicovered as way of making the optical cavity. This was important because in the other groups at GEC and MIT, polishing was pursued. Cleaving is much simpler and was still used in most laser structures in the late 1980's and early 1990's. In October, continuous wave (CW) operation was observed at 2 K, it was found that light came from the p-side of the junction, and laser operation at room temperature was observed. The competition among the groups was fierce, but the remarkable thing to emerge from the writing of these early pioneers is the strong lifelong friendship that developed amid the rivalry.

Returning once more to Rediker, by 1963 his group had also reported laser action in InAs, InSb, and InGaAs, an alloy of InAs and GaAs. PbTe followed in 1964, and lead salts of selenium and sulphur also yielded diode laser action. All these diodes operated at cryognic temperatures, for reasons which will become apparent, but Rediker's group were also involved in the significant developments that led to room temperature operation, and to the developments in diode laser technology that enabled optical fibre transmission. The diode laser had finally arrived, and today it's used in so many different areas of our daily life that it is impossible to describe them all. This book does not attempt this task, nor is it a historical account of the development. Rather it is an overview of the field of diode lasers, seen from a historical perspective. It starts with basic semiconductor physics, laser fundamentals, and the optical properties of semiconductors, before moving on to devices themselves, including the double heterostructure laser, the quantum well laser, and the vertical cavity surface emitting laser, before finishing this section with a chapter on modelling using rate equations. The order in which these are presented not only mirrors their historical development, but constitutes a logical progression through the subject. Applications are described in the next two chapters, starting with telecommunications and finishing with high power diodes lasers. In the last two chapters the development of the blue laser is described, followed by the quantum cascade laser. These are advanced devices which are still very much under development, so there is no real historical order to consider. No doubt by the next book this will have changed.

References

[1] Rediker R H 2000 *IEEE J. Select. Topics in Quant. Elect.* **6** 1355–1362
[2] Rediker R H 1987 *IEEE J. Quant. Electr.* **23** 692–695
[3] Hall R 1987 *IEEE J. Quant. Electr.* **23** 674–678
[4] Nathan M 1987 *IEEE J. Quant. Electr.* **23** 673–683
[5] Bernard M G A and Duraffourg G 1961 *Physica Status Solidi* **1** 661
[6] Sorokin P P and Stephenson M J 1960 *Phys. Rev. Lett.* **5** 557–559
[7] Sorokin P P and Stephenson M J 1961 *IBM J. Res. Develop.* **5** 56–58

Chapter 2

Essential semiconductor physics

The last chapter dealt very briefly with the origins of the diode laser. Although none of the physics was discussed in detail it was established that the laser was fabricated from a *p–n* junction. It is the purpose of this chapter to describe some of the essential concepts involved in the operation of such a junction; what do the designation *n*-type and *p*-type mean? how do electrons flow in such a structure, and how is light generated? The chapter is far from comprehensive. It is intended instead to provide a background from which the study of diode lasers can begin, and other concepts will come into play as the book progresses. It is certainly not the intention here to write a text book on semiconductor theory, as there are plenty of those you can consult [1–3] if you feel the need for a more thorough or more rigorous understanding of the topics raised here. The emphasis is placed firmly on the physics rather than the mathematics, but that does not mean that mathematics is not important. Some of the detailed concepts that arise throughout this book can only be understood properly in the light of the mathematical theories, but without a firm grasp of the physical basis for such theories much of the mathematics will be meaningless.

2.1 Free electrons in semiconductors

Semiconductors, as the term implies, lie somewhere between insulators and metallic conductors, and chemical bonding plays a part in this. In metals the atoms pack together as closely as possible, the most common form being hexagonal close-packed structure. In order to do this the valence electrons from each atom are given up to a general pool of electrons and metals exhibit very high electrical conductivity as a result. In insulators, on the other hand, the electrons tend to be bound tightly to one or other of the constituent atoms, and so require the input of a large amount of energy to overcome the strength of the bond. If that energy is supplied there is no reason why the electrons cannot then move freely through the crystal, as they do in the metal, and contribute to an electrical current.

In crystalline insulators the most common form of bonding is the ionic bond, in which one atom gives up one or more of its valence electrons to another atom. Atoms which tend to give up electrons are described as electro-positive and those which tend to accept them are described as electro-negative. The driving force for this is stability. The most stable atoms are those that have a full shell of outer electrons, for example helium, neon, argon, etc., and an electro-positive atom that doesn't possess a full outer shell can form a stable chemical compound with another electro-negative atom which also has a partially filled outer shell of electrons. For example, aluminium, with three valence electrons, can bond with nitrogen, which has five valence electrons, by donating its three electrons to the nitrogen. The nitrogen then has eight electrons (a full shell) and the aluminium has a full lower level of electrons and both are stable. Similarly, two aluminium atoms could bond with three oxygen atoms, each of which would receive two of the valence electrons.

It is easy to appreciate why such crystals tend to be very good insulators. The electro-positive atom has no more valence electrons it can give up to contribute to the conduction process, and the electro-negative atom, having achieved the desired stability that comes with a full outer shell, is not going to give up any of those electrons readily. The electrons are therefore tightly bound to the electro-negative atom and cannot move freely through the crystal. In semiconductors, the bonding tends to be covalent. That is to say, the stability is achieved not through ionisation but through co-operation. The electrons are shared between the atoms. In an elemental semiconductor, such as silicon or germanium, one atom is no different from its neighbours and the bonding is entirely covalent. The electron on average spends its time at the mid-point between the two atoms. In consequence the electron is not strongly bound to either atom and can be released from the bond with the addition of a small amount of energy. Once freed from the bond the electron can move freely through the crystal and therefore contribute to an electric current. In other semiconductors, such as gallium arsenide (GaAs), there is a slight difference between the atoms. Gallium is more electro-positive than arsenic, but not so much so that the gallium atom is persuaded to give up its electrons entirely, and the result is no more than a slight shift in the centre of the charge distribution toward the arsenic. The bonding is said to be partially ionic but in the main it is covalent. Covalency is especially important for wide band gap semi-conductors used in blue lasers and is discussed in more detail in chapter 12.

2.2 Formation of bands in semiconductors

These ideas lead on to band theory, which is the idea that instead of the well defined, sharp energy levels associated with atoms, energy levels in solids exist in bands spread over a wide range. Those electrons that are bound to atoms and which are therefore still part of a covalent bond reside in the valence band,

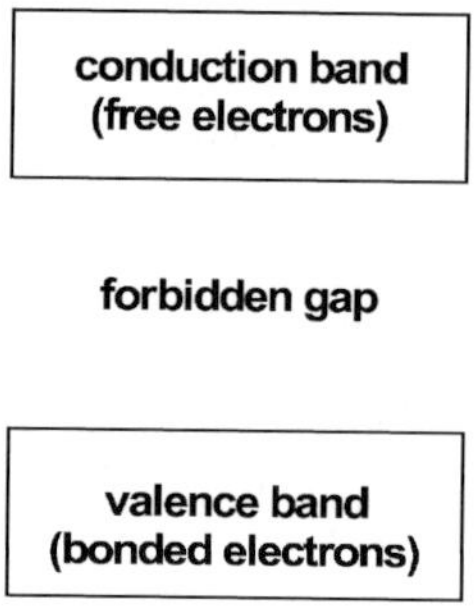

```
conduction band
(free electrons)
```

forbidden gap

```
valence band
(bonded electrons)
```

Figure 2.1. Simple picture of bands in semiconductors.

and those which have absorbed enough energy to be released from their bonds reside in the conduction band. The amount of energy needed to overcome the bond is identical to the forbidden band gap and in a perfect lattice is free of any electronic states (figure 2.1). Therefore a bound electron which absorbs energy must make a transition across the gap to a state in the conduction band.

The bands of states arise from the interactions of the valence electrons with each other. The Pauli exclusion principle states that no two electrons in a given system can have the same four quantum numbers, which is simply another way of saying that they cannot occupy the same physical space. If we consider two atoms so far apart that they do not interact with each other, then clearly the electrons from one atom have no chance of occupying the same physical space as the electrons in the other. However, if the two atoms are bound within a lattice, then by definition the atoms must be close enough for the valence electrons to interact and the possibility then arises that different electrons from different atoms could overlap in space. In order to overcome this the electronic states shift in energy slightly.

There is a close relationship between an electron's energy and its position. This is seen most easily in a simple atom, such as hydrogen, where the potential energy is simply a Coulomb term between the positive charge of the nucleus and the electron and depends inversely on the separation between the two charges. In more complicated atoms the other electrons help to shield the nuclear charge and the potential energy is not so simple to evaluate. In a solid there are not only the additional electrons in the atom but also neighbouring atoms and neighbouring electrons so the total potential energy can be quite complicated. However, the potential energy is only one aspect of the energy of the electron. The other is kinetic energy. The sum of the kinetic and potential energies is called the Hamiltonian, after the mathematician William Rowan Hamilton who developed a formulation of classical mechanics different from Newton's. Newton was concerned with forces and their effect on motion, but Hamilton was concerned with the total energy of the system. For quantum particles such as an electron, which exist within potential fields, the total energy approach is better than an approach based on forces.

The energy of an electron is found by solving the eigenvalue equation

$$H\psi = E\psi \qquad (1)$$

where ψ is the electron wavefunction, $H = K + V$ is the Hamiltonian, and E is the electronic energy level. K is the kinetic energy and V the potential energy. This is a specific form of a general principle in wave mechanics; an operator, in this case H, operating on a wave function ψ gives rise to an observable, in this case E, according to the equality in (1). Operators are not simply multipliers, as would appear to be the case above, but are specific mathematical functions such as derivatives that change the function being operated on. Equation (1) will be used in various places throughout this book and the full differential form of the kinetic energy operator will become apparent.

To return to the problem at hand, there is often a series of discrete solutions E_i, where $i = 1, 2, 3$ etc. For the hydrogen atom, for example, $E_1 = -13.6$ eV, $E_2 = -3.4$ eV, and $E_3 = -1.5$ eV. The trick in quantum mechanics is to find the appropriate potential V and the appropriate wavefunction ψ, so that the energy of the electron can be found. This is easy for the hydrogen atom. For two interacting electrons the total wavefunction will contain terms from each of the two independent electron wavefunctions and each energy E_i will have two solutions, one for each electron. These two solutions correspond to the shift in energy described earlier, and can be related back to the potential energy term. A difference in energy effectively means that the potential energy is different, and as the potential energy is an intimate function of the position of the electron within the potential field, the positions of the two electrons must be different.

The definite and discrete energy states associated with an electron in an isolated atom are therefore transformed in a solid into a series of states of closely spaced energies, with one state for each electron (or one state for two electrons of opposite spin if no magnetic field is present). Thus bands of states are formed. The conduction properties of the crystal depend crucially on the properties of these bands, and because the bands themselves arise from the interactions of neighbouring atoms, chemical bonding is again seen to be important. In a metal, the closeness of the atoms implies a strong interaction and a correspondingly large shift in the electron states such that the bands will be very wide and even overlapping. In an ionic insulator the interaction between the atoms is minimised by the transfer of the electrons and the resulting existence of full shells of electrons on both ions. The bands are correspondingly narrow with a large gap between them. In semiconductors, the bands are quite wide in energy but not wide enough to overlap, and there is a small gap between the bands which must be overcome if conduction is to occur.

2.3 Band theory and conduction

It should be obvious from the foregoing that if the valence band is completely full no electrical conduction can occur because all the electrons are bound to

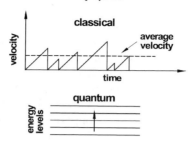

Figure 2.2. Classical and quantum models of conduction.

the atoms in the lattice. Some electrons must be promoted to the conduction band for conduction to occur, but the simple intuitive picture so far developed of free electrons permeating through the lattice is only half the picture. From a quantum point of view, electrical conduction implies a change of state (figure 2.2). An electron accelerated in an electrical field gains potential energy which in quantum terms represents a change from one electron state to another. Hence in a full valence band, where each state within the band is occupied by the maximum number of electrons allowed, there can be no conduction because there can be no change of state. It is important to realise here that the change of state is not the same as the transition from the valence band to the conduction band. For the majority of semiconductors of interest in this book, the energy involved in that transition, while moderate, is still too large to be supplied by an electric field. In normal conduction processes, there is an equilibrium between the rate of gain of energy from the electric field and the rate of loss of energy to the lattice, such that the timed-averaged amount of energy gained from the electric field is very small. The loss, by the way, represents resistance and Joule heating and in simple terms arises from collisions between the electrons and the lattice. On average then, the electrons gain only a small amount of energy from the electric field and the transitions involved in the conduction process take place between states within a band. How the electrons get into the conduction band in the first place, and what determines their number is the topic of a later section. For now just accept that a given number of electrons must exist within the conduction band for conduction to take place and that promotion to the conduction band is a prerequisite of conduction rather than a consequence.

Once in the conduction band, the electrons can of course take part in conduction. In a semiconductor the number of electrons within the conduction band is normally very small compared with the number of available states so promotion to a neighbouring state is an easy matter. These are the free electrons released from the bond permeating through the lattice in response to an electric field. However, conduction can also occur within the valence band. An electron promoted to the conduction band leaves behind it an empty state within the valence band, and an electron within the valence band can be promoted to that state. In other words, conduction can take place.

Physically, this corresponds to an electron jumping from one atom to another, which can of course be imagined as occurring relatively easily if one of the atoms in question has already lost an electron by some mechanism. This method of conduction consists of electrons moving through the crystal from atom to atom, and is distinguished from the permeation of free electrons through the lattice. More importantly, we regard this method of conduction as the transport of a positive charge in the opposite direction from the motion of the free electrons. An atom in which all the covalent bonding electrons are present is electrically neutral, but an atom that has lost an electron will effectively be positively charged. As the electron moves from atom to atom, the positive charge associated with the missing bond is effectively shunted in the opposite direction. We arrive at the conclusion therefore that although conduction in both cases takes places via the motion of electrons, the two mechanisms outlined are distinctly different. In free electron conduction, the current consists of the transport of negative charge through the lattice and the magnitude of the current depends on the number of electrons within the conduction band available for conduction. In the other mechanism, conduction consists of the transport of positive charge in the opposite direction and the magnitude of the current depends on the number of empty states in the valence band, referred to generally as 'holes'. Therefore this is called hole conduction.

As both electrons and holes can carry current in a semiconductor they are often called carriers in consequence. If one type of carrier exists in greater numbers than the other this is referred to as the majority carrier and the other is referred to as the minority carrier. Semiconductors in which electrons are the majority carriers are called "*n*-type" and semiconductors in which holes are the majority carriers are called "*p*-type".

2.4 Electron and hole statistics

It has already been mentioned that electrons must be promoted to the conduction band as a pre-requisite for conduction. The density of electrons within the conduction band in *n*-type material can range from 10^{12} cm^{-2} to over 10^{20} cm^{-3}. These sort of numbers are best dealt with statistically, and our intuitive phenomenological approach is not able to take us further on this issue.

In essence, the electrons exist within the conduction band because of thermal excitations. The probability that a state a some energy E is occupied is given, from Fermi-Dirac statistics, by the Fermi function f

$$f = \frac{1}{1 + \exp\left[\dfrac{q(E - E_f)}{kT}\right]} \qquad (2)$$

where E_f is some reference energy called the Fermi energy, q is the electronic charge, T is the absolute temperature and $k =$ Boltzmann's constant.

The relative populations f_R between the top of the valence band and the minimum of the conduction band, called E_v and E_c respectively, will therefore depend exponentially on the difference between the two energies, according to

$$f_R \propto \exp\left[\frac{-q(E_c - E_v)}{kT}\right] \qquad (3)$$

and will be independent of the actual value of E_f. This is an important result which will be referred to again. It has been assumed that $q(E_c - E_f)/kT > 3$ so that the term on the bottom line of equation (3) is much larger than 1. This is known as the Boltzmann approximation to the Fermi function.

Strictly, to find the number of electrons in the conduction band the product of the density of states and the Fermi function should be integrated over the width of the conduction band but we make the simplifying assumption that only states close to E_c are important. Then the density of electrons, n, simply becomes the product of the effective density of states at E_c ($=N_c$) and the Boltzmann probability, i.e.

$$n = N_c \cdot \exp\left[\frac{-q(E_c - E_f)}{kT}\right]. \qquad (4)$$

Similarly, the density of holes at the top of the valence band is given by

$$p = N_v \cdot \exp\left[\frac{-q(E_f - E_v)}{kT}\right] \qquad (5)$$

where N_v is the effective density of states at the top of the valence band. If equations (4) and (5) are multiplied together then

$$np = N_v N_c \cdot \exp\left[\frac{-q(E_c - E_v)}{kT}\right] = n_i^2 \qquad (6)$$

which depends only on given properties of the semiconductor. n_i is known as the intrinsic carrier density and is equal to the carrier density when both the densities of holes and electrons are equal. Equation (6) defines equilibrium in a semiconductor. It follows that non-equilibrium is defined by

$$n \cdot p \neq n_i^2 \qquad (7)$$

and in fact such conditions exist inside the semiconductor laser.

For many semiconductors

$$N_c \approx N_v \qquad (8)$$

so by definition

$$n_i = N_c \cdot \exp\left[\frac{-q(E_c - E_v)}{2kT}\right] \quad (9)$$

and by comparison with equation (4)

$$E_f = \frac{E_c + E_v}{2}. \quad (10)$$

Therefore the Fermi energy lies in the middle of the forbidden gap if $n = n_i$.

Let us examine further the concept of the Fermi energy. As described earlier, if all the electrons were bound to atoms then there would be no conduction. The valence band would be completely full and the conduction band would be completely empty. This is not physically very realistic because in a semiconductor such as silicon or GaAs there are still appreciable densities of electrons and holes even when the Fermi level lies at mid-gap. However, if we were to imagine a semiconductor at a very low temperature, close to absolute zero, we can imagine that the thermal fluctuations in the solid are considerably diminished and the distribution of energies available to the carriers is also correspondingly reduced. In these circumstances it is possible to imagine that the valence band is completely full and the conduction band is completely empty. From equation (2) it is clear that for

$$E - E_f \leqslant 3kT \quad (11)$$

$f = 1$, i.e. the probability of a state being occupied is unity. At very low temperatures, where kT is very small, we can say that states up to the Fermi energy are full and states above are empty. This is the most common definition of the Fermi energy.

It does not explain why the Fermi energy should lie in the middle of the gap, however. From the above definition the Fermi energy would be expected to lie at the top of the valence band rather in the middle of the band gap. Intuitively we would expect that the Fermi energy must coincide with a real energy in the solid, and as there are no states in the middle of the gap it is hard to imagine why the Fermi energy should lie there. However, the Fermi energy is no more than a statistical convenience that allows us to define the probability of occupancy. One of the properties of the Fermi function is that it is anti symmetric about the Fermi energy. We can most easily appreciate this if there is just one electron in the conduction band and one hole in the valence band. The Fermi energy must lie in the middle of the gap in order for the symmetry of the Fermi function to be obeyed. Even if the valence band were completely full this would also be the case, because if the Fermi energy were to coincide with E_v, i.e.

$$E_v - E_f = 0 \quad (12)$$

then irrespective of the magnitude of kT the density of holes would not be zero but equal to the density of states in the valence band according to equation (5). By similar reasoning the Fermi energy cannot lie at the bottom of the conduction band because then the electron density would be too high. It must lie in the middle of the gap.

2.5 Doping

One of the most useful properties of semiconductors is the ability to change the density of holes or electrons by the process of doping, defined as the substitution of atoms in the lattice by atoms of a lower or higher valency to bring about an increase in the carrier density.

Consider an elemental semiconductor such as silicon. Each silicon atom is surrounded by four neighbours, and shares an electron with each of those neighbours in a covalent bond. Suppose now one of the silicon atoms is replaced by an atom from group V of the periodic table, for example phosphorous. This element has five electrons, four of which will bond with the surrounding silicon atoms. The fifth electron should remain bound to the phosphorous atom in order to maintain charge neutrality, but it is fairly easy to imagine that if the phosphorous were to lose an electron it would be easier to lose this one rather than one of the four in the covalent bonds. In fact, the electron can very easily be lost and is said to be weakly bound. The density of electrons will then equal the density of dopant atoms in the semiconductor except at very high doping levels. For a very wide range of semiconductor applications the concentration will be measured in parts per billion, but in the diode laser atomic concentrations approaching 0.1% are necessary.

Increasing the electron concentration by doping moves the Fermi level closer to the conduction band, and from equation (6) the hole concentration must be reduced accordingly. Physically, the hole concentration decreases because at any time the steady state concentrations of each are maintained by transitions across the band gap in both directions. Electron-hole pairs are generated and recombine, the latter process depending on the densities of both holes and electrons. If the electron density is increased by doping the recombination rate must increase, and the steady state condition is shifted to a lower hole density.

Similar arguments can be used to explain doping by group III elements such as boron. This type of impurity can only bond with three of its neighbours. The fourth silicon atom has an unsatisfied bond, and there is a high probability that a free electron generated elsewhere in the silicon will attach itself to the bond. The hole that accompanied its generation is left behind in the valence band, and the hole density is correspondingly increased. As above, the density of holes is equal to the density of dopant atoms except for very heavy doping. Atoms which donate electrons are called donors and the material is designated *n*-type, and atoms which accept electrons are called acceptors. This type of material is designated *p*-type.

In GaAs, group III and group V atoms already exist as part of the lattice. In this case donor atoms can be either group VI elements, such as Se, sitting substitutionally for the group V element (As) or group IV elements such as silicon sitting substitutionally for the group III element (Ga). Acceptors can be a group II element, such as Be or Zn, substituting for Ga or group IV elements, again silicon is an example, sitting for the group V element. In essence, an atom of lower valency than the host acts as an acceptor and an atom of higher valency than the host acts as a donor.

2.6 Heavy doping

It is necessary in many semiconductor devices for parts of the semiconductor to be doped very heavily, of the order of 10^{19} dopants/cm^3 or more. Such high densities of dopants represent a serious perturbation of the lattice and can distort the band structure. In order to properly appreciate how this can arise it is necessary to appreciate that dopants introduce shallow localised states into the band gap. A localised state is a state localised in space in contrast to the states occupied by the free electrons, which are called extended states. The term arises from consideration of the quantum mechanics of the electron inside the crystal.

The free electron is a quantum particle and has a wave-vector k associated with it. This wave-vector is definite and well defined and consequently the momentum, $\hbar k$, is also well defined. Heisenberg's famous uncertainty principle tells us that it is not possible to know both the momentum, p, and the position, r, simultaneously, and that the uncertainties on each, Δp and Δr respectively, are related by

$$\Delta p \cdot \Delta r \approx h \qquad (13)$$

where h is Planck's constant. If p is well defined such that

$$\Delta p = 0 \qquad (14)$$

then

$$\Delta r = \infty. \qquad (15)$$

In other words the electron can be anywhere in the crystal. This is what is meant by the term "extended state"; the electron cannot be associated with any particular point in space or any particular atom and instead behaves as if it extends throughout the whole of the crystal.

A donor state on the other hand is different. The extra electron associated with the donor can be released into the conduction band but it leaves behind a positive charge, and that positive charge is localised in space on that particular atom. This charge is an attractive centre for the spare electron and the system can be modelled as a hydrogen atom modified to take into account the effective mass of the electron and the dielectric constant of the semiconductor. The effective mass will be described in more detail in chapter 1.8, but for now it is sufficient to accept that the electron within the crystal behaves as if it has a lighter mass than the free electron (which is 9.1×10^{-31} kg). This is known as the effective mass.

If the ionisation energy for the hydrogen atom is given by

$$E_H = \frac{m_o q^4}{32\pi^2 \varepsilon_o^2 \hbar^2} = 13.6 \text{ eV} \qquad (16)$$

then the ionisation energy for the donor atom, which is the amount of energy required to liberate the spare electron from the coulombic attraction of the parent atom

$$E_d = \left(\frac{\varepsilon_o}{\varepsilon_s}\right)^2 \left(\frac{m^*}{m_o}\right) E_H \qquad (17)$$

where ε_s is the relative permittivity of the semiconductor, and m^* is the effective mass.

It follows that if the donor is not ionised and has an electron attached to it, this electron must lie at an energy E_d below the conduction band. This is called a shallow state, and in GaAs experimentally measured donor energies are 0.003 eV (Te), 0.0059 eV (Se), 0.0058 eV (Si) and 0.006 eV (Ge), in very good agreement with the simple model [4]. Such low binding energies are well within a thermal fluctuation, kT, (≈ 0.025 eV), so the electron is easily lost and the donor ionised at room temperature. In wide band gap semiconductors, however, donors can lie fairly deep below the conduction band, in which case only a fraction of the donors are ionised leading to a correspondingly low electron density. Similar arguments apply to acceptors. The most common acceptors produce shallow states of the order of 0.03 eV *above* the valence band edge, and it is important to recognise that for a hole the lowest energy state appears to be the highest. Ordinarily a hole will lie at the top of the valence band so in order for the state to be lower in energy it must lie above the valence band edge somewhere within the band gap.

Heavy doping raises the carrier density and pushes the Fermi energy closer to the edge of the majority carrier band. Is there a limit to how far the Fermi level can move if the density of dopants is increased? No. It is possible to increase the density of dopants until the Fermi energy is pushed right into the band edge and beyond but in such circumstances only a fraction of the dopants are ionised. As soon as the Fermi level approaches within $3kT$ of the dopant level the Boltzmann approximation is no longer appropriate and instead the full expression for the Fermi function given in equation (2) must be used. Once the Fermi level moves above the donor state there is a greater probability that it will be occupied than not, so that at high doping densities there will always be a sizeable fraction of dopants that are not ionised, unlike the situation at moderate doping levels where effectively every dopant is ionized. This has consequences for the ionization energy of the dopants and for the optical properties of the semiconductor as electrons bound to neighbouring donors begin to interact.

The density of donors at which this interaction occurs can be estimated from the radius of the orbit of a bound electron, which can be calculated using the hydrogenic model. Similar calculations can be done with acceptors. The Bohr radius for the hydrogen atom [5]

$$a_0 = \frac{\varepsilon_o h^2}{\pi m_o q^2} = 0.053 \text{ nm} \tag{18}$$

can be modified to take into account the effective mass ($m^* m_0$) and the dielectric constant ($\varepsilon_0 \varepsilon_r$), and typically is of the order of 100 Å in GaAs because the electron effective mass is so small. At a moderate doping density of 10^{15} cm^{-3}, the average distance between the dopant atoms will be 1000 Å so there is no possibility of any interaction. Besides, the dopants would be ionised at room temperature so there would be no electron wavefunctions to overlap. For a doping density three orders of magnitude higher at 10^{18} cm^{-3}, the average distance between dopants will be correspondingly reduced to 100 Å. Clearly the dopants will now interact with each other, a situation similar to that already met in relation to the formation of the conduction and valence bands. The same thing happens with the dopant energy levels. They adjust to form bands of states centred around the ionization energy, which is itself reduced slightly by such high densities of carriers. The attractive centre of the ionised dopant is screened from the other carriers and the binding energy is reduced. The shallow level at the centre of the band therefore moves closer to the majority carrier band edge, and for very heavy doping the impurity bands will merge with, and become part of, the majority carrier band (figure 2.3). The band gap is therefore

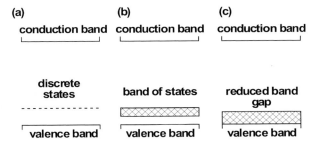

Figure 2.3. The effect of doping is to introduce shallow states at low doping densities (a), which form bands at a lower binding energy at higher densities (b), and eventually merge with the majority carrier band at even higher densities (c). *P*-type material is assumed here but the same applies to *n*-type.

reduced and though expressions for this reduction exist [6] this detail is not necessary here.

In silicon donors and acceptors also form bands that can affect device performance but the doping density has to be very high before those bands merge with the majority carrier band edge. In GaAs, however, the ionization energy is considerably smaller and at the doping level of $\sim 10^{18}$ cm^{-3} the bands merge with, and become part of, the majority carrier band edge [7]. These are, of course, averages. If the dopants are distributed non-uniformly and clump together then localised regions of the semiconductor will have a lower band gap than others. The donor band starts to merge with the conduction band at a density as low as 3×10^{16} cm^{-3} so at doping densities as high as 10^{18} cm^{-3} all that remains of the donor band is a tail of states descending into the band-gap [4]. The higher binding energy of the acceptors (by a factor of 5 or 6) means that for densities less than about 10^{18} cm^{-3} the hole impurity band is still a significant feature of the band edge, but at about 10^{18} cm^{-3} or above the impurity band will have merged with the valence band edge. The tail of states that rises from the valence band into the energy gap is much more pronounced than that descending from the conduction band, and effectively the band-gap shrinks in heavily doped p-GaAs. This is represented schematically in figure 2.4, where the dotted lines represent the band-edges in the low-doped case.

Band-gap narrowing in p-GaAs played a significant part in the operation of the earliest p–n junction lasers by altering the injection of holes across a p–n junction and by altering the refractive index of the p-GaAs (see chapter 4). The attempts made to understand these effects led directly to new ways of improving the performance of diode lasers by minimising the disadvantages and maximising the advantages, resulting in the double heterostructure device

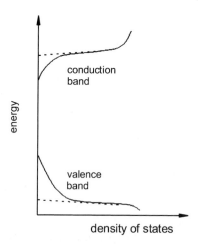

Figure 2.4. Band tails in heavily doped semiconductors. The dashed line is the band edge in low-doped material.

described in chapter 5, and from there to the quantum well laser, the vertical cavity laser, and the quantum cascade laser.

2.7 Recombination and generation

The process of recombination of excess electrons and holes with the subsequent emission of a photon lies at the heart of the diode laser. Recombination is the process whereby an electron in the conduction band occupies an empty state (hole) in the valence band, thereby removing both carriers from the conduction process. In some cases the recombination process can lead to the emission of radiation, for example if the electron drops directly from the conduction band to the valence band, and in other cases it does not, if for example the electron is first captured in a deep-lying state from which it subsequently makes the transition to the valence band. Generation is the opposite process. An electron in the valence band is promoted to the conduction band, either through thermal fluctuations or through the absorption of a photon with energy equivalent to the band-gap.

Generation and recombination occur constantly in order to maintain the equilibrium occupancy of the conduction band as required by equation (6), and in equilibrium the rate of recombination and generation are equal. There will be no noticeable luminescence as a result of the recombination, however, because the equilibrium rate is too low to be observed. An excess of carriers such,

$$n \cdot p > n_i^2 \tag{19}$$

is needed to observe significant light emission, because in attempting to restore equilibrium the rate of recombination can be very high.

Whether a recombination event is radiative or not depends on a number of factors. Recombination can be regarded as a classical collision between two particles in which momentum must be conserved. The energy of the electron and hole is used to create a third particle, the photon. In some cases the momentum of the photon is not sufficient, and a fourth particle called a phonon (figure 2.5) is involved. Phonons are lattice vibrations that are treated mathematically as travelling waves with momentum $\hbar k$ and energy $\hbar \omega$. The idea of a lattice vibration acting as a wave arises from the notion of a coupled oscillator. If an atom is displaced from its equilibrium position the interatomic forces that couple it to its neighbours will lead to energy being transported through the lattice. Detailed analysis of the mechanical properties of lattices [3] reveals that there are essentially two phonon branches. Acoustic mode phonons can have frequencies from zero to the hypersonic (100 s of GHz) and optical mode phonons have a lower frequency limit around a few THz. Acoustic mode phonons are associated with the transport of sound and optical mode phonons are associated with the transport of thermal energy.

20   *Essential semiconductor physics*

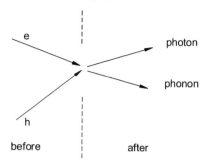

Figure 2.5. Electron-hole recombination as a four-particle collision.

Transitions involving phonons are called "indirect" and transitions that do not involve phonons are called "direct". Direct transitions have a much higher probability than indirect transitions, all other things, such as band gap and density of states, being equal, because in an indirect semiconductor the electron and hole have to interact with a phonon at the same time as they interact with each other. Of course, some of the energy of the transition can go into creating phonons in indirect materials, but as the phonon population is determined by the temperature of the lattice this will raise the temperature. In thermal equilibrium, it is better to think of a given phonon density with a given probability of interaction.

A key aspect of the semiconductor laser is therefore choosing a direct band gap material as well as designing a structure that enables strongly non-equilibrium carrier concentrations to be achieved. Direct and indirect materials are discussed further below. Non-equilibrium is achieved by means of a *p–n* junction. The earliest lasers used a simple GaAs junction in which electrons from the *n*-type are injected into the *p*-type and vice-versa, though the reduction in the band gap on the *p*-side due to the heavy doping meant that the injected hole density was considerably less than the injected electron density. In later designs the structure of the junction is modified, but in essence charge is still injected from both *n*- and *p*-type regions. The exception to this is the quantum cascade laser where quantum engineering of the material has rendered the presence of holes unnecessary.

2.8   Energy bands in real semiconductors

Electrons are quantum particles with momentum $p = \hbar k$ and energy

$$E = E_c + \frac{(\hbar k)^2}{2m^*} \qquad (20)$$

where E_c is the energy at the bottom of the conduction band and m* is the effective mass. Electrons in a laser will occupy a large range of energies from the bottom of the conduction band up. The parabolic dependence of E on k is often assumed to apply over this range because it simplifies the calculation of the electron distribution, but in fact there is often a significant deviation from parabolicity at large k. However, the assumption is accurate enough over a small range of energies close to the band edges.

Plotting E against k for real semiconductors, we find that there are two basic types: those where the minimum in E_c occurs at the same value of k as the maximum in E_v (usually but not always at $k=0$) and those for which k at the maxima and minima of the two bands differs by π/a, where a is the lattice parameter (figures 2.6a,b). These are, respectively, the direct band-gap and indirect band-gap semiconductors described above (figures 2.7).

The idea of overlapping bonds leading to conduction and valence bands has the virtue of being intuitive. In so far as it explains conduction and recombination it is also useful. However, the band structure in a real solid is often far more complicated, especially the valence band structure, and when modifications to the band structure occur, as they do in quantum confined structures, it is important to realise that this simple approach is quite limited. The band structure in real solids can be calculated quite accurately, and appendix III contains some of the detail. It is not necessary to review that detail here, but some appreciation of the problem will help.

As described in chapter 1.2 the trick is to find an appropriate wave function. A common approach is to construct the wavefunction from a linear combination of a basis set, i.e.

$$\psi = \sum_n a_n \psi_n \tag{21}$$

and there are various levels of complexity depending on the size of the basis set and the wavefunctions chosen to make up the basis set. One approach is to use eight functions corresponding to three p-orbitals and one s-orbital, each with two electrons of opposite spin. Again this corresponds to the chemistry, because these are the orbitals involved in bonding. The process of using basis

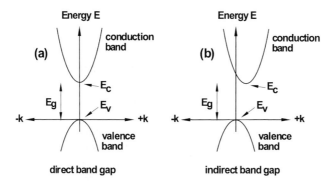

Figure 2.6. Direct and indirect band gaps.

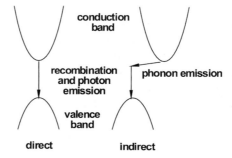

Figure 2.7. Conservation of momentum in recombination.

states in this manner is called *mixing*, and you might come across the term in other descriptions of the valence band. If the conduction band is said to be mixed with the valence band it means nothing more than basis states appropriate to both are used to derive the energy levels inside the solid. The outcome of all this is that there are in fact four bands of interest in semiconductors; three valence bands and one conduction band. The conduction band is the simplest and is essentially derived from *s*-orbitals. The three valence bands arise essentially from the three *p*-orbitals and are called respectively the light hole band, the heavy hole band, and the split-off band, as illustrated in figure 2.8.

The terms, "light and heavy hole" reflect the effective masses. It follows from equation (20) that

$$m^* = \frac{1}{\hbar^2}\left[\frac{d^2E}{dk^2}\right]^{-1} \qquad (22)$$

so where the band energy varies little with *k* the effective mass is large. It will be immediately apparent that for an electron in the valence band the effective

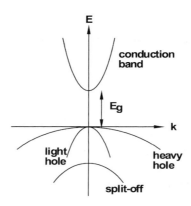

Figure 2.8. The valence bands typical of III-V laser materials.

mass is negative. It should be appreciated that the concept of effective mass is an approximation. The electron and hole have so far been treated essentially as single particles moving through the crystal but of course they are surrounded by other electrons, atoms, and ions, all of which contribute to the total potential inside the crystal and all of which therefore affect the motion. The effective mass represents the combined effect of all the other electrons and holes in the system and is essentially a correction that allows the electrons and holes to be treated as single particles. The negative effective mass of the electron in the valence band becomes the positive effective mass of the empty state, i.e. the hole. It should also be noted that the effective mass might well be anisotropic if the crystal does not have cubic symmetry.

In bulk semiconductors the distinction between the different valence bands is often not important. The light hole and heavy hole bands have the same energy at $k=0$ (states that have the same energy are said to be "degenerate"), so it makes no difference to radiative recombination. Neither is the distinction important in normal electrical conduction. The average increase in energy due to the electric field is small and hence the change in wave-vector is small. Only in the case of high field conduction, which does not generally occur in laser devices, do significant perturbations in the wave-vector occur. In general, then, most of the conduction takes place around the valence band maximum and both valence bands contribute to the total hole current. Under these conditions a single effective mass, which is a weighted average of the different masses, is often used. There are circumstances, however, when the distinction is important, particularly in thin layers where quantum confinement lifts the light and heavy hole degeneracy at $k=0$. Under these circumstances the valence band with the lowest energy dominates the recombination.

2.9 Minority carrier lifetime

Radiative recombination is the most important recombination mechanism in a laser, but it is important to appreciate that other recombination pathways exist. If these other pathways were to dominate there would be very few electron-hole pairs recombining radiatively and the laser either will not work or will be very inefficient.

Recombination pathways are characterised by a recombination lifetime τ. Restricting the discussion to low-level injection, defined as the regime where only the minority carrier density is significantly disturbed, the inequality in equation (19) is then due entirely to the presence of excess minority carriers. In this case the recombination lifetime is equal to the minority carrier lifetime. Assuming for the sake of argument p-type material, the rate of decay of the excess minority carriers (electrons) Δn depends on their density, so

$$\frac{d\,\Delta n}{dt} = -C\,.\,\Delta n \qquad (23)$$

24   *Essential semiconductor physics*

where C is some constant. This has the solution

$$\Delta n = \Delta n_0 \exp\left(-\frac{t}{\tau}\right) \quad (24)$$

where Δn_0 is the excess density at $t=0$, and τ, the minority carrier lifetime, is defined as the time taken for the excess density to decay to $\Delta n_0/e$. Equations (23) and (24) can be combined to show

$$\frac{d\,\Delta n}{dt} = -\frac{\Delta n}{\tau}. \quad (25)$$

In words, the rate of recombinations is given simply by the ratio of the excess density to the minority carrier lifetime.

Each recombination pathway, say A, B, C etc., has its own characteristic lifetime τ_A, τ_B, τ_C, which, in combination with others, gives a total minority carrier lifetime

$$\frac{1}{\tau} = \frac{1}{\tau_A} + \frac{1}{\tau_B} + \frac{1}{\tau_C} + \ldots \quad (26)$$

from which it can be seen that the smallest lifetime, i.e. the fastest pathway, dominates. The most common forms of non-radiative recombination arise from the presence of defects within the crystal lattice, and since these are usually the fastest, it is important for light emitting devices to have semiconductor crystals of high purity and high crystalline quality. In the case of heterostructure devices, where thin layers of one semiconductor are deposited on to another, special techniques have been developed to ensure that the crystalline quality of the deposited layers is sufficient for laser action.

2.10 Minority carrier diffusion

Although the precise mechanism of minority carrier injection has not been described yet, it is nonetheless possible to imagine that an excess of minority carriers is injected into the semiconductor at the *p–n* junction. We can assume that the junction constitutes a two-dimensional surface and that the excess density of minority carriers gives rise to a concentration gradient extending from the plane into the bulk of the *p*-type material. From Fick's law this must give rise to diffusion, and we can describe the flux of minority carriers F by

$$F = -D\frac{d\,\Delta n}{dx} \quad (27)$$

where D is the diffusion coefficient and $d(\Delta n)/dx$ is the concentration gradient. This eventually gives

$$\Delta n \approx \Delta n_0 \exp\left(-\frac{x}{L}\right) \tag{28}$$

where $L \approx \sqrt{(D\tau)}$ is the minority carrier diffusion length. The carriers can only diffuse as long as they exist so the diffusion length is governed by the minority carrier lifetime.

2.11 Current continuity

The processes of diffusion and recombination lead naturally onto the concept of current continuity. Under normal conditions of current flow, which we will assume for simplicity lies along the x-direction only, current continuity requires that the current is uniform throughout the semiconductor. The electron current density J, in A cm^{-2}, at some point x is

$$J = v \cdot n(x) \cdot q \tag{29}$$

where $n(x)$ is the concentration of electrons, q is the electronic charge, and v is the velocity of the electrons. If at some adjacent point, say $x + \delta x$, the concentration is different the velocities must be correspondingly different. If the semiconductor is divided into small elements of width δx then whatever electrons enter an element are swept out into the next, otherwise the charge must be stored in one of the elements. This can be expressed by

$$\nabla \cdot J = \frac{d\rho}{dt} \tag{30}$$

where the symbol ∇ represent the divergence operator (in one dimension $\nabla = d/dx$) and ρ is the charge density. The presence of recombination and generation alters this simple picture. If electrons recombine within an element, they enter that element but do not emerge from the other side so there is a divergence in the current. However, they are not stored within the element either. Similarly, if generation occurs electrons appear within the element and emerge from it as part of the current without seeming to have entered it. The current continuity equation has to be modified therefore, to take this into account.

$$\nabla \cdot J + q(G_n - U_n) = \frac{d\rho}{dt}. \tag{31}$$

Here G_n is the rate of generation and U_n is the rate of recombination and the term $q(G_n - U_n)$ constitutes the net electron current due to these processes. It should be noted that these are strictly the rates within the small elements. Similar expressions exist for the hole currents.

2.12 Non-equilibrium carrier statistics

The final topic in this chapter concerns the statistics of electrons and holes in non-equilibrium. In equilibrium the product of the electron and hole density is constant and independent of the position of the Fermi level, from which it follows that the same Fermi level applies to both carriers. If we know the density of one we can know the density of the other. This is not the case out of equilibrium, where, depending on the injection mechanism, the densities of both the minority and majority carriers can be varied. There is no reason why a Fermi level cannot still be ascribed to each carrier in such circumstances, but it clearly cannot be the same Fermi level. It is important to recognise also that such a Fermi level is purely an expedient device for calculating carrier densities and essentially describes a non-equilibrium Fermi distribution. Such a Fermi level is called a "quasi-Fermi level", and in some texts is referred to rather clumsily as an "Imref". Quasi-Fermi levels have exactly the same properties as ordinary Fermi levels. We regard states below them as full and states above them as empty. However, the fact of independent quasi-Fermi levels means that the electron density is no longer tied to the hole density.

2.13 Summary

The essential solid state physics relevant to semiconductor diode lasers has been described. Starting from the position of recognising that a p–n junction exists which allows charge to be injected into the semiconductor in order for laser action to take place, the origin of the conduction and valence bands, effective masses, doping, carrier statistics, injection, diffusion, and light generation have all been described. In particular, the concept of recombination, minority carrier lifetime, and minority carrier diffusion length have all been developed. Thus, it is now possible to understand that electrons injected into the p-type material diffuse away from the junction, recombining as they go. In the main that recombination will be radiative, so light will be emitted, but of course this does not necessarily describe a laser. A light emitting diode works very much this way, but there are crucial differences between an LED and a laser, not only in the details of the structure but also in the number of carriers injected. These concepts are described in the next chapter, which deals with laser fundamentals.

2.14 References

[1] Sze S M 1981 *Physics of Semiconductor Devices* 2nd edition (New York: Wiley)
[2] Seeger K 1985 *Semiconductor Physics; An Introduction* 3rd edition (Berlin: Springer)

[3] Kittel C 1976 *Introduction to Solid State Physics* 5th edition (New York: Wiley)
[4] Thompson G H B 1980 *Physics of Semiconductor Laser Devices* (Chichester: John Wiley & Sons)
[5] Mott N F 1972 *Elementary Quantum Mechanics* (London: Wykeham Publications) p 79
[6] Abenante L 2001 *Sol. Energy Mater. Sol. Cells* **67** 491–501
[7] Willardson R K and Beer A C (eds) 1966 *Semiconductors and Semimetals* Vol 1 (London: Academic Press)
[8] Kuok M H, Ng S C, Rang Z L and Lockwood D J 2000 *Phys. Rev. B* **62** 12902–12908

Problems

1. The effective densities of states in the conduction and valence bands of GaAs are 4.7×10^{17} cm^{-3} and 7.0×10^{18} cm^{-3} respectively, and the band gap is 1.424 eV at room temperature. Calculate the densities of electrons and holes when the Fermi level lies; (i) 0.3 eV below the conduction band edge, (ii) at the conduction band edge, and (iii) at the valence band edge. Note: $q/kT = 38.69$ eV^{-1} at room temperature.
2. Calculate the Bohr radius of a typical donor in GaAs ($m_e^* = 0.067$, $\varepsilon_r = 13.1$) and hence estimate the average density at which some interaction between donors can be expected.
3. Silicon is an indirect band gap semiconductor with a lattice parameter of 5.43 Å. Calculate the momentum of the electron in the conduction band minimum. Compare this with the momentum of a band gap photon ($E_g = 1.12$ eV). Take Planck's constant, $h = 6.63 \times 10^{-34}$ m^2 kg s^{-1}, the velocity of light $c = 3.0 \times 10^8$ ms^{-1}, and the electronic charge to be 1.602×10^{-19} C.
4. Assuming the longitudinal optical phonon energy in silicon to be 63 meV calculate the frequency of the phonon. Hence calculate the wavevector and the momentum assuming the velocity of the phonon to be 8350 ms^{-1} [8]. Compare this with the momenta calculated in Q3 above. Is the momentum of the phonon larger than the electron momentum?

Chapter 3

Laser fundamentals

This chapter is concerned with the basic physics of the homo-junction laser. Although some important concepts in semiconductor physics have already been described it is not yet possible to apply them directly to the semiconductor laser without first looking at some fundamental aspects of lasers in general.

All laser oscillators have certain basic features in common. The term "oscillator" arises from the fact of having a medium exhibiting gain in conjunction with feedback. The output from such a system is self selecting; of the frequency components fed back into the system that which undergoes the largest amplification will, after several cycles, be the largest component in the system, and eventually the output will be dominated by this component. Before considering a laser oscillator consider the simplest electrical analogue, a basic series LCR circuit with a resonant frequency ω_o, as shown in figure 3.1. Suppose at first the input to the circuit comprises a signal at a single frequency ω. The output from the circuit can be the voltage across one of the components, say the resistor. We know that the impedance of the circuit is given by

$$Z = \sqrt{\left(\omega L - \frac{1}{\omega C}\right) + R^2} \qquad (1)$$

so that, at resonance,

$$Z = R. \qquad (2)$$

Hence, the current flowing in the circuit is a maximum and the voltage dropped across the resistor will also be a maximum.

Expressed as a differential equation for the motion of the charge in the circuit, we have, for a sinusoidal input

$$V = V_o \sin(\omega t) = L\frac{d^2 q}{dt^2} + R\frac{dq}{dt} + \frac{q}{C}. \qquad (3)$$

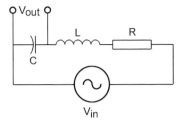

Figure 3.1. A simple resonant LCR circuit.

Mathematically, this is identical to a range of problems going under the description of a damped, forced harmonic oscillator with the solution for the amplitude of the motion

$$A = \frac{K}{\omega_o^2 - \omega^2 + i\frac{R}{L}\omega} \tag{4}$$

where K is some constant. In short, the amplitude is constant for low frequencies such that $\omega_o^2 \gg \omega^2$, and at resonance, $\omega = \omega_o$, the amplitude is a maximum limited by the size of the resistance. For $R = 0$ the amplitude would be infinite. Equations similar to (4) will be met in the treatment of refractive index in semiconductors in chapter 4 and in a treatment of the small signal modulation characteristics of diode lasers.

Continuing with the electrical analogy, it is instructive to ask what exactly is meant by the term "amplitude" here. These equations have been formulated for the specific case of charge flowing round a LCR cicuit, but, the general form of the equations applies to a very wide range of physical systems. The same equations with different coefficients can be used to describe the motion of a mass on a spring, or a pendulum, for example, where it is clear that the amplitude is the amplitude of the motion of the body. In the present case, the oscillatory motion is the physical motion of the charges around the circuit as the energy is alternately stored in the capacitor and then the inductor. The amplitude refers to the size of the packet of charge moving round the circuit, so when the amplitude is a maximum at resonance, the current is also a maximum. Equation (4) makes it clear that there is also a complex component of the amplitude, which in the present context means that the current is complex. This refers to the well known fact that it is in phase with the voltage in the resistive element, which is of course 90° out of phase with the voltage in the inductive or capacitive components. For frequencies higher than the resonant frequency the amplitude decreases with increasing frequency until it is zero. Under these circumstances, there is very little current flowing and the output from the circuit is reduced.

Suppose now that instead of a single sinusoidal input, a spectrum of frequencies were used to drive the circuit. Such a spectrum might arise from a periodic, but anharmonic, driving signal such as a square wave. Fourier's theorem tells us that this periodic signal can be sub-divided into a series of harmonic components, and each of these can be applied to equation (4). Components at frequencies much lower than resonant will be amplified by a small but frequency independent amount. Components at or near the resonant frequency will be amplified greatly, and components at frequencies much higher than resonance will be damped. If part of the output is tapped off and fed back into the input – positive feedback – then these components will be amplified yet again. As the signal passes around the circuit repeatedly the components will continue to be amplified according to the gain characteristics of equation (4). Obviously, the high frequency components, will continue to be damped and will eventually die out. The low frequency components will continue to be amplified by a low amount. The components at or near resonance will be amplified greatly and each time the output is fed back into the circuit these components will grow relative to all the others. Eventually they will dominate the output of the system, and from a square wave input a near-sinusoidal (or harmonic) output will have been selected.

In a laser, it is not an electrical signal but the optical field which is amplified. Optical gain occurs by the process of stimulated emission (hence the acronym LASER – Light Amplification by Stimulated Emission of Radiation). Optical feed-back occurs when the amplified optical field is fed back into the active medium simply by a process of reflection. Reflection will occur at both ends of the system so light bounces back and forth within an optical cavity. In this way the light makes repeated passes through the active material so that the intensity of the optical field is built up. Only one of the mirrors need be fully reflecting. The other can be designed to be partially transmitting in order to let out a constant fraction of the radiation, which then forms the output of the device, i.e. the laser beam. Eventually a steady state will be achieved in which the optical gain is balanced by the losses from the mirror.

The analogy with the electrical signal is instructive. Optical gain can occur over a wide range of frequencies, all of which have the potential to be amplified. The starting signal for the amplification is not fed in from an external source, however, but originates from within the gain medium itself via spontaneous emission. An electron in an excited state decays to the ground state, emitting a photon. Photons can be emitted over a wide range of frequencies, and if these are fed back into the system, then like the electrical signal, the different components will undergo selective amplification according to the gain spectrum. However, unlike the electrical system not all emitted photons will be fed back. Two parallel mirrors constitutes a Fabry–Perot cavity and the possible wavelengths that can propagate between the two mirrors are

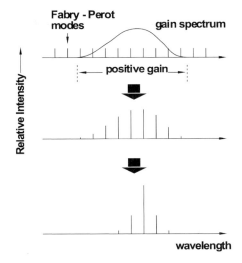

Figure 3.2. Amplification of Fabry–Perot modes to produce a self-selected narrow output spectrum. The amplitude of the final stage will clearly be much greater than the initial and intermediate stages.

restricted to those that will allow a whole number of half wavelengths to fit between the mirrors, i.e.

$$m \cdot \frac{\lambda}{2} = n \cdot L \tag{5}$$

where m is an integer, λ is the wavelength, n is the refractive index, and L is the length of the cavity. These modes, as they are referred to, will be evenly spaced in wavelength and will therefore constitute a spectrum of optical signals that can be fed back into the active medium. Those nearest in wavelength to the centre of the gain curve will be self selected after a number of cycles, and the output of the laser will consist of a single line, or at most a few modes, of very narrow spectral width. The process is illustrated in figure 3.2.

3.1 Stimulated emission

Stimulated emission, the process behind optical gain, is not ordinarily observed in nature, although it does occur. It is the process by which an electron in an excited state can be made to relax to the ground state by the the interaction of the electron with an incident photon. There are then two photons, both of which occupy the same quantum state. There is no difficulty about this since photons obey Bose–Einstein statistics, in which the number of particles occupying a given energy state is not limited. Indeed, one of the features of Bose–Einstein statistics is the tendency for particles to accumulate in a particular energy state [1] and if stimulated emission were to occur enough times the number of photons occupying that particular energy state would be enormous. Macroscopically, this is manifested in the phenomenon of coherence, which is defined here simply as the condition required for light beams to interfere. This somewhat simple definition nonetheless implies a definite phase-relationship between two different light beams. Light from a normal source does not normally exhibit coherence, but radiation from a laser does, a property often exploited in the fabrication of holograms, for example.

Stimulated emission obviously competes with absorption (figure 3.3). If the electron is in the ground state when the photon is incident the photon will be absorbed and the electron excited. Hence the number of photons comprising the optical field will be reduced. In thermal equilibrium the number of electrons in the ground state far exceeds the number in the excited state so absorption dominates. As will be shown below, the populations of the two states have to be inverted in order to observe stimulated emission, i.e. there must be more electrons in the excited state than in the lower. Such systems are very far removed from thermal equilibrium, which is why stimulated emission is not normally observed.

A full mathematical treatment of absorption, stimulated emission, and spontaneous emission was developed by Einstein. The relative populations of two energy levels E_2 and E_1 in thermal equilibrium is governed by Fermi–Dirac statistics, but under certain conditions it is possible to use the Boltzmann approximation. The two population densities are then related by

$$n_2 = n_1 e^{(E_1 - E_2/kT)} \tag{6}$$

and equilibrium is maintained by spontaneous and stimulated relaxation from the excited state to the ground state, with the emission of radiation. As the name implies, spontaneous relaxation is random and there is no definable phase relationship between emitted photons.

Now consider a black body placed within a furnace at some elevated temperature. The rate of absorption of radiation at a frequency ω_{12} which causes transitions between levels 1 and 2 in the body is

$$R_{12} = B_{12} n_1 \rho(\omega_{12}) \tag{7}$$

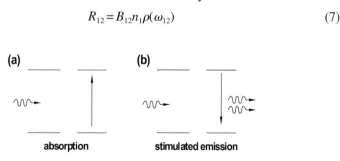

Figure 3.3. The competing processes of absorption and stimulated emission.

where B_{12} is a coefficient and $\rho(\omega_{12})$ is the density of radiation at the frequency ω_{12}. The rate of emission has both stimulated and spontaneous components, i.e.

$$R_{21} = A_{21}n_2 + B_{21}n_2\rho(\omega_{21}) \qquad (8)$$

where A_{21} is the coefficient for spontaneous emission and B_{21} is the coefficient for stimulated emission. In a simple two-level system the energy of the upward transition is identical to the energy of the downward transition so

$$\rho(\omega_{12}) = \rho(\omega_{21}). \qquad (9)$$

In thermal equilibrium then,

$$R_{12} = R_{21} \qquad (10)$$

hence

$$\rho(\omega_{12}) = \frac{A_{21}}{B_{12}e^{\hbar\omega_{12}/kT} - B_{21}}. \qquad (11)$$

Since the radiation arises from the furnace, $\rho(\omega_{12})$ must be identical to the black body spectrum derived by Planck,

$$\rho(\omega) = \frac{\omega^3}{\pi^2 c^3}\left[\exp\left(\frac{\hbar\omega}{kT}\right) - 1\right]^{-1}. \qquad (12)$$

Comparison between equations (10) and (11) yields

$$B_{12} = B_{21} = B. \qquad (13)$$

The coefficients for stimulated emission and absorption are identical so the population determines which process dominates. In thermal equilibrium $n_1 > n_2$ so absorption dominates but in a laser $n_1 < n_2$ so stimulated emission dominates. The coefficient for spontaneous emission is

$$A_{21} = B\hbar \frac{\omega^3}{\pi^2 c^3}. \qquad (14)$$

In general population inversion, i.e. $n_2 > n_1$, cannot be achieved using the simple two-level system so far described. At best the populations will be equal so at least three levels are required to realise inversion. The ruby laser utilises three levels, but other materials, such as Nd-based lasers, utilise four levels (figure 3.4). In an optically pumped four-level system, absorption occurs at ω_{12} and radiation at a frequency ω_{34} is emitted. The electron raised to state 2 is non-radiatively transferred to state 3 accompanied by phonon emission, and relaxation to state 4 is accompanied by the emission of radiation. As state 4 is empty to start with it is necessary to excite only a few electrons to state 2 before the population of state 3 is raised above that of state 4 and population inversion is achieved. It is important, of course, that the rate of arrival of electrons at

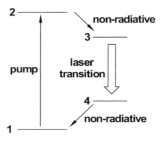

Figure 3.4. A four level system with population inversion between levels 3 and 4.

state 3 be fast compared with the rate of relaxation to state 4, otherwise it will be difficult for electrons to accumulate in the upper state (level 3). Similarly electrons must transfer rapidly from state 4 back to state 1. If not, and electrons arrive at a similar or even slightly faster rate, electrons will accumulate in 4 and population inversion may be lost. As the dominant mechanism for the relaxation from state 3 to 4 is stimulated emission the rate of stimulated emission will also increase as the intensity of radiation in the laser increases.

3.2 Population inversion in semiconductors

A semiconductor with an inverted population behaves like a four-level system with the main difference that the pumping is supplied by the current. Charge is fed in at the Fermi level because current conduction takes place within the states around this energy. From the quantum point of view conduction implies a change of state, and it is only at the quasi-Fermi level that electrons and empty states coincide. States well below the quasi-Fermi level are full so it is not possible for electrons to be scattered into these states, and states well above the quasi-Fermi level are essentially empty so there are no electrons there to be scattered. However, the states close to the quasi-Fermi level are partially filled so there are plenty of electrons to be scattered and plenty of empty states within an accessible energy range for the electrons to be scattered into. Similar arguments apply to hole conduction.

The quasi-Fermi levels, E_f^p and E_f^n, therefore define the pumping levels of the semiconductor laser. Comparison with a four-level system indicates that

$$E_f^n - E_f^p > E_c - E_v \qquad (15)$$

and indeed this is the condition derived by Bernard and Duraffourg (see Appendix 1). In short, the quasi-Fermi levels are pushed into the band edges so that all valence band states above E_f^p are filled with holes (i.e. empty of electrons) and all conduction band states below E_f^n are filled with electrons. Physically this means that the semiconductor is now transparent to band-gap

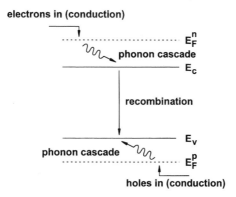

Figure 3.5. An equivalent four-level scheme for a diode laser with phonon cascade for the electrons and holes.

radiation because the minimum energy for an electron to be excited into the conduction band from the valence band by optical absorption is

$$E_{\min} = E_f^n - E_f^p. \quad (16)$$

The onset of population inversion is therefore also called the onset of transparency.

If it is assumed that the only electrons that recombine are those at the bottom of the conduction band then electrons injected at the quasi-Fermi level must lose kinetic energy in order to reach the band edge. This is achieved by the emission or absorption of phonons so that both energy and momentum are lost to the lattice. The electrons therefore "cascade" through the states to occupy those made empty by recombination (figure 3.5). Similar arguments apply to the hole current. This "cascade" process is extremely fast, of the order of a pico-second, in the extended states of the semiconductor. It has been shown by experiments on photo-pumped semiconductor lasers that the process of relaxation through the states is also extremely fast in the band-tails [2] which form the band edges in heavily doped semiconductors. In consequence there are always electron-hole pairs available for recombination and it is possible to get very high optical energy densities in semiconductor lasers.

3.3 The *p–n* homojunction laser

The *p–n* homojunction laser is relegated to history, but it serves to illustrate some of the principles enunciated above. The highly non-equilibrium conditions defined by equation (15) are achieved by means of a forward-biased *p–n* junction. The junction is needed because *n*-type semiconductors conduct electrons and *p*-type semiconductors conduct holes, but neither can simultaneously conduct large amounts of both. Carrier densities of at least 10^{18} cm^{-3} are needed to push the Fermi levels into the band edges. In fact the lower density of states in the conduction band for GaAs means that the electron density at which the Fermi level enters the conduction band is slightly lower than the hole density at which the Fermi level enters the valence band, but in a laser equal numbers of electrons and holes must be injected or else the laser is effectively charged. For very heavily doped n-type GaAs, then, the equilibrium hole density will be very small, at less than 1 cm^{-3}. Similarly, with very heavily doped p-GaAs the electron concentration will be very small, so population inversion cannot be achieved in either n-type or p-type material alone. However, if a junction between the two exists then under the right conditions enough electrons can be injected into the p-type material so that in the vicinity of the junction population inversion can be achieved.

If the Fermi levels on either side of the junction have been pushed into the band edges by doping on either side of the junction, then population inversion will always be achieved if enough current flows. The quasi-Fermi levels define the pumping levels, so if their separation is insufficient it is simply necessary to increase the current. These lasers required threshold currents of the around 10^5 A cm^{-2} before lasing occurred and so were operated at cryogenic temperatures. Imagine a piece of semiconductor 1 cm × 1 cm in cross-section with 10^5 A flowing through it. If even one volt is dropped across the whole of that semiconductor then 100 kW of electrical power would be dissipated, and without cryogenic cooling the lasers would simply melt. Alternatively, the lasers could be operated under pulsed conditions.

One of the consequences of such high currents is that the contacts to the semiconductor need to be very good. Fabricating good contacts to semiconductors is as much an art as a science, but successful methods have been established [3]. Ideally the electrons in the metal, which reside at the Fermi level, will lie at the same energy as the conduction band in the semiconductor so that a transition from one to the other is easy. However, the metal Fermi level usually lies somewhere in the semiconductor band gap (figure 3.6), so electrons moving from the metal to the semiconductor must gain energy. There is said to be a potential barrier. Thermal equilibrium requires ultimately that the Fermi levels within the two materials align, but this does not eliminate the barrier. Instead the potential is lowered on the semiconductor side of the contact to form a triangular barrier, the base of which is inversely proportional to the square root of the doping density [4]. Two approaches to low resistance contacts are therefore possible. One is to diffuse a metal in to make a graded junction so that sharp potential barriers do not exist, but finding the correct metal, or combination of metals, and the right processing conditions is the art. The second approach is to increase the doping density in the contact region so that the potential barrier is very thin, which increases the probability of a direct transition across the barrier, via quantum mechanical tunneling. Similar ideas apply to p-type contacts.

Injecting contacts of this sort are usually called "ohmic", but a better term is "low resistance". There is always some resistance associated with a contact,

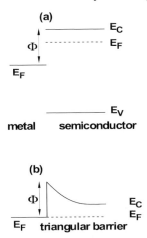

Figure 3.6. Formation of a barrier at a metal-semiconductor junction.

and the resistance might even vary with the voltage applied, so in this sense it is not ohmic. The resistance is made as small as possible, but even a moderately high resistance is not the reason for such high threshold currents. This is more to do with what happens to the electrons when they have been injected across the p–n junction, but clearly, in the presence of such large currents poor contacts will cause a problem. Indeed, one of the problems that always besets a new material system is the development of a technology for low resistance contacts.

The p–n junction itself is generally made by diffusing Zn in from the surface before any other processes such as the formation of contacts. It is a fact that group II elements (acceptors) generally diffuse through III-V semiconductors much faster than group VI elements (donors), so in fabricating a p–n junction the substrate is n-type and acceptors are diffused in to form a p-type layer on the surface [5,6]. The doping densities will typically be around 10^{20} cm^{-3} at the surface falling almost linearly for a deep diffusion (≈ 5 μm) to below 10^{19} cm^{-3} at the junction. Doping densities as high as this are needed to push the Fermi level into the band, so the semiconductor is not merely doped but "doped-plus". This is denoted as n^+ and p^+ respectively.

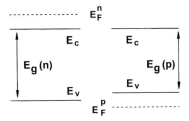

Figure 3.7. Band diagrams of heavily doped p and n-GaAs.

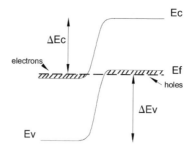

Figure 3.8. A heavily doped *p–n* junction in equilibrium.

The effect of heavy doping on the band-gap of *p*-GaAs has already been described; the valence band is lifted slightly as the dopant levels themselves form a band and join with the valence band, leaving deep-lying band-tails. In isolation the two sides of the junction will be as illustrated in figure 3.7 where the conduction and valence bands (E_C, E_V), and the respective Fermi levels are all shown. In contact, the Fermi levels must equilibrate by the diffusion of electrons from the n^+ to the p^+ and *vice versa* to establish thermal equilibrium (figure 3.8). A space charge of ionised dopants is exposed by this diffusion, leaving so-called "depletion regions" at the boundary in which there are effectively no carriers, and which, in consequence, are highly resistive. Moreover, the space charge gives rise to an electric field which establishes a drift current opposing the diffusion, and an alternative way of looking at the junction is to say that equilibrium has been established when the two currents are equal.

The existence of highly resistive depletion regions means that any voltage applied to a *p–n* junction is dropped across the junction itself. Usually the reverse bias is the main focus of attention in semiconductor physics texts, for the simple reason that the depletion regions extend under an applied voltage and the capacitance of the junction changes. This makes the *p–n* junction an important element in an integrated circuit but is of no consequence in a diode laser. Rather the forward bias, in which the voltage on the *n*-type is negative with respect the *p*-type, is more important. In this configuration charge is injected across the junction. The voltage dropped across the junction reduces the barrier, so the diffusion current exceeds the drift current. In an ideal case, the current I flowing in response to an applied voltage V_A is

$$I = I_O \left[\exp\left(\frac{qV_A}{kT}\right) - 1 \right] \quad (17)$$

where the negative term takes account of the current flowing in the opposite direction due to drift. Ideally this current is independent of bias but in practice may vary slightly. In forward bias the current is essentially exponentially dependent in V_A, which leads to very large densities of injected carriers.

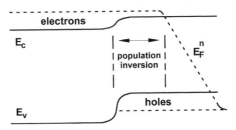

Figure 3.9. The junction under forward bias showing the active region.

However, the diffusion of holes and electrons is not necessarily equal, as the reduced band gap in the *p*-type leads to a larger barrier to hole injection than electron injection.

Equation (17) arises because of the barrier lowering under bias. Ordinarily, one would expect carriers of both types to be injected across the junction but it should be clear that the density of carriers available for injection depends on the doping density on either side. Thus, if one side of the junction, say *n*, is much more heavily doped than the other, say *p*, the density of electrons injected will far exceed the density of holes injected, and in fact the hole current can be ignored. This is the one-sided step junction. If this were not the case, the current-voltage relationship might be more complicated than equation (17) implies. Of course, in the *p–n* junction laser both sides are similarly doped and it would ordinarily be necessary to take into account hole injection as well as electron injection, but the larger barrier at the valence band means that to a first approximation the holes are confined to the p^+ side of the junction and the hole current is ignored.

The active region, i.e. the region of semiconductor over which population inversion exists (figure 3.9) is confined to the *p*-type semiconductor and is defined by the behaviour of the electrons once they are injected into the p^+ region. One side of the active region is formed by the valence band barrier at the junction itself, where the holes are confined, but the other side is defined by the diffusion of electrons away from the junction. The concentration of electrons will decrease with increasing distance due to recombination until eventually the electron population reaches the equilibrium population in the p^+ material. At this point the quasi-Fermi level for the electrons becomes identical to the Fermi level for the holes. Long before this point, however, the quasi-Fermi level drops below the conduction band edge and population inversion stops.

So far only two of the conditions for lasing have been met. There is a pumping source (the current) and an active medium (population inversion at the junction) but no optical feedback. Given that the semiconductor wafer is typically ~0.5 mm thick from top to bottom, aligning external mirrors such that optical radiation can be fed back into the semiconductor without substantial loss would seem to be very difficult, if not impossible. Fortunately,

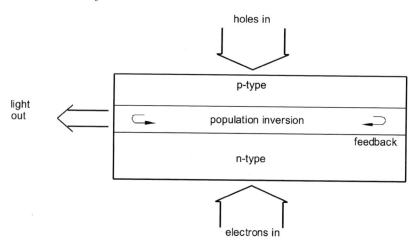

Figure 3.10. The basic *p–n* junction laser.

it is possible to fabricate the mirrors out of the semiconductor itself. Material of the right crystal orientation can be cleaved along crystal planes which are perfectly parallel to each other. The cavity is therefore formed in the plane of the junction and the device emits from the edge (figure 3.10). A semiconductor such as GaAs typically has a refractive index of about 3.5 at the wavelengths of interest so a cleaved facet has a reflectivity of the order of 30–40%. This is not as high as might be found in other lasers with external mirrors but it's high enough. For increased feedback one of the facets can be coated with aluminium.

The gain spectrum spans a broad range of wavelengths consistent with recombination between states of higher energy but with identical wave-vectors. In addition it is possible for states to violate the k-selection rule, as the conservation of momentum is sometimes called, thereby allowing other transitions and an even broader emission spectrum. Strictly, k-selection is not violated in the sense that no fundamental laws of physics are broken, but as some of the tail states at the band edges are localised there is a corresponding uncertainty in the momentum and strict k-selection cannot apply. In the lasers that superceded the homo-junction device the active region is not heavily doped and the issue does not arise. However, the choice or otherwise of k-selected transitions affects the density of states function used in modelling the gain, so the topic will be discussed further in chapter 4.

3.4 The active region and threshold current

It is necessary now to begin to quantify some of the ideas that have been described in order to appreciate the failings of the homo-junction device and the improvements that were made in 1968 with the development of the double heterostructure laser. We start with the threshold current density, in A cm^{-2}, required for the onset of laser action.

Recalling the discussion in chapter 2 on current continuity, and restricting the argument to the electron current only, but bearing in mind that similar expressions apply to the hole current,

$$\nabla \cdot J + q(G_n - U_n) = \frac{d\rho}{dt} \tag{18}$$

where $\rho = q \cdot n$ is the charge density. Therefore

$$\frac{1}{q}\nabla \cdot J + G_n - U_n = \frac{dn}{dt}. \tag{19}$$

At steady state $dn/dt = 0$, so

$$\frac{1}{q}\nabla \cdot J = U_n - G_n. \tag{20}$$

The gain per unit length, g, in the active region is related to the net number of recombinations across the whole of the active region,

$$g = c \int (U_n - G_n) \, dx \tag{21}$$

where c is some constant here. Therefore,

$$\frac{J}{q} = \int (U_n - G_n) \, dx = \frac{g}{c}. \tag{22}$$

This expression takes no account of other sources of loss in the electrical current such as that caused by a lack of electrical confinement. This is the singlemost important reason for the high threshold currents in these lasers. The active region is defined by diffusion and recombination, but even when the carrier density drops below the transparency density and the material becomes absorbing there is still a very high carrier density that just diffuses away and recombines. Intuitively, then, not only do we expect the gain to be proportional to the current, as above, but we also expect an offset current, because the current needs to be high enough to achieve population inversion before the gain is non-zero. Therefore

$$g = A(J - J_t) \tag{23}$$

where A is a constant. In fact, it can be shown that this relationship is exact [7] for an ideal semiconductor for $J > J_t$, where J_t is the offset current. J_t is not the threshold current of the laser, but is the current required to bring about non-zero gain in unit volume of the semiconductor. It is also called the transparency current. The condition for threshold requires not only that the gain be non-zero but also that the round-trip optical gain within the cavity is sufficient to overcome any optical losses. Assuming for the present that the only losses are those caused by reflection from the Fabry–Perot mirrors, with power reflection coefficients R_1 and R_2, then after a complete round trip of length $2L$, where L is the cavity length, the loss in the optical field is balanced by gain and

$$\exp(g \cdot 2L) = \frac{1}{R_1 R_2} \qquad (24)$$

and

$$g_{th} = \frac{1}{2L} \ln\left(\frac{1}{R_1 R_2}\right). \qquad (25)$$

This is a standard condition for any laser. Therefore

$$J_{th} = J_t + \frac{g_{th}}{A} = J_t + \frac{1}{2LA} \ln\left(\frac{1}{R_1 R_2}\right). \qquad (26)$$

The width of the population inversion region is defined by the diffusion length. However, it is not the spontaneous recombination lifetime but the stimulated recombination lifetime that determines the diffusion length over the active region. Stern [5] has provided a simple method of estimating the length of the active region.

The diffusion length, d, is

$$d = \sqrt{D\tau} \qquad (27)$$

where D is the diffusion coefficient and τ is the carrier lifetime. Bearing in mind that electrons are injected into p-type material, and that the hole concentration varies substantially away from the junction in line with the diffusion profile, Stern assumed that the lifetime is inversely proportional to the average hole density p_{av}, which is given by the product of the impurity concentration gradient across the junction and the diffusion distance, i.e.

$$\tau \approx \frac{1}{Bp_{av}}. \qquad (28)$$

However,

$$C = \frac{p(d) - p(0)}{d} \qquad (29)$$

from which

$$p_{av} = p(0) + \frac{C \cdot d}{2} \approx C \cdot d \qquad (30)$$

and

$$d \approx \left(\frac{D}{AB}\right)^{1/3}. \qquad (31)$$

Stern quotes a value of B at 77 K to be 4×10^{-10} cm^3/sec, and a diffusion coefficient for electrons of about 15 cm^2/sec. The concentration gradient is about 2×10^{23} cm^{-4} as the concentration in diffused junctions can change by about an order of magnitude over 5 μm. In some cases, if the diffusion is shallower, the gradient will be higher. Using these values $d \approx 1$ μm and $\tau \approx 1$ ns, which agrees with the value of ≈ 1.5 μm quoted by Stern from several experimental sources. As the condition for population inversion is essentially a condition on the electron density, it is necessary to inject a large number of electrons at a high rate in order to achieve a sufficient density over such a large volume. This is one of the principal reasons why the homojunction laser requires a threshold current of the order of 10^4–10^5 A cm^{-2}.

The lack of a well defined active region is one of the major reasons for the poor efficiency of the homojunction laser. The extent of the active region depends entirely on the diffusion properties of the electrons, rather than on any feature built into the device. Even when the electron density has decayed to a point where population inversion ceases there is still a significant density of electrons that must diffuse away and recombine spontaneously. It would be nice to be able to quote a single figure for the spontaneous recombination lifetime of GaAs but it is impossible to do so because quoted minority lifetimes vary greatly from device to device and from wafer to wafer. There are so many different forms of non-radiative recombination that can be introduced during processing that separating the intrinsic spontaneous radiative lifetime from others in the total minority carrier lifetime is simply not possible. What is clear, though, is that these excess electrons do not contribute to the laser output and are therefore a source of some of the excess current that has to be applied. Furthermore, they are a source of noise because any spontaneous emission arising from the recombination of these electrons has a random phase with respect to the stimulated emission.

3.5 Optical properties of the junction

It was discovered fairly early on the research into GaAs diode lasers that there exist small, but significant, changes in refractive index across the junction, due to various effects such as the variations in both the carrier density and the band gap. The origin of these refractive index changes is dealt with in chapter 4. Here it can simply be accepted that they occur and that the refractive index in the active region is higher than in the surrounding n^+ and p^+ material. An optical wave-guide is therefore formed in the plane of the junction which helps to guide the light along the active region between the cleaved facets. The refractive index changes are small (a few parts in a thousand), and the waveguiding effect is correspondingly weak, so the optical field spreads out considerably into the material on either side of the junction. Nonetheless it is advantageous to the operation of the laser because the intensity of the optical field inside the active region is enhanced, leading to a greater rate of stimulated emission.

Waveguiding in lasers is most often treated using the idea of a slab of material with lower refractive index on either side because it is the easiest to solve mathematically. In fact there are very few refractive index profiles that can be solved analytically and more often than not numerical methods are used to solve for the waveguiding properties of a particular layer structure. Appendix II contains a brief description of the ray propagation model of the slab waveguide, and shows that rays undergoing total internal reflection propagate at specific angles (modes) within the guide. It is convenient to think of total internal reflection occurring at the interface but in fact the ray penetrates some way into the cladding before being turned around.

An alternative approach is to imagine an optical field with a particular, well-defined lateral intensity distribution propagating down the guide, as shown in figure 3.11. The intensity distribution illustrated here corresponds to the fundamental mode of the guide, but higher order modes will have bimodal, trimodal, and other distributions. In general the fundamental mode will not be centred in the core unless the refractive index in both the cladding layers is identical. If one side has a much lower refractive index than the other the mode will be shifted over towards this side and away from the high refractive index difference on the other side. Essentially both the electromagnetic and ray pictures lead to the same end but the ray approach is more intuitive whilst the optical field model allows the intensity distributions to be more easily calculated, and in particular the distribution of light within the cladding layers. The mode field diameter is the same in both cases is defined as the limit where

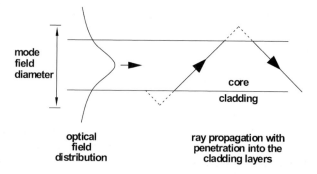

Figure 3.11. The electromagnetic model of the fundamental mode of a slab waveguide alongside the propagating ray model.

the intensity drops to 1/e of the intensity at the core/cladding interface, and is γ^{-1} and δ^{-1} respectively. These are none other than the ray penetration depths defined in appendix II.

The homojunction laser is weakly guiding so the light penetrates quite far into the cladding. In itself this is a source of inefficiency. Which ever way you look at it, either as an optical field with a fraction of the light propagating in the cladding, or as a ray which spends a fraction of its time within the cladding, there is a substantial fraction of light that is not in the active region and which cannot stimulate emission. Intuitively we might expect the gain to be diminished as a result of the loss of confinement so if the optical confinement factor Γ is defined as the ratio of integrated intensity within the core to the integrated intensity across the guide, the relationship between the gain and the threshold current (equation 23) can be modified as follows.

$$g = \Gamma A (J - J_t) \qquad (32)$$

from which

$$J_{th} = J_t + \frac{1}{2\Gamma L A} \ln\left(\frac{1}{\sqrt{R_1 R_2}}\right). \qquad (33)$$

The threshold current is therefore linearly dependent on the width of the guide and inversely dependent on the optical confinement. For strong guiding where Γ is close to unity, the threshold current is linearly dependent on the width of the active region, but for systems where Γ is reduced, the threshold current rises dramatically.

It is difficult to separate out the contributions of optical confinement and carrier confinement to the threshold current in the homojunction device because the active region is so ill-defined. What is certainly true is that the carriers are not confined and that the waveguide is weak. For both these reasons a large current needs to be supplied to establish transparency and to raise the photon density. The gain in the central region, the refractive index changes, and the width of the guide all depend on how hard the laser is driven, rather than on any specific features built into the device. The double heterostructure laser, on the other hand, has very definite carrier confinement and changes of refractive index built into it, which makes it much more efficient.

It remains to consider how optical losses, in particular absorption in the cladding regions, will affect the laser. The only real loss of intensity so far considered is the reflection loss at either end of the cavity. The loss of optical confinement may be a source of inefficiency but it is not strictly a loss. In principle, there should not be any other source of losses in the laser because the nature of population inversion shows clearly that the active region is transparent to the emitted radiation. However, it is an assumption that the gain is uniform along the cavity. Enhanced recombination at the facets might well lead to a loss of transparency as the carrier density is reduced. Moreover, radiation propagating within the cladding layers is not necessarily propagating within transparent material, and clearly a loss of intensity here reduces the intensity of the wave propagating down the guide.

The power in an electromagnetic wave propagating in the z-direction is given by the Poynting vector, $S_z = E \times H$. In a guide the power flow, P, is given by the integral over the guide cross-section, i.e.

$$P = \int_{-\infty}^{\infty} S_z \, dx = \frac{1}{2} \int_{-\infty}^{\infty} Re[E \times H^*]_z \, dx. \qquad (34)$$

It can be shown [8] that the total power is

$$P = \left(\frac{\beta}{2\omega\mu_o}\right) \frac{A^2}{2} \left(\frac{K^2 + \delta^2}{K^2}\right) \left[d + \frac{1}{\gamma} + \frac{1}{\delta}\right]. \qquad (35)$$

The last term in (36) is simply the effective width of the guide. The total power can therefore, be regarded as the sum of the powers flowing in each of the separate regions of the guide,

$$P = P_1 + P_2 + P_3 \qquad (36)$$

and if the total attenuation is αP then

$$\alpha = \frac{P_1\alpha_1 + P_2\alpha_2 + P_3\alpha_3}{P} \qquad (37)$$

which is what might be expected intuitively; the attenuation coefficients in each section of the guide α are simply weighted according the respective fractional powers in the different regions of the guide.

The attenuation coefficient in each section of the guide is defined by the complex refractive index, which is described in detail in the next chapter. Here it will simply be recognised that a complex refractive index $n_j + ik_j$ exists in each section of the guide (denoted by $j = 1, 2, 3$) where k_j is called the extinction coefficient. The attenuation coefficient is then

$$\alpha = 2k_j k_0 \qquad (38)$$

where k_0 is the wave vector. Given transparency in the active region, you might think the extinction coefficient in the core, k_1, should be zero, but in fact it is negative. Gain can be considered as negative absorption as the light intensity grows with distance travelled. In the cladding region, however, the radiation corresponds to the band edge or slightly higher energy on the p-side of the junction. The extinction coefficient will be small, at around 0.06, corresponding to an absorption coefficient of $10^4 \, cm^{-1}$, which is typical of the absorption coefficient at the band edge. In the homojunction laser, then, this can be understood as a further source of loss. Light must propagate the entire length

3.6 Output characteristics of the homojunction laser

The output of a diode laser will usually consist of several discrete wavelengths (figure 3.12). These wavelengths constitute the longitudinal modes of the laser, or the modes of the Fabry–Perot cavity, as the arrangement of two parallel reflectors is called. Self-selection of the Fabry–Perot modes lying closest to the centre of the gain curve has been described at the beginning of this chapter, but additionally the gain spectrum narrows in a diode laser as steady state is approached. In fact, it is one of the features of homojunction lasers that not only does the spectrum narrow but the emission wavelength changes with the driving current as the gain spectrum itself changes. For example, Nathan [9] reports laser emission at shorter wavelengths than the spontaneous emission spectrum below threshold.

In order to understand this phenomenon, consider two electrons at differing energies; one that has just been injected at the quasi-Fermi level and the other at the bottom of the conduction band. Either of these can undergo spontaneous or stimulated emission. When the laser is switched on and the carrier density rises to its maximum value over a period of time defined by the RC time constant of the device, and while it is increasing the photon population is low. The photon population lags behind the carrier population and has no serious damping effect on the electron density. Spontaneous emission occurs, and is amplified, and the photon population begins to rise. The very nature of spontaneous emission means that the spectrum will contain a wide range of

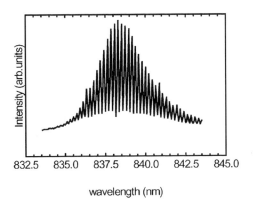

Figure 3.12. Spectrum of a laser diode at relatively low forward power (after Nathan *et al.* [9]).

wavelengths corresponding to all the possible transitions. The high carrier density means that there will be high gain within the system and entire spontaneous emission spectrum can be amplified. Our two electrons at the different energies can both make stimulated transitions.

With the passage of time, however, the electron injected at the quasi-Fermi level can take one of two paths to recombination. It can recombine directly from the quasi-Fermi level or it can cascade through the conduction band states to replace the electron at the bottom of the conduction band which has also recombined. The rate at which electrons can move through the states by phonon emission is exceedingly fast – it takes a few picoseconds at most – so unless there is a very, very high density of photons at this energy (the separation of the quasi-Fermi levels) to create a rate of stimulated emission which is comparable, the electrons will tend to decay by phonon cascade. This of course feeds back. There will be fewer electrons recombining by stimulated transitions so the photon density at this energy will be smaller, and hence the rate of stimulated recombination decreases. The phonon cascade process therefore becomes even more favourable. By this process emissions at the periphery of the gain spectrum decay away with time and the output spectrum of the laser narrows considerably. Moreover, the carrier density is reduced by the tendency for the upper levels to depopulate rapidly as the photon density rises. The steady state carrier density is much lower than the maximum and in fact tends to be pinned at threshold during the operation of the laser.

In relation to the density of states, figure 3.13 shows the carrier densities in the conduction and valence band as products of the density of states (ignoring band tails) and the Fermi function. The initial current injection causes

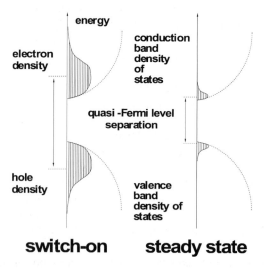

Figure 3.13. Contraction in the carrier density from switch-on to steady state.

the quasi-Fermi levels to penetrate both bands so that the maximum carrier density occurs well away from the band edge. The gain over an interval of energy, dE, can be expected to be proportional to the joint carrier densities,

$$g(E) \cdot de \propto \rho_c(E) f_c(E) \rho_v(E)(1 - f_v(E)) \, dE \qquad (39)$$

and the maximum gain will also occur well away from the band edge. High energy transitions from the extreme carrier densities at the exponential tail of the Fermi function are possible but the gain will be small. As described above, such transitions will occur immediately after the laser has been switched on, but they will not be sustained in competition with transitions at higher gain. Moreover, as the density of carriers is depleted through stimulated emission and steady state is achieved the carrier density profile will be compressed into a smaller range of energies and the gain maximum will lie closer to the band edges.

The carrier density overshoot actually causes oscillations in the light output over a period of time. There is a delay between switch-on and the build up of charge within the laser, as described above, but there is a further delay between the build-up of charge and the photon density within the Fabry–Perot cavity. There is a time delay before the photon density is high enough to reduce the carrier density by stimulated emission so, as described above, the carrier density momentarily exceeds the steady state density. There is a correspondingly large gain so when the photon density begins to increase the maximum also exceeds that found in equilibrium. The carrier density is thus depleted, and actually falls below the threshold level. The low gain means that the photon density also begins to fall. The carrier density then begins to rise again, and both the carrier and photon densities oscillate until steady state is achieved and the two are balanced. This is illustrated schematically in figure 3.14, where, instead of the photon density, the light output is shown. Note the light output is a maximum when the carrier density is a minimum. These oscillations in light output are known as "relaxation oscillations".

You might be forgiven for thinking that these ideas are not relevant to modern devices because, as mentioned elsewhere, the homojunction laser was rapidly superceded by the double heterostructure laser. The disorder in the band structure within the homojunction device undoubtedly contributes to its complicated output characteristics, but the double heterostructure is no less susceptible to relaxation oscillations during which the output spectrum is broad. Of course, in steady state the spectrum is narrow, but diode lasers are not always operated in steady state. In optical communications, for example, the successive increases in modulation frequency over the years mean that today's diode lasers may be modulated at tens of GHz. The diodes are on for such a short time that steady state is not established, and even for modulation at 100s of MHz the pulse length coincides with the duration of the relaxation oscillations. The output spectrum of the device is correspondingly broad, and,

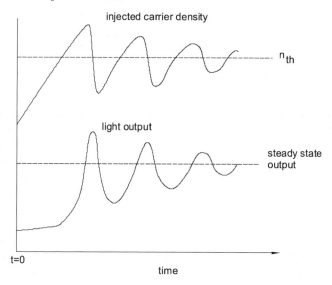

Figure 3.14. Oscillation of the carrier and photon densities at switch-on.

as discussed in chapter 9, a great deal of research has taken place to improve the spectral characteristics of lasers for communications.

The diode laser is generally a much broader source than other laser types. The diode laser illustrated in figure 3.12 has a linewidth

$$\Delta \nu = -\frac{c}{\lambda^2} \Delta \lambda \tag{40}$$

so taking the wavelength span to be approximately 5 nm, $\Delta \nu \approx 2$ THz. It is possible to reduce the linewidth by driving the laser harder, and indeed for this particular device the authors showed that the output reduced to a single mode at ~ 837 nm, but even a single mode device will have a large bandwidth relative to other types of lasers. The Schawlow–Townes equation describes the fundamental limits on the linewidth,

$$\Delta \nu = \frac{h\nu}{(2\pi \tau_c^2 P)} \tag{41}$$

where $\Delta \nu$ is the linewidth in Hertz, P is the output power, and τ_c is the cavity decay time. The cavity decay time may be estimated from the loss in the cavity and at the mirrors. The total loss is usually expressed in dB/cm and is simply the sum of the mirror loss and the cavity loss

$$\alpha_{tot} = \alpha_c + \alpha_m \tag{42}$$

where the mirror loss can be calculated from the power reflection coefficients at each end. If these are R_1 and R_2 then after one complete round trip the loss due to the mirrors is simply the product of these two. The loss (in cm^{-1}) is

$$\alpha_m = -\frac{1}{2L} \ln(R_1 R_2) \quad (43)$$

the characteristic decay length is simply $1/\alpha_m$, and the decay time can be taken approximately as the time to traverse this distance (equation chapter 8, 31 gives the exact expression for the cavity decay time). For example, for a cavity 300 μm long with power reflection coefficient of 0.35 at each end, and a refractive index ≈ 3.5 inside the cavity, the decay time is ≈ 3 ps. For a 5 mW output power at 830 nm the minimum linewidth is 850 kHz. For an equivalent solid state laser the cavity decay time may be a factor of greater 10^3, leading to a theoretical linewidth of about a Hz.

The linewidths quoted above are very much theoretical. In fact, the linewidth of a diode laser is enhanced by a factor $(1 + \alpha^2)$, where α is known as the linewidth enhancement factor and is typically in the range 2–4. Chapter 10 describes the linewidth enhancement factor in greater detail. Practical linewidths of a few THz are much greater than the Fabry–Perot mode spacing, which is why, in figure 3.12, the Fabry–Perot modes are superimposed onto a broad (~ 5 nm) spectrum.

The Fabry–Perot mode spacing is defined by equation (5), which can be differentiated to give

$$\Delta \lambda = \frac{\lambda^2}{2 \cdot n \cdot L}. \quad (44)$$

The mode numbers are of the order of a few thousand for near infra-red lasers with cavity lengths of a few hundred micrometres, and mode spacings are fractions of a nanometre.

3.7 Summary

The fundamentals of laser oscillators have been described. The process of mode selection has been described as a result of the interaction between feedback and gain. Within this process the gain spectrum is fixed, but in the diode laser the gain spectrum also shrinks as a result of the interplay between the photon and carrier densities just after the laser is switched on. This is an entirely general result for diode lasers that operate by radiative recombination into a Fabry–Perot cavity, and is not just applicable to the p–n homojunction laser. The basic physics of homojunction devices has been described in order to illustrate the practical reality of population inversion, threshold currents, and linewidth in lasers. The mechanisms contributing to the very high threshold current have been identified as; the lack of electrical confinement, the lack of optical confinement; and losses in the waveguide caused by the propagation of light in the narrower band gap *p*-type GaAs.

3.8 References

[1] Feynman R P, Leighton R B and Sands M 1965 *The Feynman Lectures on Physics* vol III, ch 4–7 (Addison Wesley)
[2] Holonyak N and Lee M H 1966 *Semiconductors and Semimetals* vol 1 eds R K Willardson and A C Beer (London: Academic Press)
[3] Murakami M and Koide Y 1998 *Critical Reviews in Solid State and Materials Sciences* **23** 1–60
[4] Sze S M 1981 *Physics of Semiconductor Devices* 2nd edition (New York: John Wiley & Sons) chapter 5
[5] Stern F 1966 *Semiconductors and Semimetals* vol 2 eds R K Willardson and A C Beer (New York: Academic Press) pp 371–411
[6] For details of diffusion processes see S M Sze 1988 *VLSI Technology* 2nd edition (McGraw-Hill)
[7] Thompson G H B 1980 *Physics of Semiconductor Laser Devices* (Chichester: John Wiley & Sons) p 83
[8] Marcuse D 1974 *Theory of Dielectric Optical Waveguides* (NY: Academic Press)
[9] Nathan M I, Fowler A B and Burns G 1963 *Phys. Rev. Lett.* **11** 152–154

Problems

1. Assuming a refractive index of 3.4 and a cavity length of 500 μm, calculate the mode number corresponding to emission at around 830 nm, and calculate also the exact emission wavelength of this mode.
2. Calculate the mode spacing for the laser described above.
3. Taking the power reflection coefficient to be 0.3 at each facet calculate the effective mirror loss and the cavity decay time. Hence calculate the Schawlow–Townes linewidth for a power output of 1 mW.
4. If one of the mirrors is coated with Al and has a reflectivity of 98% as a consequence, calculate the effect on the mirror loss and cavity decay time.
5. Thirty percent of the light propagating in a homojunction cavity is propagating in a material with an average extinction coefficient of 0.06. If the wavelength is 830 nm calculate:
    i. the absorption coefficient of the material;
    ii. the contribution of this to the cavity loss;
    iii. the total cavity loss, assuming the mirror loss is the same as in (3).
6. Assume an average electron density of 8×10^{17} cm^{-3} is needed to establish transparency over an active region 1.5 μm wide. If the stimulated lifetime is 2 ns calculate the transparency current density. Neglect the effects of carrier diffusion beyond the active region.

7. Assuming the cavity loss in (5) above calculate the threshold current for the laser in (6). Take $A = 0.045$ cm/A (equation 23) and ignore the effects of optical confinement.
8. Taking the same average transparency density and recombination time as in (6), assume that the carriers are confined to an active region 200 nm thick and that there is no loss in the cavity other than the mirror loss. Calculate the transparency and threshold current densities.

Chapter 4

Optical properties of semiconductor materials

4.1 A model of the refractive index

As chapter 3 showed, the refractive index is not a given property that remains constant under changing conditions of wavelength, current flow, carrier density, or even optical intensity, and the purpose of this chapter is to provide some insight into these phenomena. A detailed theoretical treatment of the subject is not necessary; instead the essentail physics of the problem is described in a phenomenological way, leading finally to a quantum treatment of optical transitions and absorption.

Let's take as the starting point radiation incident upon a surface. For non-normal incidence the electric vector of the incident light can either lie parallel to the interface or in the plane of incidence (figure 4.1). Electrostatics determines the continuity conditions at the interface. For a parallel electric vector the electric fields immediately adjacent to the boundary on either side must be equal. For an electric vector lying in the plane there will be a component perpendicular to the interface and for this component the displacement vector D is given by the continuity equation

$$\nabla \cdot D = \rho. \tag{1}$$

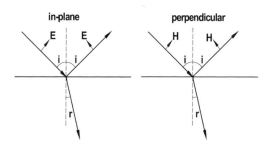

Figure 4.1. Electric and magnetic vectors lying in the plane of incidence or parallel to the interface.

A model of the refractive index

In short, D is continuous unless there is an interfacial charge density, in which case D is discontinuous by this amount. Mathematically, $\nabla = d/dx$ in one dimension, but it is important to distinguish physically between the *grad* and *div* operators. Both are vector operators, but *div* acts on a vector (in this case D) to produce a scalar (in this case ρ) whereas the *grad* operator acts on a scalar, for example potential V, to produce a vector such as the electric field E. Physically, this reflects the idea that an electric field either originates or terminates on charge, so if there is no charge present the displacement vector remains unchanged on crossing the interface. However

$$D = \varepsilon_0 \varepsilon_r E \tag{2}$$

where ε_0 is the permittivity of free space, and ε_r is the relative permittivity so the electric field changes across the interface in proportion to the respective dielectric constants.

If the boundary is between vacuum (or air) and a metal in which there are a lot of free electrons we might expect a large surface charge to exist. If, on the other hand, the boundary is between vacuum and an insulator (dielectric) then there will be virtually no charge except that due to polarisation. Polarisation of the dielectric will lead to the existence of some charge at the surface but there will be considerably less than in a metal. To all intents and purposes the displacement vector can be taken to be continous and the electric vector to be discontinuous. These boundary conditions determine the amount of light transmitted and so we might expect intuitively that they also determine the amount of light reflected from a surface. If so, then ε_r must be related to the refractive index.

The wave propagating through the medium will have an electric vector which is perpendicular to the direction of propagation oriented at some angle to the crystal planes. If the material is anisotropic we should consider the propagation in terms of tensors, but we will make the simplifying assumption that the material is homogeneous and isotropic, and therefore dispense with all considerations of directionality. The electric field will polarise the material and from classical electrostatics

$$D = \varepsilon_0 E + P \tag{3}$$

where, for uniform isotropic material

$$P = \varepsilon_0 \chi E \tag{4}$$

is the polarisation. The constant, χ, is the macroscopic polarizability, which is simply the product of the microscopic polarizability α and the number of polarizable charges, N.

$$\chi = N\alpha. \tag{5}$$

From this we obtain the result that

$$\varepsilon_r = 1 + N\alpha \tag{6}$$

which links the dielectric permittivity to the microscopic response of the material to the electric field.

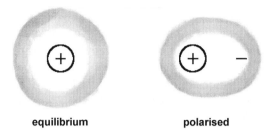

Figure 4.2. Schematic illustration of the polarisation of the electron cloud around an atom.

It remains to show the relationship between ε_r and n. Let us assume that the microscopic polarizability is entirely electronic, that is we have atoms consisting of positive nuclei and surrounding charges only. In an ionic crystal this will not necessarily be true. A polarisation could arise from the oscillation of ions about their equilibrium position, and indeed there are frequencies within the electromagnetic spectrum where such modes of lattice vibration are excited. However, these are mainly in the infra-red, well below the energy corresponding to band-gap radiation. We will assume instead that the only polarisation of interest lies within the atom; that the distribution of charges around the atom is perturbed by the electric field (figure 4.2). Since the electric field oscillates the electronic motion should also oscillate. An electron orbiting a nucleus is a bound system, and it is possible to show by expanding the potential energy in a Taylor series that any such system will undergo simple harmonic motion (SHM) if subjected to a small perturbing force.

The problem of the electron motion is therefore reduced to a problem of SHM, the solutions to which are well known. You may question whether this is physically realistic. SHM requires a force that increases linearly with the displacement from the central position and is directed towards the centre. However, the attractive force between the nucleus and the electron is coulombic and depends inversely on the square of their separation. At first sight it appears that the condition for SHM cannot be met, but the attractive force is offset by the screening effect of the other electrons between the outermost electron and the nucleus. These will repel the electron if it is forced out of its equilibrium position toward the nucleus. Thus, over a limited range of motion the net effect of these two forces approximates to the linear restoring force that SHM requires. The oscillating electric field due to the electromagnetic wave will drive the electron sinusoidally, and in the general case some mechanism will exist to damp down the motion. Then the equation of motion corresponds to that of a forced, damped, harmonic oscillator

$$m \frac{d^2 y}{dt^2} + m\gamma \frac{dy}{dt} + m \cdot \omega^2 y = qE_y \cos(\omega t) \qquad (7)$$

A model of the refractive index

where E_y is the electric field in the y direction, γ represents the magnitude of the damping force, and m is the mass of the electron. This is a classical approach and means that the quantum nature of the atom is ignored for the moment.

For an isolated atom, the atomic polarisability, α, is just the dipole moment, qy, induced by the electric field, where y, the displacement, is the solution to equation (7)

$$x = \frac{q_e}{m\varepsilon_0} \frac{1}{-\omega^2 + i\gamma\omega + \omega_0^2} \quad (8)$$

hence [1]

$$\alpha = \frac{q_e^2}{m\varepsilon_0} \frac{1}{-\omega^2 + i\gamma\omega + \omega_0^2}. \quad (9)$$

The relative permittivity now becomes

$$n^2 = 1 + \frac{Nq_e^2}{m\varepsilon_0} \frac{1}{-\omega^2 + i\gamma\omega + \omega_0^2}. \quad (10)$$

This expression contains nearly all the physics needed to understand the essence of the refractive index in semiconductors. It's not in the final form that would be used in practice because in a real solid there are charges other than those bound to the atoms, but whilst this modifies slightly the right hand side of equation (10) the left hand side is still essentially the resonance condition of a damped harmonic oscillator. Therefore nothing new is gained from making these adjustments apart from some calculational accuracy. Sticking with equation (10), then, the permittivity (n^2) is complex, arising from the term in $i\gamma\omega$. This is a damping term and limits the magnitude of the electronic displacement at the resonant frequency ω_0, as can be seen in equation (11), which has been rationalised to bring out more clearly the complex nature of n^2.

$$n^2 = 1 + \frac{Nq_e^2}{m\varepsilon_0} \frac{(\omega_0^2 - \omega^2) - i\gamma\omega}{(\omega_0^2 - \omega^2)^2 + (\gamma\omega)^2}. \quad (11)$$

The relative permittivity, $\varepsilon_1 + j\varepsilon_2$, and its square root, refractive index $n_r - jk$, of a dielectric with a resonant frequency $\omega_0 = 10^{15}$ and $\gamma = 3.5 \times 10^{14}$ s^{-1} is shown in figures 4.3 and 4.4. The meaning of γ will become clear later, but it has the dimensions of frequency and has to have value close to the resonant frequency in order to have any damping effect. If γ is too low then ε_1 will be very large at resonance. Equation (10) is strictly valid for isolated atoms, as described above, and if realistic atomic densities ($N \approx 10^{29}$ m^{-3}) are included in the pre-factor it becomes too large to be representative of a solid. Therefore a reduced density $N \approx 10^{27}$ m^{-3} is chosen, which, along with these values of γ and ω_0, give something reasonably representative of a solid.

The resonant nature of the model is clearly shown by the peak in ε_2 at the resonant frequency, and by the appearance of both normal dispersion, where n_r

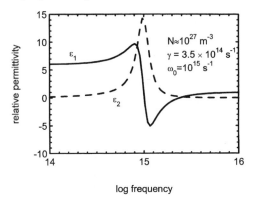

Figure 4.3. The real and imaginary parts of the relative permitivity for a set of independent oscillators with a single resonance.

increases with increasing frequency below the maximum, and anomalous dispersion, where n_r decreases with increasing frequency above the maximum. Resonances such as these are a strictly classical phenomenon, but they correspond to quantum transitions between levels [2]. That is, whereas energy is absorbed by the oscillators from the driving electric field, in quantum terms a photon is absorbed and the electron makes a transition to a higher energy level. In so far as the resonant model is the classical analogue of a quantum model of the atom, it is valid. However, in the quantum model an electron can make several transitions to higher lying energy levels and therefore more than

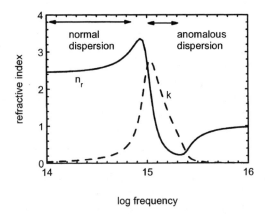

Figure 4.4. The real and imaginary parts of the refractive index corresponding to the relative permitivity in figure 4.3.

one resonance must be included. Each transition m has its own resonant frequency ω_{0m}, and a corresponding strength factor f_m, so that [1]

$$n^2 = 1 + \frac{Nq_e^2}{m\varepsilon_0} \sum_m \frac{f_m}{-\omega^2 + i\gamma_m\omega + \omega_{0m}^2}. \tag{12}$$

As described above, the permittivity is complex, and the refractive index, which is also complex, is given by

$$n_r = \left[\frac{|\varepsilon| + \varepsilon_1}{2}\right]^{1/2} \tag{13}$$

and

$$k = \left[\frac{|\varepsilon| - \varepsilon_1}{2}\right]^{1/2} \tag{14}$$

where

$$|\varepsilon| = [\varepsilon_1^2 + \varepsilon_2^2]^{1/2}. \tag{15}$$

Conversely,

$$\varepsilon_1 = n_r^2 - k^2 \tag{16}$$

and

$$\varepsilon_2 = 2n_r k. \tag{17}$$

The damped harmonic oscillator forms the basis of nearly all theoretical models of the refractive index. For example, figure 4.5 shows the real and imaginary parts of the relative permittivity of silicon [3] together with the results of one such model [4]. A model such as this is relatively easy to programme and provides a good description of the refractive index over a wide range of photon energies. The similarity to figures 4.3 and 4.4 is immediate, but clearly there is more than one transition, which reflects in part the extra complexity of equation (12), but also the fact that in a solid the transitions occur not between discrete energy levels but between bands and that several direct band gaps exist in solids at higher energies.

The agreement between the theory and the data is very good over a wide range of photon energies except in the imaginary part of the permittivity at low energy, and this seems to be a general feature not only of this model but theoretical models in general. Forouhi and Bloomer, for example, formulated a model for the refractive index directly in terms of the damped harmonic oscillators [5], and the results for GaAs are shown in figure 4.6, together with experimental data [6]. The model appears to match the data better at small k but it should be borne in mind that the values are nearly zero here and quite large

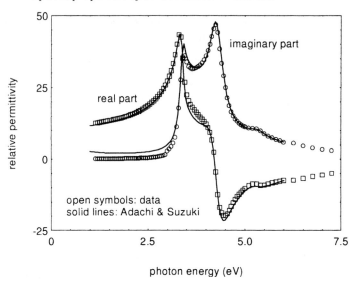

Figure 4.5. Comparison of the modelled dielectric function [3] against experimental data [4] for silicon.

discrepancies will not be apparent on this scale. The Forouhi–Bloomer model does in fact perform better than other models in this respect but it is not accurate enough, or perhaps appropriate, to model the absorption, and conversely the gain, at the band gap. A quantum mechanical approach to these

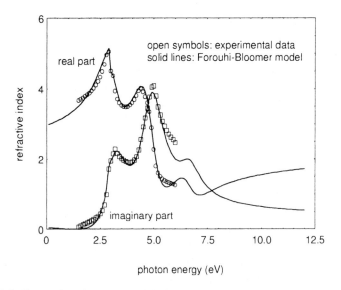

Figure 4.6. Comparison between model [5] and experiment for GaAs [6].

transitions will be developed later in this chapter. Before doing so, the relevance of the foregoing to the Fabry–Perot cavity in a laser will be discussed.

4.2 The refractive index of a semiconductor laser cavity

Equation (10) contains all the essential physics necessary to understand the changes in refractive index occurring in a semiconductor under population inversion. Three things need to be considered; the resonant frequency, the role of free carriers, and the effect of gain.

As figures 4.5 and 4.6 show, there are several resonances, but only the resonance at band gap, defined by

$$E_g = \hbar \omega_o \tag{18}$$

is important here. For sub-band gap radiation ($\omega < \omega_0$) an increase in the resonant frquency ω_0 will cause a decrease in the extinction coefficient k. This corresponds to the physical reality that band filling shifts the transition energy upwards and the semiconductor becomes transparent. A small change in k causes a change in the real part of the permittivity of $-2 \cdot k \cdot \delta k$, and a change in the imaginary part $2 \cdot n \cdot \delta k$. For $k \ll n$ and close to zero the change in the imaginary part far exceeds the change in the real part. For $\delta k < 0$, i.e. a decrease, the modulus of the permittivity will be slightly reduced and the real part of the refractive index will therefore decrease slightly in consequence. This interplay between the real and imaginary parts of the refractive index is hard to imagine if the two are seen in isolation, but as they are derived from the complex permittivity it should be clear that a change in one part of the refractive index cannot occur without a change in the other.

It is a general rule that the refractive index decreases with increasing bandgap, which is very useful for the double heterostructure device in which the active region is sandwiched between two layers of wider band gap material. Figures 4.7 and 4.8 show the refractive index of GaAs and AlGaAs alloys [7]. The first resonance corresponds to band gap radiation. It is not pronounced because it sits in the normal dispersion tail of the second resonance which clearly moves to higher energy as the aluminium content (and with it the band gap) is increased. Normal dispersion is associated with a decrease in the refractive index with increasing resonance energy for a given photon energy. For example, at 1.5 eV, corresponding roughly to the band gap of GaAs, the refractive index changes from ≈ 3.6 in GaAs to ≈ 3.2 in $Al_{0.5}Ga_{0.5}As$. Embedding the active region between layers of wide band gap AlGaAs will therefore result not only in large refractive index differences, but from figure 4.8 the extinction coefficient will be zero and the waveguide cladding layers will be transparent. In principle, then, the only losses of significance within the laser cavity should be the mirror losses.

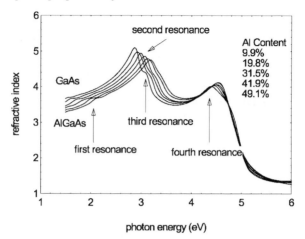

Figure 4.7. Real part of the refractive index for GaAs and AlGaAs alloys over a wide range of photon energy [7].

Contrast this with the *p–n* homojunction device in which absorption in the *p*-type layer will be significant. Not only this, but there is a reduction in band gap in going from *n* to *p* and we would expect, therefore, the refractive index on the *n*-side of the junction to be slightly lower than that on the *p*-side. Indeed this is the case, so the waveguide, apart from being weakly guiding, is also asymmetric. In the junction itself, the onset of threshold, also known as the onset of transparency, will cause the resonant frequency to be increased, which

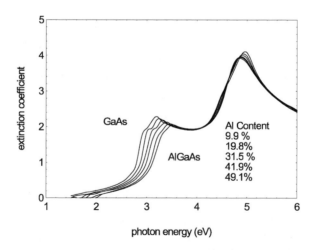

Figure 4.8. Imaginary part of the refractive index of GaAs and AlGaAs alloys over a wide range of photon energy [7].

will tend to decrease n_r. At first sight, then, this decrease would seem to rule out the possibility of waveguiding, but in fact the situation is more complicated than this. The effect of free carriers has to be considered.

Both the *p*- and *n*-sides of the junction are heavily doped and therefore have high carrier densities, almost to the point of rendering the semiconductor metallic. In addition, the high currents required for operation of the laser means that the carrier density will be further increased and we want to know the effect of this on the laser. There are two things to note in the case of high carrier densities; first, the correction to the electric field made for a dense dielectric does not need to be applied because the charges in a metal are mobile and will average out; and second, the electrons are not bound to atoms so by definition there is no restoring force and no resonant frequency. Therefore equation (10) can be used with the following modifications,

$$n^2 = 1 + \frac{Nq_e^2}{m\varepsilon_0} \frac{1}{-\omega^2 + i\gamma\omega}. \tag{19}$$

At first sight it would appear that increasing N will increase the refractive index, but that will only be the case if $\gamma\omega > \omega^2$. This term needs to be examined more closely.

The term in γ represents damping, and for a free electron this must be related in some way to the resistance. In a damped harmonic oscillator, the resistive force on the motion responsible for the damping normally is proportional to the velocity, i.e.

$$F_r = \gamma m \frac{dx}{dt} \tag{20}$$

which is equal to the accelerating force due to the electric field, i.e.

$$F_r = qE. \tag{21}$$

This arises because in normal conduction the electrons travel with an average velocity and there is no net acceleration, and hence no net force, on the electron. We can say therefore, if the mean time between collisions is τ, then the drift velocity of the electrons is v_{drift},

$$v_{drift} = \frac{dx}{dt} = \frac{qE}{m}\tau \tag{22}$$

and

$$\gamma = \frac{1}{\tau} \tag{23}$$

which gives us a physical insight into the meaning of γ in materials with high densities of free carriers.

The conductivity is given by

$$\sigma = Nq\mu \tag{24}$$

where N is the concentration of electrons and the mobility, μ, is

$$\mu = \frac{v_{drift}}{E}. \qquad (25)$$

The combination of equations (23), (24), and (25) therefore leads to the result

$$\sigma = \frac{Nq^2}{m}\tau \qquad (26)$$

and hence

$$n^2 = 1 + \frac{\sigma/\varepsilon_0}{i\omega(1+i\omega\tau)}. \qquad (27)$$

For the situation that $\omega\tau > 1$, which will be the case at the optical frequencies of interest to us, equation (27) approximates to

$$n^2 = 1 - \frac{\sigma}{\varepsilon_0 \omega^2 \tau} \qquad (28)$$

which shows that for an increase in σ the refractive index will decrease and in a metal it can even be negative.

A semiconductor is not a metal, so the refractive index will lie somewhere in between the two extremes of a dielectric and a metal. We can expect contributions from dielectric components (equation (10)) and metallic components (equation (28)). Hence we can expect that an increase in carrier density will reduce the refractive index of the semiconductor. This has been confirmed by experimental observation and is an important contribution to the refractive index changes across the junction. The very high density of free electrons on the *n*-side of the junction therefore depresses the refractive index, and even on the *p*-side the effect of decreasing the band gap is offset somewhat.

It is worthwhile at this point discussing further the effects of free carriers, which are most most noticeable in the infra-red, first through the free-carrier absorption, and second, at much longer wavelengths, through the surface plasma oscillations, known as plasmons. This is not strictly relevant to the Fabry–Perot cavities under discussion but does have implications for other lasers. Free carrier absorption is seen mostly in long wavelength vertical cavity devices (chapter 7) where thick, highly conducting layers form both the cavity mirror and the contact. Plasmons are used to good effect in long wavelength quantum cascade lasers (chapter 12), where the refractive index associated with them is important for the design of the waveguide. In a metal a plasmon is the quantum of a collective longitudinal excitation of the conduction electron gas, i.e. $\hbar\omega_p$, where ω_p is the plasma oscillation frequency, defined as

$$\omega_p^2 = \frac{Nq^2}{\varepsilon_0 m}. \qquad (29)$$

In a dielectric, the entire valence electron sea oscillates back and forth relative to the ion cores [8]. Substituting (29) into (28) gives

$$n^2 = 1 - \frac{\omega_p^2}{\omega^2} \tag{30}$$

and if, as before, there is some background dielectric constant from the ion cores that extends well out to optical frequencies, say $\varepsilon(\infty)$, then

$$n^2 = \varepsilon(\infty) - \frac{\omega_p^2}{\omega^2}. \tag{31}$$

For frequencies close to the plasma oscillation frequency the permittivity becomes small and may even become negative for long wavelengths.

The free carrier absorption can be treated by re-arranging (27) to give

$$n^2 = \varepsilon(\infty) - \frac{\omega_p^2}{\omega^2 + \gamma^2} - i\frac{\omega_p^2 \gamma}{\omega(\omega^2 + \gamma^2)}. \tag{32}$$

The first two terms are recognisable as the real dielectric constant (31) for $\gamma = 0$ and the third term is the imaginary dielectric constant which gives an extinction coefficient,

$$k = \frac{\omega_p^2 \gamma}{2\omega^3 \sqrt{\varepsilon(\infty)}} \tag{33}$$

and hence the absorption coefficient

$$\alpha = \frac{4\pi k}{\lambda} \propto \lambda^2. \tag{34}$$

The absorption coefficient increases with the square of the wavelength, which is why the effect is most noticeable in the infra red.

The final effect to consider is gain. An electromagnetic wave propagating through the semiconductor will have the following form

$$\psi_x = \psi_0 e^{i(\omega t - nk_0 x)} \tag{35}$$

where n is the refractive index and k_0 is the wave-vector. Expansion of the refractive index into real and imaginary parts gives

$$\psi_x = \psi_0 e^{-kk_0 x} e^{i(\omega t - n_r k x)} \tag{36}$$

which shows that the imaginary part of the refractive index is associated with attenuation of the wave, i.e. absorption. Under the conditions of stimulated emission, however, the wave is not attenuated but grows exponentially, so gain is "negative absorption". The fact of gain, i.e. a change in the magnitude and sign of k, must affect n_r. It is not immediately obvious from any of the foregoing precisely what the effect will be, and in fact it is necssary to use a different treatment to calculate the changes. This is the Kramers–Kronig transformation, in which a change in the real part of the permittivity, $\delta\varepsilon_1(\omega)$, can be evaluated from the change in the imaginary part, $\delta\varepsilon_2(\omega)$, over the range $\omega - x_0$ to $\omega + x_0$ [9]

$$\delta\varepsilon_1(\omega) \approx -\frac{1}{\pi} \int_{\omega-x_0}^{\omega+x_0} \frac{[\delta\varepsilon_2(x) - \delta\varepsilon_2(\omega)]}{x - \omega} \, dx. \tag{37}$$

However, this does not add to our physical understanding of the problem. As before, a decrease in k leads to a decrease in n_r through the effect on the modulus of the permittivity. This reaches a minimum at $k=0$ and as k becomes negative the real part of the refractive index will increase. The net effect is that under conditions of gain the refractive index in the active region is greater, albeit by a small amount, than it is on either side of the junction and a waveguide is formed. It should be emphasized, however, that these are small changes; much smaller than the effect of increasing the band gap on either side of the junction.

4.3 Gain in semiconductors

To a great extent, the preceding treatment is as far as we need go with the optical properties of semiconductors. The output characteristics of the laser depend directly on the refractive index in the Fabry–Perot cavity, and it is now possible to understand the effects of changes in the band gap, the presence of large densities of free carriers, and even gain on the refractive index. However, you might want to go further and model the gain, and although the gain coefficient is just the negative absorption coefficient of the semiconductor the preceeding treatment is insufficient for the purpose. In order to model the gain properly it is necessary to look at the detailed quantum mechanics of the transitions between the bands. This will be examined using the wave mechanics of Schrödinger, but a word of qualification is necessary.

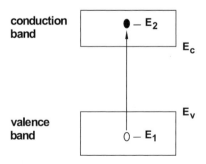

Figure 4.9. An absorption transition in a semiconductor.

Much of the early work on modelling gain applied to the heavily doped semiconductors described in chapter 3 which are no longer relevant to today's devices. This does not alter the general approach, but it modifies the outcome. Heavy doping perturbs the band structure and in so doing permits a range of transitions that would otherwise not occur because of momentum conservation. The densities of states in both band available for the transitions is therefore much greater, and much of the early work concentrated on finding the appropriate form of the density of states [10]. The active region of the DH laser is lightly doped and therefore has a different density of states function. The quantum well is much simpler still, but the density of states changes with the width of the well. This makes it quite different from the bulk material of the DH laser, which need only be modelled once. Optical and electrical confinement may alter the performance of the laser, but the material properties remain fixed. Therefore the remainder of this chapter can be understood as being most relevant to quantum well lasers, and though these will not be dealt with until chapter 6 an appreciation of the physics is best dealt with here in the context of the optical properties.

The photon density is related to the Einstein coefficients by equation (11) of chapter 3 for a two-level system, but for a semiconductor band structure must be taken into acount. For transitions between a state 1 in the valence band and 2 in the conduction band, the photon density is

$$\rho(\hbar\omega_{12}) = \frac{A_{21}f_2(1-f_1)}{B_{12}f_1(1-f_2) - B_{21}f_2(1-f_1)} \tag{38}$$

where $\hbar\omega_{12}$ is the energy of the transition. The absorption rate can be found by taking the difference between the rate of stimulated emission and the rate of upward transition, and is

$$r_{12}(abs) = B_{12}\rho_c(E_2 - E_c)\rho_v(E_v - E_1)(f_1 - f_2)\rho(\hbar\omega_{12}) \tag{39}$$

where ρ_c and ρ_v are the densities of states. The absorption coefficient is independent of the photon density and given by

$$\alpha(\hbar\omega_{12}) = \frac{n}{c} B_{12}\rho_c(E_2 - E_c)\rho_v(E_v - E_1)(f_1 - f_2) \tag{40}$$

so to proceed further it is necessary to know more about the probability B_{21}. First, however, some background on the nature of quantum states is necessary. Schrödinger's formulation wave mechanics is preferred here, but in fact it is but one of nine formulations of quantum mechanics [11], not all of which use a wave function. There is no good reason, therefore, to ascribe to the wave function a physical reality. As with most wave phenomena, the measurable reality lies not in the amplitude of the wave but in the intensity, $\psi^* \cdot \psi$, where ψ^* is the complex conjugate of ψ. Formally,

$$|\psi^2| = \psi^*\psi \tag{41}$$

4.3.1 The vector potential and the interaction Hamiltonian

For absorption or stimulated emission, the interaction Hamiltonian can be determined from consideration of the electrodynamics via Maxwell's equations. These can be found in any text book on electromagnetism and will not be given again. However, it should be realised that there is a symmetry between magnetic and electric phenomena which is revealed when the solution is expressed in the form of a vector potential. In their most fundamental form Maxwell's equations allows the measurable quantities E and H to be calculated, from which the forces acting on an electron, respectively $q \cdot E$ and $q \cdot v \times B$, where $B = \mu_0 H$, can also be calculated.

As discussed, forces are not so important in quantum mechanics and the scalar and vector potential solution of the electromagnetic equations is particularly suited. That is to say, the electric and magnetic fields can be written as functions of a vector potential A and a scalar potential φ such that

$$B = \nabla \times A$$

$$E + \frac{\partial A}{\partial t} = -\nabla \varphi. \tag{55}$$

The vector potential is discussed at length by Feynman [ref 1, chapter 15]. Unlike the scalar potential, which is no more than the potential of our common experience that has magnitude only and results from the integrated effect of applying a force, the physical origin and significance of the vector potential A is not so easy to comprehend. The differential of the scalar potential leads to the electric field via the force $q \cdot E$, and in so far as the curl operator ($\nabla \times$) is a differential, but with directional properties, the vector potential plays a similar role in the derivation of the magnetic field. Given the highly directional nature of the magnetic field it should not be surprising that the potential from which the magnetic field is derived has both direction as well as magnitude, and hence is a vector.

For many problems in electrodynamics the vector potential A can be regarded as no more than a mathematical convenience, it being just as easy to work with B and E as A and φ. However, it has a particularly useful property of direct relevance to the quantum problem at hand. The scalar potential can always be chosen such that

$$E = -\frac{\partial A}{\partial t} \tag{56}$$

so if we consider the force in relation to Newton's second law

$$F = qE = \frac{dp}{dt} \tag{57}$$

we arrive directly at the conclusion

$$p = -qA. \tag{58}$$

In other words, the momentum gained by the quantum particle due to the time-dependent electric field is directly porportional to the vector potential A. Therefore, if this momentum is added into the Hamiltonian we have

$$H = \frac{1}{2m}(p - qA)^2 + V. \quad (59)$$

Expanding the brackets and neglecting the term in $(qA)^2$ as being of no significance, the Hamiltonian becomes

$$H = \frac{1}{2m}(p^2 - 2qA \cdot p) + V \quad (60)$$

so the interaction Hamiltonian is just

$$H' = \left(-\frac{q}{m}\right) A \cdot p. \quad (61)$$

4.3.2 Fermi's golden rule

It is fairly straight forward to show that a harmonic (sinusoidal) perturbation results in a transition rate

$$W = \frac{\pi |H'_{mn}|^2}{2\hbar^2} g(\omega) \quad (62)$$

where H'_{mn} is known as the matrix element for the transition from state n to m and $g(\omega)$ is the density of pairs of states separated by an energy $\hbar\omega$ over which the transition can occur [12]. This is Fermi's golden rule and is used to calculate the rate of scattering from one state to another.

Formally the matrix element is written as

$$H'_{mn} = \int \psi_m^* H' \psi_n \, d\tau \quad (63)$$

where $d\tau$ is a volume element, and arises because the expectation value of a quantum measurement is given by

$$\langle E \rangle = \int \psi^* H \psi \, d\tau. \quad (64)$$

The expectation value reduces to the eigenvalue in the case where the quantum state is known, but for the case where the state is unknown then ψ has to be constructed from known states in a manner analogous to Fourier's principle for classical waves, i.e.

$$\psi = \sum_n c_n \psi_n. \quad (65)$$

The states ψ_n are known as the basis states, and are said to be orthogonal to each other. That is, they are independent from each other, and mathematically this is represented by the property $\psi_m^* \psi_n = 0$, so expansion of the expectation value equation leads to the elimination of all cross terms of this type. Hence

$$\langle E \rangle = \sum_n |c_n|^2 E_n \tag{66}$$

and the probability that a particle is in state k is simply given by $|c_k|^2$.

The essence of the perturbation method is that the details of the states are not known, but can be expressed in terms of the states of the system before the perturbation is applied through an expansion in the basis states. Hence,

$$j\hbar \frac{\partial}{\partial t} \sum_k c_k u_k \exp(-j\omega_k t) = (H_0 + H') \sum_k c_k u_k \exp(-j\omega_k t) \tag{67}$$

where $u_k \exp(-j\omega_k t)$ is one of the basis states with the time dependence explicitly included. Multiplying through by the complex conjugate of a state u_m^* and integrating over all space is equivalent to calculating the expectation value of the state m. Orthogonality can be used to eliminate all states that do not interact with m, leading to the transition probability expressed by the golden rule. Hence, B_{21} can be determined.

Recognising that

$$\frac{E}{\hbar} = \omega \tag{68}$$

the photon density can be expressed as a function of energy rather than frequency and Fermi's golden rule becomes

$$W = \frac{\pi |H'_{mn}|^2}{2\hbar} g(E). \tag{69}$$

The Einstein transition probability is therefore

$$B_{12} = \frac{\pi |H'_{mn}|^2}{2\hbar} \tag{70}$$

so it remains to determine the transition matrix. Recall that the electric field is given by

$$E_x = E_{0x} \exp[j(\omega t - kz)] \tag{71}$$

so that

$$A = \frac{jE_{0x}}{\omega} \exp[j(\omega t - kz)] \tag{72}$$

and

$$|A|^2 = |A \cdot A^*| = \frac{E_{0x}^2}{\omega^2}. \tag{73}$$

Matching the energy flux, $\hbar\omega c/n$, to that derived from the Poynting vector [10] allows E_0 to be expressed in terms of known parameters and

$$|A|^2 = \frac{2\hbar}{\varepsilon_0 n^2 \omega}. \tag{74}$$

After some manipulation,

$$B_{12} = \frac{\pi q^2}{m^2 \varepsilon_0 n^2 \omega} |M|^2 \tag{75}$$

where

$$M = \int_v \psi_1^* p \psi_2 \, d\tau \tag{76}$$

is known as the dipole matrix element.

It is but a short step from here to the final result. First, though, it is necessary to look at the density of states function in Fermi's golden rule. This is not just the density of states in the semiconductor, but the density of pair of states over which the transition occurs. There are essentially two circumstances to distinguish. States in the valence band can communicate either with all the states in the conduction band at the energy, $\hbar\omega$, or can communicate with only a limited set through the restriction of conservation of momentum. The first case corresponds to heavy doping and will not be considered further. The second case corresponds to high quality bulk material or quantum wells and the reduced density of states ρ_{red}, must be used. Consideration of a few limiting cases will help to arrive at a sensible expression for the reduced density of states.

Suppose first that only one conduction band state exists but that every valence band state can communicate with it. The density of pairs of states will then be equal to the density of valence band states. This situation is unlikely to occur. Much more likely, a smaller subset of valence band states will communicates with the conduction band, but we don't know how many. The density of states will therefore be smaller than the valence band density but larger than the conduction band density of states. Similarly, if the situation is reversed and only one valence band state exists we would expect a reduced density somewhere between the two. If the densities were identical and a one-to-one correspondence existed between the states so that one state in the valence band communicated with only one state in the conduction band we would expect the density of pairs would be equal to the density of either of the two. We could simply average over the two densities but, counter-intuitively, this would actually overestimate the density of states in the case of equal densities.

If we define the density of joint states to be

$$\rho_{red} = \frac{dN}{dE} \tag{77}$$

where N is the volume density of joint states, then for the limiting case that

$$N = N_c = N_v \tag{78}$$

the Fermi level will penetrate equally into the valence and conduction bands. Therefore the photon energy is given by

$$E = \hbar\omega = E_g + \Delta E_c + \Delta E_v = E_g + 2\Delta E_c \tag{79}$$

where ΔE_c is the penetration of the Fermi level into the conduction band. In terms of the conduction band density of states,

$$\rho_{red} = \frac{dN_c}{dE_c}\frac{dE_c}{dE} = \frac{\rho_c}{2}. \tag{80}$$

Although the volume density of pairs of states is the same as the volume density of conduction band states, in terms of the energy density we find that the density of pairs of states is, quite surprisingly, halved. This arises because the photon energy is split between the conduction and valence bands so in effect only half the states are accessible. In fact the assertion that the volume densities of conduction band and valence band states are equal is not really an assumption. Each arises from an atom and there is a fixed number of atoms in the solid, so recognising in general that

$$dE = dE_c + dE_v \tag{81}$$

and approximating the density as N/dE (the number of states within a small energy interval is assumed to be constant), then

$$\frac{dE}{N} = \frac{dE_c}{N} + \frac{dE_v}{N} \tag{82}$$

and

$$\frac{1}{\rho_{red}} = \frac{1}{\rho_c} + \frac{1}{\rho_v} \tag{83}$$

giving

$$\rho_{red} = \frac{\rho_c \rho_v}{\rho_c + \rho_v}. \tag{84}$$

Gain modelling in bulk semiconductors is most accurate close to the band extrema because the bands are assumed to be parabolic so that the density per unit energy varies as $\Delta E^{1/2}$ [13], where ΔE is the energy from the band edge. This should not be confused with the effective density of states, which is in effect an integral of equation (85) over a small energy range close to the band edge. This gives a reduced density of states for a transition at energy E

$$\rho_r(E) = \frac{1}{2\pi^2}\left(\frac{2\mu}{\hbar^2}\right)^{3/2}(E - E_g)^{1/2} \tag{85}$$

where μ is the reduced effective mass

$$\mu = \frac{m_e^* \cdot m_h^*}{m_e^* + m_h^*}. \tag{86}$$

Non–parabolicity is particularly a problem in the III–V materials where the density of states in the valence band is much higher than in the conduction band, principally because of the heavy hole contribution. Charge neutrality is assumed to exist within the active region so in order to achieve population inversion the electron quasi-Fermi level often penetrates much further into the conduction band than the hole quasi-Fermi level penetrates into the valence band. At high injection levels in particular, deviations from parabolicity become obvious.

4.3.2 The matrix element and densities of states

The matrix element (strictly the oscillator strength $2|M|^2/m$, which has the units of energy) can be determined experimentally [14] but it can also be estimated using accurate measurements of the effective mass. It is possible to show [15] that for light polarized in a particular direction

$$|M_{avg}|^2 = \frac{1}{6}\left[\frac{m_0}{m^*} - 1\right]\frac{(E_g + \Delta)}{\left[E_g + \frac{2\Delta}{3}\right]} \quad m_0 E_g = \frac{M}{3} \tag{87}$$

where Δ is the split-off energy for the valence band, E_g is the band gap, and m_0 is the free electron mass. The polarisation properties of the laser can reduce the average oscillator strength, and though it is relevant to quantum wells, it is a complication that need not concern us here.

This formula is not exact but it serves as a good guide for materials that have not been characterised fully. However, its origins lie in the quantum mechanical derivation of the valence bands, which is only accurate if all the possible bands of the semiconductor are considered together. It is common to consider only four; the three valence bands and the conduction band, as described in appendix III where it is shown that the conduction band effective mass, m^*, falls out naturally from these calculations. In essence, each band makes a contribution to each other band, especially for values of the electron wavevector $k \neq 0$. Hence the $E - k$ dispersion of the conduction band changes with more contributions and for a truly accurate picture of the effective mass, it is necessary to include in the calculation valence bands lying deeper than the three normally considered.

The result is that the effective mass derived from the four-band models are underestimates. For example, Yan et al. [15] quote the effective mass of electrons in GaAs derived from four-band models to be $0.053m_0$ whereas the

Table 4.1. Important III–V materials and matrix elements.

Material	E_p (eV)
GaAs	28.8
AlAs	21.1
InAs	21.5
GaP	31.4
AlP	17.7
InP	20.7

experimentally measured effective mass, which includes the effect of all the higher bands, is $0.0665m_0$. As the matrix element in equation (87) is a product of only four bands (the inclusion of deeper lying bands would lead to additional complexity) an effective mass of 0.053 should be used, otherwise the matrix element is underestimated. A comprehensive summary of the most important opto-electronic III–V compounds is provided by Vurgaftman *et al.* [16] from which the data in Table 4.1 has been selected. Ternary and quaternary compounds of the above can be treated using a parametric model for the band gap, effective mass, and split-off energy and the reader is referred to Vurgaftman for details.

Finally, it remains to modify equation (40) to derive the gain. Following Coldren and Corzine, [14, p 128].

$$g_{21} = g_{max}(E_{21})(f_2 - f_1) \qquad (88)$$

where

$$g_{max}(E_{21}) = \frac{\pi q^2 \hbar}{\varepsilon_0 nc \cdot m_0^2} \frac{1}{E_{21}} |M(E_{21})|^2 \rho_r(E_{21}). \qquad (89)$$

In GaAs quantum wells about 10 nm wide $g_{max} \approx 10^4$ cm^{-1}. The actual gain is modified by the Fermi levels at the two levels involved in the transition, but it should be noted that these are electron Fermi levels in both cases, so that for a hole quasi-Fermi level pushed deep into the valence band $f_1 \approx 0$ at the valence band edge.

4.4 Summary

In this chapter some of the essential physics behind optical transitions in semiconductors has been reviewed. A damped harmonic oscillator model of the electron bound to an atom can be used to explain the origin of the refractive index, and in particular such a model can be used to explain why decreases in the real part of the refractive index can be expected as the band-gap reduces and the carrier density increases, and why increases in the refractive index can be expected as threshold and gain are achieved.

This can be applied to the *p–n* homojunction laser. In equilibrium, bandgap narrowing on the *p*-side of the junction and high dopant concentrations on both sides of the junction lead to a reduction in the refractive index of the semiconductor, but the index is lower on the *p*-side than on the *n*-side. In the junction itself, the effect of carrier depletion negates the decrease due to high carrier concentrations and a slight increase in n_r is apparent in the centre of the junction. Therefore there is a wave-guide present, although it is asymmetric, and this will affect the propagation of the modes down the guide, as discussed in the next chapter.

In operation, the increase in carrier density due to the high current density required for lasing will bring about a general decrease in the refractive index, but this is more than offset by the increase in resonant frequency caused by transparency, and the increase due to gain. Under operating conditions, therefore, the wave-guiding structure is enhanced slightly, though it must be emphasized that confinement in this type of structure is by no means large.

Moving on from the refractive index, the quantum mechanics of optical transitions has been developed to allow an understanding of absorption and gain. The matrix element has been defined and related to the Einstein coefficient for stimulated emission and the gain function for a semiconductor in which *k*-selection occurs has been developed.

4.5 References

[1] Feynman R P, Leighton R B and Sands M 1975 *The Feynman Lectures on Physics* vol II, ch 32 (Addison Wesley)
[2] Fedak W A and Prentis J J 2002 *Am. J. Phys.* **70** 332–344
[3] Palik E D 1985 *Handbook of Optical Constants of Solids* (Orlando: Academic Press)
[4] Suzuki T and Adachi S 1993 *Jap. J. Appl. Phys. Pt 1,* **32** 4900–4906
[5] Forouhi A R and Bloomer I 1988 Optical Properties Of Crystalline Semiconductors And Dielectrics *Phys. Rev. B* **38**(3) 1865–1874
[6] Aspnes D E and Studna A A 1983 *Phys. Rev. B* **27** 985–1009
[7] Aspnes D E, Kelso S M, Logan R A and Bhat R 1986 Optical Properties Of AlXGa1-XAs *J. Appl. Phys.* **60**(2) 754–767
[8] Kittel C 1976 *Introduction to Solid State Physics* 5th Edition (New York: John Wiley & Sons) p 293
[9] Thompson G H B 1980 *Physics of Semiconductor Laser Devices* (Chichester: John Wiley & Sons) p 535
[10] Casey H C Jr and Panish M B 1978 *Heterostructure Lasers Part A: Fundamental Principles* (New York: Academic Press) p 111
[11] Styer D F, Balkin M S, Becker K M, Burns M R, Dudley C E, Forth S T, Gaumer J S, Kramer M A, Oertel D C, Park L H, Rinkoski M T, Smith C T and Wotherspoon T D 2002 *Am. J. Phys.* **70** 288–297

[12] Rae A I M 2002 *Quantum Mechanics* 4th Edition (Bristol: IoP Publishing Ltd)
[13] Sze S M 1982 *Physics of Semiconductor Devices* 2nd Edition (Wiley)
[14] Coldren L A and Corzine S W 1995 *Diode Lasers and Photonic Integrated Circuits* (John Wiley & Sons) pp 490–494
[15] Yan R H, Corzine S W, Coldren L A and Suemune I 1990 *IEEE J. Quant. Electr.* **QE-26** 213–216
[16] Vurgaftman I, Meyer J R and Ram-Mohan L R 2001 *J. Appl. Phys.* **89** 5815–5875

Problems

1. Using the damped harmonic oscillator model, $n^2 = 1 + N\alpha$, calculate the dielectric constant and the complex refractive index at: (a) resonance corresponding to 1.5 eV; and (b) at $\gamma/2$ below resonance for a system of 10^{27} oscillators with $\gamma = 5 \times 10^{14}$ Hz.
2. The onset of transparency shifts the resonant frequency upward by 0.05 eV. Recalculate the refractive index to show that both the real and imaginary parts decrease.
3. Calculate the change in refractive index in the above example due to a change in the carrier density of $\Delta N = 10^{18}$ cm^{-3}. Hint: use the identity

$$\frac{dn^2}{dN} = 2 \cdot n \cdot \frac{dn}{dN}$$

to calculate a total change $\Delta n = dn/dN \cdot N$, and assume an effective mass of 0.068 corresponding to GaAs.

Chapter 5

The double heterostructure laser

5.1 Introduction

The use of heterojunctions in lasers was first demonstrated in 1968 and showed immediately that a significant reduction in threshold current ensues [1] (see figure 5.1). The threshold current is one of the principal figures of merit in semiconductor laser technology. A lower threshold current means significantly reduced power losses through heating, and the introduction of the heterojunction therefore made the prospect of room temperature operation much closer.

A heterostructure, or sometimes simply a heterojunction, is a junction between two different materials. A semiconductor structure in which the chemical composition changes with position in this manner is usually achieved by deposition of a thin film from its constituent components onto a substrate. Atoms arriving at the substrate surface from a vapour or liquid source adhere

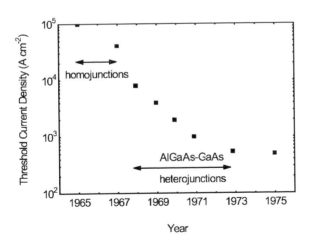

Figure 5.1. Progress in the lowest threshold current over time. (After Kressel and Butler.)

79

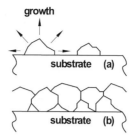

Figure 5.2. Schematic development of polycrystalline material from island growth.

to the surface, but normally there is not a unique position at which the film is nucleated. Nucleation can occur at several places at once and a random distribution of nucleation sites over the substrate surface will result in islands of material which spread laterally until eventually they join to make a continuous covering. Within the separate nucleation centres the material may be crystalline or amorphous, depending on the arrival rate of atoms at the surface and their ability to take up an equilibrium lattice position. The lateral mobility at the surface has to be high compared with the rate of arrival. If the material within the nucleation centres is crystalline it does not necessarily follow that the orientation of one crystallite will be identical to the orientation of another crystallite. Crystallites of different orientation meeting at some point must form a grain boundary, which is a region of material over which some atomic disorder occurs as the orientation changes. A film so constructed is said to be polycrystalline as it is made up in effect from myriad smaller crystals (figure 5.2b).

The term heterojunction does not necessarily imply therefore that the two materials making up the junction are single crystal, but for most opto-electronic devices single crystal material is essential and semiconductor lasers are no exception. It should already be clear that a high crystalline quality is vital. The growth of single crystal material requires that individual nucleation centres take up a particular orientation so that when separate islands meet there is complete registry of the lattice from one crystallite to the next and no grain boundary is formed. Some of the techniques for achieving single crystal films will be described within this chapter.

The first heterojunction laser used only a single heterojunction, however. The junction consisted of p^+-AlGaAs/p-GaAs on n^+-GaAs and was introduced specifically to contain the electrons injected into the p-material. Band-gap narrowing in p^+-GaAs resulting from heavy doping leads to a higher barrier in the valence band than in the conduction band, so electrons were injected into the p^+-GaAs while the holes were confined somewhat by the valence band barrier. The introduction of a wide band-gap material such as AlGaAs served to confine the electrons, thereby restricting the depth over which they could diffuse (figure 5.3), and defined more precisely the width of the active region.

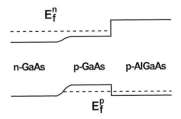

Figure 5.3. Schematic band diagram of the single heterostructure (SH) laser.

The carriers that would otherwise have diffused beyond the active region and recombined without producing stimulated emission were utilised more effectively in producing coherent radiation. In addition the introduction of the heterojunction allowed the photons generated by stimulated emission to be used more effectively. AlGaAs has a larger band-gap than GaAs and hence a smaller refractive index at wavelengths corresponding to near band-gap emission in GaAs. The waveguiding effects described chapter 3 are much enhanced thus confining the radiation to the active region.

The single heterostructure (SH) laser, as this particular design was named, was rapidly succeeded by the double heterostructure (DH) laser, in which the active layer was low-doped *n*- or *p*-type GaAs sandwiched between two AlGaAs layers. So rapid was this development that the SH laser was rendered obsolete before it could be fully characterised, and now it is little more than a historical curiosity. The advantage of the DH laser was two-fold; first, though the overwhelming need for improved confinement lay within the *p*-side of the junction, the hole confinement was by no means complete, and a second heterojunction improved the confinement dramatically; second, the SH structure gave rise to an asymmetric wave-guide, with the refractive index difference at the heterojunction far exceeding that at the *p–n* homojunction. The significant differences in the propagation of TE and TM modes that occurred as a result were corrected by the addition of a second heterojunction to make a symmetrical waveguide. Improved wave propagation, particularly at small core (active layer) thicknesses, in conjunction with the better optical confinement meant that the DH laser exhibited lower threshold currents than the SH laser.

5.2 Materials and epitaxy

The heterojunctions described above require very careful preparation in order to ensure crystalline perfection across the junction. A large number of defects will provide a clear route for non-radiative processes and effectively prevent laser action, but a high quality, defect free interface will allow effective confinement and efficient radiative recombination. Epitaxy is the term used to describe the gowth of one material on top of another such that the thin film is single crystal with a lattice structure matched to the substrate. Hetero-epitaxy further describes the production of a junction between two different materials, for example AlGaAs-GaAs, whereas homo-epitaxy describes the growth of like upon like, e.g. GaAs on GaAs. While hetero-epitaxy is clearly important, so too is homo-epitaxy because it is often necessary to grow an initial buffer layer on to the substrate.

Homo-epitaxy is straightforward to understand, as the thin film and substrate materials are identical and there is no question of the thin film taking any structure other than the substrate structure, provided the conditions at the surface exist to allow the incoming atoms to migrate across the surface and find the correct position relative to the substrate atoms. Hetero-epitaxy is more difficult to envisage because different materials have different lattice parameters. Not only must the correct surface conditions exist, but in taking up a lattice position dictated by the substrate the thin film might well be taking up an artificial structure in so far as the equilibrium position of atoms within the bulk single crystal is often different from that assumed in the epitaxial thin film.

It is not strictly necessary for the two materials in question to have the same lattice structure. It is possible for example to grow thin films of single crystal silicon, which has a zinc blende structure, on top of sapphire, which has a hexagonal structure. However, the sapphire has to be oriented such that it appears cubic to the silicon. In conventional III-V materials such as arsenides, phosphides and antimonides, no such problem exists. The constituents of any particular heterojunction system of technological interest have the same crystal structure and only the lattice parameters are different. GaAs, for example, has the zinc blende structure with a lattice parameter of 5.65359 Angstrom [2] while AlAs has an almost identical lattice parameter at 5.6605 Angstroms [3]. InAs, however, has a much larger lattice parameter of 6.05838 Angstroms [4]. Solid solutions of any two compounds will have both a lattice parameter and a band-gap intermediate between the two. Hence a solution of GaAs and AlAs, forming the ternary compound AlGaAs, will have a lattice parameter very nearly identical to that of GaAs over the whole range of composition from 0% Al to 100%. On the other hand, InGaAs will have a much larger lattice parameter than GaAs and forcing the ternary compound to take the same lattice parameter will cause a significant compressive strain to develop within the film. Conversely, thin films can be grown that have a smaller equilibrium lattice parameter than the substrate, which will lead to tensile strain in the film plane, as illustrated in figure 5.4.

There is a limit to the elastic strain that can be incorporated in such structures. An atomic displacement of a few percent or less represents an enormous mechanical deformation and there comes a point when such deformation can no longer be sustained. For InGaAs grown on GaAs, the

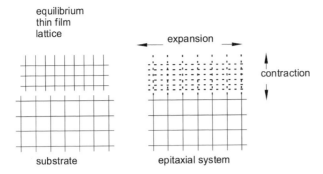

Figure 5.4. Schematic deformation of the lattice in hetero-epitaxy. This particular example shows tensile strain in the plane of the film but compression is equally possible.

mismatch in lattice parameters is too large to allow reasonable growth of high quality semiconducting layers. InGaAs and InGaAsP are important materials for longer wavelength lasers in the near infra-red so it is necessary to use an alternative substrate material. Fortunately InP provides a very good lattice-matched substrate for which the materials technology is sufficiently well advanced to produce InGaAs lasers emitting at 1.6 μm, and InGaAsP lasers emitting at both 1.3 μm and 1.55 μm. These are both important wavelengths in fibre communications

Strained layer systems do have a place in diode laser technology, but it is usually limited to quantum well systems in which the active layer is much smaller than in the conventional double heterostructure laser. Quantum well lasers will be dealt with in a separate chapter. For the DH laser the layer thickness can be quite large, and if strain exists it is important in such systems to ensure that the critical thickness of the layer is not exceeded. The critical thickness effectively defines the upper limit of film thickness for a given lattice mismatch. The mechanical energy stored within the deformation increases with film thickness until at a certain point the structure is no longer able to support the deformation. It is possible to show [5] that the energy per unit area increases linearly with film thickness, so at the critical thickness it becomes energetically more favourable to form dislocations rather than to continue to deform the crystal mechanically. A dislocation is essentially a line of atoms which terminates (or originates) within the lattice, so a dislocation allows an extra row of atoms and hence a smaller lattice spacing (figure 5.5).

The presence of dislocations gives rise to undesirable electrical effects, particularly the presence of alternative, non-radiative pathways for recombination. For a misfit with the substrate of 10^{-2} on GaAs the critical thickness is calculated at 5 nm, rising to 0.92 μm at a misfit of 10^{-3} [5]. For a DH laser with an active layer thickness up to a few hundred nanometres the maximum

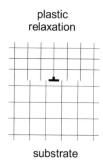

Figure 5.5. Strain relief by plastic (non-reversible) deformation.

permissible mismatch is therefore of the order of 10^{-3}, which is of the order of the mismatch between GaAs and AlAs. For ternary ccompounds of AlGaAs, however, the mismatch will be smaller since the lattice parameter will be intermediate between the two. The situation is further complicated by the fact that impurities effectively change the lattice constant by as much as 10^{-3} Å [6]. Epitaxial growth on p^+-GaAs is therefore different from epitaxial growth on n^+-GaAs, and both are different in turn from growing on intrinsic GaAs.

An additional complexity is introduced when the thermal expansivity is considered. Epitaxial layers are always produced at elevated temperatures, though for some methods of epitaxial growth the temperatures are considerably higher than for others. The lattice parameters deduced at 300 K will not apply at the growth temperature and indeed in some cases the mismatch might change sign. Cooling to room temperature can then induce compressive stresses where consideration of the lattice constants at room temperature might indicate tensile stresses, and vice versa.

Strain at an interface, then, is invariably accompanied by the creation of dislocations which relieve the strain. This is called plastic relaxation since the deformation is permanant, and is illustrated in figure 5.5. The thin film is still single crystal but with a different lattice parameter from the substrate. If the relaxation is complete the lattice parameter of the layer will be the equilibrium lattice parameter of the bulk solid. From the foregoing, a misfit of 10^{-2} or greater will produce dislocations at the interface in GaAs, and these will effectively destroy the operation of the laser because recombination at the dislocations, whether radiative or not, and usually it is not, will compete with, and dominate, the radiative band to band recombination. Lattice matching is therefore vitally important in the DH laser, but for a ternary compound at a given composition fixed by the lattice parameter the band gap is also fixed and this will determine the wavelength of the emission (figure 5.6).

The materials of greatest interest are GaAs/AlAs and InAs/GaAs/GaP/InP, and InAs/InAsSb for mid-infra-red devices. The In-Ga-As-P combination is

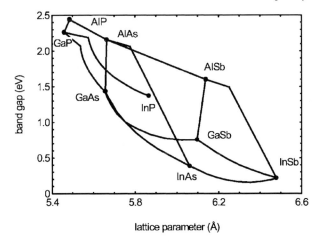

Figure 5.6. The band gap vs. lattice parameter for III–V compounds historically used in lasers.

used extensively in lightwave communications technology where quaternary compounds of InGaAsP are lattice matched to InP. The ternary compound InGaAs, by its nature, will always have a larger lattice parameter than GaAs. The composition lattice matched to InP, $In_{0.53}Ga_{0.47}As$, corresponds to an emission wavelength of 1.67 μm, which is longer than either of the two most desirable wavelengths for fibre communications, 1.31 μm and 1.55 μm at the loss minima in silica fibres. The lack of a simple lattice matched ternary with suitable band-gaps for these wavelengths has driven research to understand and control the quaternary system, which has the advantage of having four components and therefore independent control over lattice parameter and band-gap. Solid solutions of InAs/GaAs/GaP/InP can lie anywhere within the lines joining the pure compounds whereas ternary compounds are constrained to lie on those lines. The disadvantage of the four-component system is that it is much harder to obtain the desired ratio of four components. However, the growth technology is well developed and GaInAsP lasers are an established component in high bit rate long haul fibre communications systems.

Several methods of depositing these alloy materials have been used over the years; liquid phase epitaxy (LPE), molecular beam epitaxy (MBE), and several vapour phase techniques, including organo-metallic vapour phase epitaxy (OMVPE), sometimes also called MOCVD (metallo-organic chemical vapour deposition). MBE and MOCVD are the dominant growth technologies today. LPE was common certainly around the late 1980s and possibly still has its place today as a technique for fabricating high quality epitaxial layers, but there are problems with the surface morphology of the layers, and the technique is not suited to large area, high yield, high throughput production.

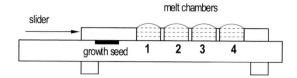

Figure 5.7. Schematic of a sliding LPE system using four chambers.

LPE will not be described in any detail here. Undoubtedly it has its place in the history of the diode laser, but the technology has been largely superceded. Moreover, a full understanding of the technique requires a knowledge of phase diagrams, which are graphical representations of the stability of mixtures of solids and liquids at various temperatures over a range of compositions. As such they are essential to a detailed understanding of growth from the melt, but such detail is a distraction here. Rather a brief overview will be given so that the limitations of the technology can be appreciated.

Historically, LPE was one of the first commercially viable techniques for producing high quality thin films, having been established in the 1960s. The principle is illustrated in figure 5.7, where four chambers containing melts of different compositions are shown. However, for complicated structures six may be used [7]. The substrate, or growth seed, is slid underneath to make contact with the melt and at each chamber a film is deposited. This simple statement of principle hides a number of difficulties, not least because this is a near equilibrium technique. The deposition has to take place close to the melt temperature and for any molten solution in contact with a solid there is a tendency either for material to be deposited or for solid material to be dissolved.

Ideally the composition of the melt is designed to be in equilibrium with a solid of known composition at a certain temperature, so that there is no net transfer of material across the melt–solid interface. This is actually very hard to achieve, and in some cases impossible. A solution of a ternary, such as AlGaAs, or quaternary, such as InGaAsP, cannot be in equilibrium with a binary solid, such as GaAs, so some etching of the surface on contact with the melt must occur. For example, suppose AlGaAs with 1% Al on GaAs is desired. It can be shown (with the aid of the phase diagram) that the only solid composition in equilibrium with this solution at 900°C is $Al_{0.63}Ga_{0.37}As$. That is, a solid of this composition can co-exist with a melt containing 1% Al without etching or deposition onto the solid. Thus, when a solid GaAs substrate is brought into contact with the melt containing 1% Al the solid will react and if left would dissolve completely.

Etching occurs whenever the composition of the melt is different from the substrate, i.e. whenever a heterojunction is grown, but it is only temporary while the temperature is lowered to induce supercooling and force growth. It has the unfortunate consequence, though, that the surface is roughened. An alternative method is to cool the melt to a lower temperature before contact is made with the substrate, so the driving force for crystal gowth exists prior to the driving force for dissolution of the solid. The amount of cooling that can be induced is only a few degrees (maximum 10°C in GaAs) otherwise nucleation occurs in the melt itself rather than on the substrate. This type of growth tends to result in smoother surfaces but the initial growth rate is rapid, and very thin layers are difficult to produce. Growth rates can be as high as a few micrometers per minute, so accurate control of the time and temperature are needed. Nonetheless the very thin layers needed for quantum well lasers cannot be grown

The technique is a near equilibrium method, with the crystal growth occurring close to the melting temperature. This has the disadvantage that defects will be present in the film in quite high densities. The formation energy of the Ga vacancy in GaAs has recently been measured at 3.2 eV, which means that close to the melt temperature there may be something in the region of 10^{17} vacancies cm^{-3} [8]. The number of As vacancies may be even higher. If the material is cooled rapidly this vacancy concentration can be frozen in. Non-equilibrium growth techniques such as MBE use much lower temperatures and the rate of arrival of atoms at the surface can be varied independently, so the concentration of defects is correspondingly reduced.

One of the big developments in LPE was the growth of InGaAsP on InP. The advantages of the quaternary over the ternary have already been discussed, and this material system allowed the growth of DH laser structures that could be used at wavelengths of 1.31 μm and 1.55 μm in both loss minima of silica fibres [9]. The growth of InP by LPE was developed in the 1970s and because P is so volatile special precautions had to be taken against its preferential loss from the melt. Usually a tight fitting lid or an InP cover slice had to be used to maintain a P overpressure. Failure to do so leads to etching of the InP surface to replenish the P lost to evaporation.

LPE growth technology proved very successful for the growth of high quality luminescent material, i.e. material with long minority carrier lifetimes, and also for the growth of complicated layer sequences. For example, in 1979 the threshold current of 1.3 μm emitting lasers grown by LPE was reported to be 670 A cm^{-2} [10], but that grown by CVD was 1500 A cm^{-2} [11]. By contrast the MOCVD technique in 1983 [12] allowed a threshold current of 800 A cm^{-2} and in 1982 the MBE technique [13] gave rise to a device with a threshold current of 1800 A cm^{-2}. LPE was clearly producing better devices, but two essential difficulties led ultimately to its demise as a favoured technique. First, as mentioned, the high growth rates made the technique unsuitable for quantum well structures, and second, the lack of control over the surface morphology, particularly over large areas, made large scale production a problem. MBE and OMVPE do not suffer from these problems, and these two techniques will be described in some depth.

5.2.1 Molecular beam epitaxy

Molecular beam epitaxy (MBE) is a vacuum deposition technique utilising thermal beams of atoms impinging on a substrate. It is in essence an advanced form of evaporation but differs in important respects. First, the base pressure is in the range known as ultra-high vacuum (UHV) in order to allow the crystal surface to be cleaned of overlying oxides and impurities so that a crystalline surface can be presented to the incoming atoms. The low pressure also minimises the incorporation of impurities from the background gas into the growing film, and cryogenic panels around the substrate help to reduce further the impurity incorporation by providing a cold surface on which impurities can condense. Second, elemental sources, such as Ga and As, rather than a single compound source, e.g. GaAs, are usually used, but compound sources that evaporate congruently, i.e. where the constituents of the compound are vapourised together, are sometimes used in conjunction with other sources to control the composition of the film. Differences in sticking coefficient and re-evaporation from the surface of the growing film often lead to significant densities of vacancies within the film, especially if that element has a relatively high vapour pressure. Phosphorous is a particular problem. An overpressure of the element can provide an increased flux of atoms impinging on the surface to replace those lost. Ternary compounds of differing compositions can be grown simply by varying the flux of the third element relative to the other two. Quantum wells and multiple quantum well structures can be grown using shutters to close off the flux periodically.

Figure 5.8 shows a schematic of a typical MBE growth chamber. Three evaporation (effusion) cells are shown, along with the substrate holder mounted on a rotary drive, the cryogenic panelling, and a RHEED gun. RHEED – Reflecting High Energy Electron Diffraction – is used to monitor the quality of

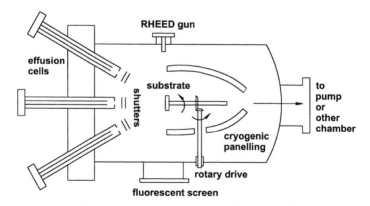

Figure 5.8. Schematic of an MBE growth chamber with three effusion cells.

the growing surface. The angle of incidence is very shallow so the electrons barely penetrate more than a few atomic layers and a highly ordered surface gives rise to strong diffraction peaks. The rotary motion on the substrate helps to ensure uniform deposition across the wafer, as effusion cells are highly directional and there is a strong likelihood of non-uniform growth across a wafer otherwise.

This equipment is quite complicated, as well as expensive, which is the principal disadvantage of MBE over other growth techniques. In many ways, though, the technique is one of the best for producing high quality heterojunctions with abrupt interfaces. Growth occurs at a slow rate and layer by layer growth can be monitored using RHEED. Moreover, the growth temperature can be very low compared with OMVPE or LPE so doped layers can be grown without fear of the dopants diffusing across interfaces, so that not only can atomically abrupt heterojunctions be grown but so also can p–n junctions. The use of molecular beams ensures not only that uniform composition can be achieved across a layer, but also graded compositions can be achieved if desired. Last, the surface morphology of epitaxial layers is usually very good. For these reasons MBE has been the preferred technique for experiments in heterojunction and quantum well physics, but the expense and low throughput have limited the commercial applications of the technique. Nevertheless, commercial MBE systems do exist and are used to produce complex heterojunction devices, including lasers.

UHV conditions are not easily achievable when pumping directly from atmospheric pressure – the vacuum chambers have to be baked at temperatures in excess of 100°C for many hours while being pumped, and if oil based pumps are used the oils are specially formulated to have a low vapour pressure – so the important functions of substrate preparation and growth are done in chambers maintained under UHV conditions and only rarely opened to the atmosphere for cleaning or replenishment of the evaporation sources. Therefore the elemental sources have a large capacity which enables them to operate for many months without interruption. The substrates are introduced into the system via a load lock – a small chamber which can be isolated from the main system and which has its own pumps – and then manoeuvred through the various chambers by a complicated magnetic or mechanical transport system. In this way the substrate preparation is often done in a separate chamber so that the growth chamber is not contaminated.

The key to epitaxial growth lies in achieving as clean a surface as is physically possible while at the same time presenting the crystal structure to the incoming atoms. In some cases the cleanliness can be achieved simply by thermal desorption of a surface oxide, but this might have to be prefaced by sacrificial oxidation of the surface prior to loading so that transition metals, and also ambient contaminants such as lead, sulphur, carbon, etc., are removed with the oxide. In other cases the cleanliness is achieved by ion beam cleaning of the surface followed by annealing to restore the lattice. Ion beams physically knock atoms off the surface and so impurities are removed along with a surface layer of material, but the process is damaging to the surface crystalline structure. This obviously needs to be restored before epitaxial overgrowth can be achieved. The precise details of the procedures employed vary from material to material. The preparation chamber would normally be equipped with a variety of diagnostic tools, among them Auger/ESCA spectroscopy, and RHEED. ESCA (electron spectroscopy for chemical analysis) is a technique whereby electrons liberated from the surface atoms following excitation are analysed according to their energies to determine their chemical origin. Auger is similar, so the two are often lumped together. Using these techniques the trace elements present on the surface can be detected, and the success of the cleaning process evaluated.

In GaAs technology, however, such elaborate processes are now redundant. It is possible to buy factory packed and processed clean wafers with only a protective oxide, which still has to be desorbed at about 600°C but it is no longer necessary to clean the wafer. The only diagnostic that is required is RHEED to ensure that the oxide has fully desorbed and that what is left is an atomically ordered and abrupt surface on which an epitaxial film can be deposited. InP is also available in epi-ready form so neither is it not necessary to clean these substrates. For InP desorption of the oxide presents some problems, because the temperature required for desorption is higher than that required for incongruent evaporation, and loss of P from the surface can occur. If InGaAs devices are being grown then desorption can be done under an As overpressure, which has been shown to be effective in reducing the loss of P. Alternatively a P overpressure can be applied.

Surface preparation is only one aspect of high quality epitaxial growth. Growth rates must be calibrated accurately to achieve high quality material and the flux from each source accurately controlled. The use of independent sources of material provides the grower with tremendous flexibility, not only to change the alloy composition but also to change the properties of binary layers such as GaAs, by controlling for example the number of point defects created. Although a material such as GaAs is nominally stoichiometric – that is to say it has equal numbers of Ga and As atoms – in reality there is a limit to the precision with which this occurs and vacancies of one element or the other can be created in sufficient quantity to alter the electrical or optical properties of the material. Independent control of the fluxes of each element arriving at the surface provides a means whereby the concentrations of such defects can be controlled accurately.

5.2.1.1 MBE of aluminium gallium arsenide

A typical MBE system might have five effusion cells. Not only are Ga, Al, and As required but also dopants. Shutters over the effusion cells are operated to turn on and off the flux from a particular cell and create heterojunctions. The evaporation rate is often measured using a variation of an ionisation gauge which measures the beam equivalent pressure (BEP). An ionisation gauge is a fairly simple device in concept, consisting of a hot filament which emits electrons by thermionic emission, and an anode. The electrons are accelerated towards the anode and any residual gases within the path of the electrons are ionised. Extra electrons are liberated which contribute to the total current reaching the anode. If such a device is placed within the beam of material evaporated from an effusion cell then there will be an increased ionisation current and an equivalent pressure associated with that current. It is not a real pressure – the atoms do not occupy the whole of the chamber but are localised to the molecular beam – but the measurement is nonetheless proportional to the flux from the cell.

Beam equivalent pressure provides a direct measurement of the rate of arrival of atoms at the surface, but this is not always directly related to the growth rate. The rate of growth also depends on the rate at which atoms are re-evaporated from the surface. The difference between the two is the so-called sticking coefficient. For GaAs, Ga atoms have a near unity sticking coefficient for temperatures up to ~650°C, and for Al on AlGaAs the sticking coefficient is near unity even above 700°C. Arsenic tends to desorb from the surface at temperatures used for growth (typically ~580°C < T < 650°C) so an As overpressure is used to prevent the surface becoming Ga-rich. It is an empirical observation that the As_2 and As_4 sticking coefficients increase with the presence of Ga adatoms on a GaAs surface. In short, as many As atoms as are required for stoichiometry tend to stick at the surface, so, provided the Ga atoms stick, which they do with near 100% probability at the growth temperatures, the As will also stick. The growth rate, and alloy composition, is therefore determined entirely by the rate of evaporation from the cells provided there is an As overpressure, which is determined in turn by the temperature of the cells. In the case of As a valved cracker source is often used. A solid source of As is heated to provide an As vapour, and the flux is controlled by means of a valve. The As species are passed over a hot filament – usually an inert material such as boron nitride – in order to crack the species and produce As_2 from As_4. However, the As_4 can be also be used directly.

The surface mobility of the group III atoms plays a crucial, but not exclusive, role in the surface morphology and quality of the layers. Ga tends to be mobile at typical growth temperatures and is able to move from islands to steps and hence fill in the missing parts of an atomic layer. Al is not so mobile, so while it is possible to grow smooth layers with good properties, the control of the conditions for growth has to be much tighter. RHEED can be used *in situ* during growth to monitor the surface quality via the strength and clarity of the diffraction pattern. Growth of thin films is ultimately an art in so far as it depends very much on the skill of the operator, though there are, of course, strong scientific principles behind it.

5.2.1.2 MBE of indium gallium arsenide phosphide

The ternary compound InGaAs is included in this material system as well as the quaternary InGaAsP. The principal difficulty with this system is the volatility and high vapour pressure of P, which causes problems both for the growth of phosphorus containing layers and for indium phosphide substrates. InP starts to sublime incongruently above 360°C but growth temperatures over 100°C higher than this are often required to achieve good growth of the lattice matched ternary $In_{0.53}Ga_{0.47}As$. As an elemental source, solid P is usually a mixture of allotropic forms of P, each of which has a different vapour pressure and precise control of the P flux is therefore difficult. Elemental source (ES) MBE, as the technique described above for the AlGaAs system is called, is therefore not suited to the growth of phosphide based materials. Variants can employ hydride source (HS) MBE in which the effusion cell is replaced by a gas source of AsH_3 and PH_3 which are mixed in the right ratio by the gas handling system outside the MBE chamber, and a thermal cracker is used to decompose the hydrides to produce As_2 and P_2.

Elemental sources for the group III elements can be used in conjunction with gas sources, but an important variation employs metallo-organic (MO) sources. MOMBE, as the technique is called, has the important advantage over ESMBE that the flux of group III elements is controlled outside the vacuum chamber as the uncracked MO sources are mixed outside in the gas handling system. The ratios of group III fluxes is therefore fixed and there is no need to rotate the substrate. Substrate rotation is only important whenever elemental sources are used because the output from an effusion cell is highly directional, and if the substrate were not rotated uniformity over a relatively large area would not be possible. In MOMBE this geometric constraint is removed. The metallo-organic sources – compounds such as tri-methyl gallium (TMG), tri-ethyl gallium (TEG), trimethyl indium (TMI), etc. – react on the hot surface of the substrate and generate the group III elements for epitaxy. Thus for a difficult quaternary system such as InGaAsP four sources are replaced by two – a single hydride source for As and P and a single gas source for In and Ga – and the complexity of the system is reduced enormously. The control of the composition is then reduced to a problem in controlling the gas handling and mixing, which, though still complicated, is easier than independent control of elemental fluxes via precise control of the temperature.

5.2.2 Chemical vapour phase epitaxy

Chemical vapour deposition techniques can be roughly divided into two; hydride/halide CVD and metallo-organic (MO) CVD. MOCVD, sometimes referred to as OMVPE, utilises essentially the same reaction as described in the MOMBE of InGaAsP. For the deposition of AlGaAs trimethyl aluminum $((CH_3)_3Al)$, trimethyl gallium $((CH_3)_3Ga)$, and arsine AsH_3 react together to form the compound, and to first order the composition is determined by the compositional mixture of gases. Organo-metallics other then TMA and TMG may be used but these are the most common.

The difference between the MOMBE and MOCVD processes lies entirely in the operating conditions. CVD processes occur under conditions of viscous flow, where the mean free path between collisions in the gas phase is much shorter than the dimensions of the reactor vessels and tubes. A molecular beam is completely different; the mean free path between collisions is very much longer than the dimensions of the vessel – the condition of so-called molecular flow – and the molecular beam impinges on the target essentially unimpeded by the ambient (figure 5.9). The low pressures used in MBE are therefore not so much concerned with transport of atoms and molecules but with the purity of the films via the incorporation of background species. CVD, on the other hand, is entirely concerned with transport. It is essentially a diffusive process. There will always exist a boundary layer of gas next to the substrate which, by the laws governing fluid flow, will be static with respect to the substrate. The reactive gases must therefore diffuse through this layer and react at the surface to deposit the layers.

5.2.2.1 Hydride chemical vapour deposition

The hydride process, along with the trichloride process, has been in existence a long time because of its success in the fabrication of GaP and GaAsP light emitting diodes. Here I shall concentrate on the InGaAsP system, since that is the more general. The hydride sytem is a hot wall epitaxial process, i.e. it takes place in a quartz tube placed inside a furnace. Gaseous sources of HCl, AsH_3, and PH_3 are introduced at low partial pressures in a hydrogen carrier gas. The arsine and phoshine decompose on entry into the hot vessel and produce As and P vapours. The HCl reacts with solid In and Ga sources at 700°C to produce InCl and GaCl vapour which is then transported to a mixing region held at 800°C. Here they combine with As and P vapours. The mixture flows onto a polished substrate wafer held at 700°C where InGaAsP is deposited. The mole

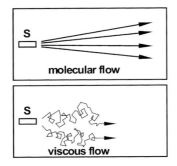

Figure 5.9. Illustration of molecular and viscous flow.

fraction of Ga in the layer is determined to first order by the ratio of GaCl to InCl and the mole fraction of As is similarly determined by the ratio of As to P.

The reactions involved in the hydride process are reversible, for example,

$$As_4 + 4GaCl + 2H_2 \rightleftharpoons 4GaAs + 4HCl \qquad (1)$$

so the deposition is actually competing with the reverse process, i.e. etching. Controlled deposition of very thin layers is possible through the slow net deposition rates consequent upon the controlled etching process, and this is one of the principal advantages of this technique. One of the disadvantages is the reactivity of HCl gas in the presence of the slightest trace of moisture, and background doping concentrations better than 10^{15} cm^{-3} are difficult to achieve.

5.2.2.2 The trichloride process

The trichloride process is essentially a variant on the above, but has the advantage that the hydrogen chloride gas is produced *in situ* and so is very pure, not having been transported through pipes where contamination can occur. The reactions

$$4AsCl_3 + 6H_2 \rightarrow As_4 + 12HCl \qquad (2)$$

and

$$4PCl_3 + 6H_2 \rightarrow P_4 + 12HCl \qquad (3)$$

are promoted by introducing AsCl$_3$ and PCl$_3$ vapours at low partial pressures in a hydrogen carrier gas from their respective liquid sources. The reaction occurs on heating and essentially goes to completion, the solid element (As or P) being condensed out. The HCl reacts with the metal source to produce a volatile chloride, as before, but also the the arsenic and phosphorus chlorides react with the metal sources to produce metal chlorides and P and As allotropes. Again, the growth processes are reversible so controlled etching is possible.

5.2.2.3 MOCVD

As described, MOCVD uses essentially the same reactions as occur in MOMBE. Various metal alkyls can be employed as sources of In or Ga and phosphine and arsine are common sources of group V elements. The reactions are not reversible, so controlled etching is not possible. Control of layer thickness has to be achieved entirely through control of the gases and through the reactor conditions. Further, preparation of the substrate surface has to be much more stringently controlled since it is not possible to etch the surface *in situ*.

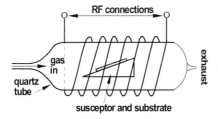

Figure 5.10. Schematic of an RF heated CVD system.

The technique is generally a cold wall technique, i.e. the vessel itself is not heated, only the substrate holder. This is usually achieved via RF induction heating (figure 5.10), which is not normally accurate to more than 10° so mass transport limited growth is desirable. Mass transport limited means in this context that the reactive gases are assumed to react instantaneously on the surface so the arrival of the reactive species at the surface limits the growth rate. This is only weakly dependent on the temperature of the gases and is preferred to the alternative; a surface reaction rate limited regime. In the surface reaction rate limited regime mass transport is considered to supply species at an abundant rate and the reactivity at the surface limits the growth. Surface reactions are activated processes and therefore depend exponentially on temperature. If these were to dominate the growth process then the control of the surface temperature would have to be much more accurate in order to achieve reproducibility from run to run.

Mass transport is hindered by a high vapour pressure so ideally growth would occur at atmospheric pressure rather than under low pressure conditions. However, the cold wall/hot substrate arrangement produces convective eddy cells which can disturb the mass transport and result in less abrupt interfaces. For this reason low pressures are preferred, though the systems can become very complex in their operation. It is absolutely essential, for example, to avoid gas phase reactions lest a low density film is formed. MOCVD is perhaps the hardest of the CVD techniques to control but it holds out the greatest promises for growth of complicated structures, particularly for InGaAsP lasers for lightwave communications. As will be shown, these structures require low capacitance insulating regions in order to confine currents to particular areas of the device and a simple reverse biased *p–n* junction, which can be grown by any of the techniques described, is not suited. A reverse biased junction certainly is insulating enough but unfortunately it has a high capacitance. What is desired is the deposition of insulating material, and with MOCVD this can be achieved through the use of organo-transition metal compounds, such as ferrocene to introduce iron into the deposited layers. Transition metals produce deep centres near mid-gap which pin the Fermi level and render the material as near insulating as is possible, so the current confinement is achieved without the expense of a high capacitance. The versatility afforded by MOCVD makes this technique a strong commercial technology.

5.3 Electronic properties of heterojunctions

Although the term heterojunction does not in itself say anything about the crystalline quality of materials, the term is usually reserved for abrupt epitaxial interfaces in which traps play a negligible role. Traps at interfaces arise from a variety of sources but especially from a disruption of the crystal lattice and the lack of long range order. In a system where lattice matching has not occurred interface states will exist as a result of the disorder and these might be quite different from the states associated with specific defects such as dislocations, vacancies, and foreign atoms. The last can arise from the cross diffusion of atoms in material systems of unlike chemical nature, for example, group II elements acting as acceptors and group VI elements acting as donors in III–V's. In III–V/II–VI heterojunctions, e.g. GaAs-ZnSe, cross diffusion of the chemical species will change the electrical properties of the junction. There may even exist a transitional region of intermediate chemical composition such as Ga_2Se_3.

In the heterojunctions of interest here interfacial compounds will not form. Interface states will still exist because of the change in chemical species and hence the change in potential, but they are considered to play an insignificant part in the electrical properties of the junction. In high-quality AlGaAs-GaAs heterojunctions, for example, it has been found that the interface is essentially atomically abrupt and the transition from one material to another occurs over the space of one atom. It depends very much on the growth method, of course. In LPE the etching of the surface prior to growth can lead to roughness and some gradation in the chemical composition across the junction. In MOCVD the abruptness of the junction depends very much on how quickly the gas mixtures can be changed over, and again some gradation in the interface is possible. In MBE the interfaces are usually abrupt, especially in elemental source MBE where shutters over the sources can shut off the flux completely.

The central feature of a heterojunction is that the bandgaps of the constituent semiconductors are usually different. The difference in energy gaps means that there must be a discontinuty in at least one of the bands, and usually in both. These discontinuities are the origin of most of the useful properties of heterojunctions and the reason why heterostructure devices attract so much interest so the question of how the discontinuities are distributed between the valence and conduction bands has been a subject of intensive study for well over forty years. The first, and simplest, model of heterostructure line-up was Anderson's electron affinity rule [14] based on experimentally measured electron affinities. The electron affinity is defined as the energy required to

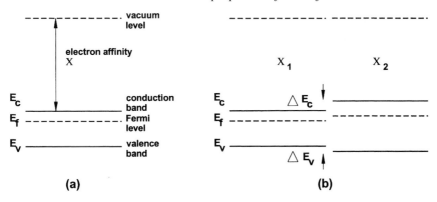

Figure 5.11. Anderson's electron affinity rule for heterojunction line-up.

remove an electron at the bottom of the conduction band to a point outside the material (figure 5.11a) where it is said to have zero energy (the vacuum level). The electron affinity is therefore analogous to the workfunction in metals, and is used in semiconductors because the workfunction changes with doping density as the Fermi level moves.

According to this model when two materials are brought together to form a heterojunction the vacuum levels align (figure 5.11b) and the conduction band and valence band offsets are then defined. The Fermi levels in the two materials must align in order to achieve thermal equilibrium and some band bending will occur, as in the p–n homojunction described in chapter 3. It has been found experimentally that the *a priori* relation between the band-edge energies implied by the electron affinity rule does not exist. This model is not a very accurate description of heterojunctions, but it stood until the early to mid-1970's after a new approach to the problem of band line-ups was adopted. That approach involves a self consistent quantum mechanical calculation of band structure on either side of the junction using pseudopotential methods and is therefore beyond the scope of this book. For our purposes it is sufficient to recognise that the band-edge energies can be regarded as fundamental properties of the semiconductors so it is possible to calculate them from first principles and hence determine the band offsets at the heterojunction. Experimentally, band offsets are measured using a variety of approaches that vary in accuracy. Estimates of band offsets are continually revised in the light of new experimental findings, as the history of the AlGaAs-GaAs system shows. In this system the band gap of $Al_xGa_{1-x}As$ varies with x as

$$E_g(x) = 1.424 + 1.266x + 0.26x^2 \qquad (4)$$

for $x < 0.45$. In this compositional range the band gap is direct. The absolute band offsets will vary with energy gap but the percentage of the total offset appearing in the conduction band appears to be constant. Initially this was thought to be 85% but in 1984 new experiments revealed a figure of 57%, and the most recent results indicate a figure of 60%. This is the commonly accepted value for the offset in this system.

For practical purposes it is not necessary to know the detailed arguments behind heterojunction theories. It's enough to know that it is possible to predict offsets theoretically. The test of the theory lies in the experiment, and therein lies the uncertainty. In the following paragraphs, the band offsets of some important heterojunctions will be presented. Many band offsets have been taken from the tabulations by Yu *et al.* [15]. In some cases, where there is conflicting data in the literature, more than one value of the offset is presented.

1. $Al_xGa_{1-x}As$ – GaAs

Yu *et al.* take the band gap of AlGaAs to first order in x only and the offsets correspond to $\Delta E_c = 0.616 \, \Delta E_g$ and $\Delta E_v = 0.384 \, \Delta E_g$ (figure 5.12). Furtado *et al.* [16] measure the conduction band offset between alloys of differing aluminium content to be $\sim 0.6 \, \Delta E_g$.

2. InGaAs – InP

This system has been measured by Guillot *et al.* [17,18] using capacitance-voltage (C-V) profiling. At room temperature the conduction band offset is reported to be 200 meV but decreases to 120 meV at temperatures less than 150 K due to a complicated effect concerned with the filling of traps located at the interfaces. At room temperature $\Delta E_v = 0.4$ eV, but Yu *et al.* reported $\Delta E_c = 0.26$ eV and $\Delta E_v = 0.34$ eV.

3. $In_{1-x}Ga_xAs$ – $In_{1-y}Ga_yAs_zP_{1-z}$

This system has been measured by Hall *et al.* [19] using a quantum well (QW) structure, i.e. a thin layer (9nm) of $In_{1-x}Ga_xAs$ sandwiched between barriers of $In_{1-y}Ga_yAs_zP_{1-z}$ 540 nm thick on the bottom (substrate side) and 50 nm thick on top (toward the surface). The fractions of As and Ga varied slightly from run to run, but the average values were 0.49 and 0.23 respectively, so chosen as to give a nominal transition energy (band gap) of 1.03 eV in the barrier layers. The principle variable was the fraction of Ga, x, in the quantum well, and consequently the strain varied as well. Five structures were used in all and the

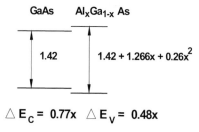

Figure 5.12. Band offsets for GaAs-AlGaAs. (After Yu [15].)

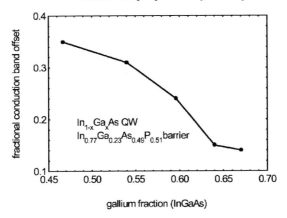

Figure 5.13. The conduction band offset as a function of InGaAs composition [19].

fractional offsets shown in figure 5.14 were obtained. This is a system susceptible to accurate modelling.

4. $In_{0.53}Ga_{0.47}As - In_{0.52}Al_{0.48}As$
Investigated by Yu *et al.*, the band lineups are $\Delta E_c = 0.47$ eV, $\Delta E_v = 0.22$ eV

5. InAlAs – InP
This system has also been investigated by Abraham *et al.* [20] using a very similar composition of 46% aluminium. The offsets reported ($\Delta E_c = 0.35$ eV, $\Delta E_v = -0.27$) differed considerably from those presented by Yu *et al.* [15] ($\Delta E_c = 0.25$ eV, $\Delta E_v = -0.16$ eV), where the negative sign means a type II alignment as illustrated in figure 5.14.

5.3.1 Band bending at heterojunctions

Band offsets are just one of the important electrical properties of heterojunctions. These give rise to a discontinuity in the potential across the junction, at

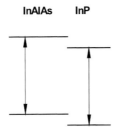

Figure 5.14. A type II alignment in InAlAs-InP.

least on a macroscopic level. The charge distribution in the vicinity of the junction is the key determinant of the potential profile, commonly known as the band bending. Charge neutrality across the junction must exist in equilibrium since there are no externally applied voltages. Maxwell's equation for the continuity of the displacement vector applies, i.e.

$$div D = \rho \qquad (5)$$

where

$$D = \varepsilon_r \varepsilon_o \xi \qquad (6)$$

and ε_r is the relative permittivity, ε_o is the permittivity of free space, ξ is the external electric field, and ρ is the charge density. In the presence of interface states, which represent a sheet of charge localised at the interface, D will be discontinuous. However, interface states are not considered to play a significant part in these heterojunctions, so D is continuous across the interface. It can be shown that this leads to the condition that the total charge on either side of the heterojunction is equal in magnitude, but opposite in sign. If this charge is distributed uniformly in space a parabolic potential is superimposed on the band structure. Thus, for n-GaAs/p-AlGaAs, for example, the band diagram at equilibrium is similar to figure 5.15.

The depletion width on either side of the junction is determined by the condition for the equality of the total charge on either side of the junction, i.e.

$$N_d w_1 = N_a w_2 \qquad (7)$$

where $N_{d,a}$ are the doping densities for donors and acceptors respectively and $w_{1,2}$ are the depletion widths. Hence, if $N_d \gg N_a$ then $w_1 \ll w_2$. Doping densities may differ by orders of magnitude across the junction, for example p^+-AlGaAs doped to greater than 10^{19} cm^{-3} might form a heterojunction with undoped GaAs. Undoped GaAs is in fact a misleading term since it implies intrinsic material. It should be understood that "undoped" means that dopants have not been introduced intentionally but a background density of 10^{13} impurities cm^{-3} is common. The density of states in the conduction band of GaAs is such that

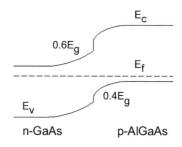

Figure 5.15. Band bending at an isotope heterojunction.

the intrinsic carrier density is very low, $\approx 2 \times 10^6$ cm^{-3}, and it is not possible to achieve background impurity levels comparable to this. A doping density of 10^{13} cm^{-3}, represents an atomic purity of 1 part in 10^{10}. MOCVD and gas source MBE utilise very pure starting materials but even so the material will still be very low doped *n*- or *p*-type, usually *n*-type. This should not be confused with semi-insulating (SI) GaAs produced by doping with transition metal ions such as iron or vanadium, both of which pin the Fermi level close to mid-gap by means of deeps states. The depletion region of "intrinsic" or semi-insulating GaAs will correspondingly be many orders of magnitude larger than in the AlGaAs, and may extend many tens or even hundreds of microns.

The GaAs region in a DH laser does not, however, extend this far. For reasons which will become apparent, it is usually no more than ~ 200 nm wide, after which another heterojunction with oppositely doped AlGaAs occurs. The two possible types of heterojunction – isotype, where the doping is of the same type, and anisotype, where the doping is of opposite type – will exist together in the DH laser. Assuming that the GaAs layer is residually doped *n*-type, the n^+-AlGaAs/GaAs junction is isotype and the p^+-AlGaAs/GaAs junction is anisotype. Figure 5.16 shows a typical isotype junction in this system. Equilibration of the Fermi levels leads to the formation of severe band bending in the GaAs close to the junction, such that the conduction band can dip below the Fermi level. In this case a two dimensional electron gas (2-DEG) will exist at the surface of the GaAs. The electrons originate on the donors present in the AlGaAs but reside in the GaAs, their presence at the interface ensured by the coulombic attraction between the ionised donors and the electrons. Ordinarily, the appearance of a 2-DEG will effectively terminate the electric field arising from the ionised donors but in the case of the DH structure the field terminates instead within the p^+-AlGaAs. The system is effectively a *p–n* junction in AlGaAs with a thin, insulating layer in between (figure 5.17).

Comparing the above diagrams with those of the *p–n* homojunction laser, the differences in the positions of the Fermi level are immediately apparent. In the homojunction laser a pre-requisite for population inversion is doping heavy enough to push the Fermi levels into the bands edges. In the DH laser, large densities of electrons and holes can be injected without this initial condition, which is an advantage because it is to some extent counter productive to try to dope the AlGaAs cladding layers to this extent. In common with many wide

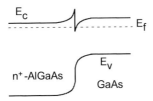

Figure 5.16. An isotype junction showing a 2-DEG at the interface.

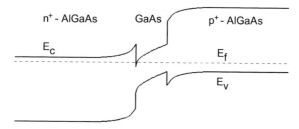

Figure 5.17. Band diagram of the AlGaAs-GaAs DH laser in equilibrium.

band gap semiconductors, both acceptors and donors can be deep lying with relatively high ionisation energies in AlGaAs. The ionisation energies vary with Al mole fraction [21], so for Ge doped p-AlGaAs the ionisation energies varies linearly from 4 meV in GaAs to ~110 meV at 55% Al. In Te doped n-AlGaAs the behaviour is more complicated. The ionisation energy remains constant at 6 meV up to just under 30% Al, then rises dramatically to a maximum of ~150 meV at just under 40% Al, and then drops linearly to 50 meV in AlAs. These large ionisation energies cause the ionisation fraction to decrease dramatically as the Fermi energy moves towards the band edge in accordance with Fermi–Dirac statistics, so pushing the Fermi energy into the band doesn't actually achieve much. For this reason most early depictions of the band diagram of AlGaAs-GaAs DH lasers show the Fermi level to be in the conduction band but not in the valence band, but the precise doping densities are rarely mentioned in detailed discussions of the physics of the DH laser. If the Fermi level is pushed into the conduction band the 2-DEG at the interface is accentuated, but a corresponding hole gas is unlikely to exist.

5.4 The double heterostructure under forward bias

Under forward bias the total voltage dropped across the junction will decrease. Figure 5.18 illustrates several features of the heterojunction under forward bias. First, the injected carrier densities in the GaAs active region are seen to be higher than either carrier density in the AlGaAs regions. Second, the quasi-Fermi levels are shown as constant across the active region, which will be the case if the active region is much shorter than the diffusion length. This normally applies but there are some laser designs employing large optical cavities where the active region may be 2 or 3 μm wide. Third, the bands are depicted as flat within the active region, indicating that there is no net electric field inside this region. The presence of such large densities of carriers means that the effects of the space charge regions on either side of the junction, which

The double heterostructure under forward bias   103

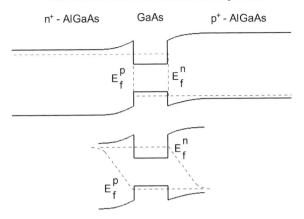

Figure 5.18. The quasi-Fermi levels under forward bias.

are in any case diminished because of the forward bias, will be screened out. Moreover, equal densities of free electrons must be injected otherwise charge would build up. For thin active regions, then, the carrier density can be assumed to be uniform across the region and equal to the hole density, but in general numerical calculations which take into account a variety of transport phenomena are necessary. For example, electrons diffusing across the active region will be almost entirely reflected off the opposite barrier owing to their wave-like nature [22], and therefore generate a reverse diffusion current that virtually cancels out the forward current except for the effects of recombination in the bulk and at the interfaces. The recombination rates themselves are concentration dependent (heavy injection) and are further affected by the onset of laser action.

Figure 5.18 also illustrates two possible variations of the quasi-Fermi levels with the barrier regions. The first shows the quasi-Fermi levels changing discontinuously at the barrier and the second shows the quasi-Fermi levels changing smoothly with distance inside the barriers. In any junction under forward bias the quasi-Fermi levels must eventually meet as the excess carrier density decays through recombination. A discontinuous change can only occur if the confinement is absolute and there are no carriers injected over the barrier. In such a case the excess electron and hole densities inside the barriers are zero and the quasi-Fermi levels drop abruptly to meet the majority carrier Fermi levels. However, in the realistic case that some carriers are injected, these will diffuse away and recombine and the quasi-Fermi level will decay with distance.

Considering electrons only, the diffusion current away from the junction and into the barrier is given by

$$J_n = qD_n \left.\frac{\partial n}{\partial x}\right|_{x=x_n} \tag{8}$$

where D_n is the diffusion coefficient for electrons and x_n is the position of the junction. The electron density is

$$n(x) = n(x_n)\exp\left(\frac{x_n - x}{L_D}\right) \tag{9}$$

where L_D is the diffusion length of the electrons. Therefore

$$\frac{\partial n(x)}{\partial x} = \frac{n(x_n)}{L_D}\exp\left(\frac{x_n - x}{L_D}\right) \tag{10}$$

and at $x = x_n$

$$\frac{\partial n(x_n)}{\partial x} = \frac{n(x_n)}{L_D}. \tag{11}$$

Therefore

$$J_n = qD_n \frac{n(x_n)}{L_D}. \tag{12}$$

The density of electrons can be evaluated from the interfacial band offsets. Figure 5.19 shows a schematic diagram of the conduction band under the condition that forward bias across the junction is sufficient to eliminate any band bending. Regions (1) & (3) correspond to the confining layers and region (2) to the active region. The electron density in region (3) is

$$n_3 = N_c \exp\left(\frac{-F_{n3}}{kT}\right) \tag{13}$$

where

$$F_{n3} = \Delta E_c - F_{n2}. \tag{14}$$

This can be written in terms of the total band gap difference and the valence band offset as

$$-F_{n3} = F_{n2} - \Delta E_g + \delta E_v. \tag{15}$$

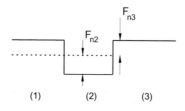

Figure 5.19. Confinement parameters in the conduction band.

The valence band offset can be written in terms of the hole densities, where, to a first approximation,

$$p_3 = p_2 \exp\left(\frac{-\delta E_v}{kT}\right) \qquad (16)$$

assuming both that the hole quasi-Fermi level lies close to the band edge, which can be justified in GaAs where $N_v \gg N_c$, and that the barrier valence band is similar to that in the active region. Hence

$$kT \ln\left(\frac{p_2}{p_3}\right) = \delta E_v \qquad (17)$$

and

$$n_3 = N_c \left(\frac{p_2}{p_3}\right) \exp\left(\frac{F_{n2} - \Delta E_g}{kt}\right). \qquad (18)$$

Therefore

$$J_n = \frac{qD_n N_c}{d} \left(\frac{p_2}{p_3}\right) \exp\left(\frac{F_{n2} - \Delta E_g}{kt}\right) \qquad (19)$$

where d is the diffusion length of the electrons in region (3) or, the width of the layer, if smaller, on the assumption that anything reaching the GaAs substrate recombines, and p_2 is the total density of holes in the active region.

This form of the electron leakage current [23] or similar [1, p 74] is used because it depends only on the total band gap difference and not on the detail of the distribution between the conduction band and valence band. As discussed previously, these parameters may be known in some cases but not all, and even where experiments have been performed there may be uncertainty over the results. The hole leakage current can be formulated similarly but holes generally have much smaller mobilities, and hence smaller diffusion coefficients, and so hole leakage is not as important as electron leakage.

Looking in detail at equation (19) the term in the exponential offsets the coefficient, and both must be considered. F_{n2} (in eV) can be calculated from the Fermi integral and is approximately [1, p 73]

$$F_{n2} = 3.64 \times 10^{-15} \frac{m_0}{m_e^*} n^{2/3} \qquad (20)$$

where m_e^* is the effective mass and n is the injected electron density. In GaAs $m_e^* = 0.068 m_0$ so for an injected electron density of 10^{18} cm^{-3} $F_{n2} \approx 0.05$ eV at $T = 300$ K. This is quite small compared with a total band offset of say 0.3 eV, which is typical, and means that the term in the exponential will always be negative for any reasonable heterojunction design. The magnitude of the exponential is determined by the magnitude of this coefficient, so if this is large the band offset must be large to reduce the total leakage current. The worst case occurs if region (3), the p-doped AlGaAs layer is both relatively low doped and narrow. It is straightforward to show that for the above example ΔE_g must be ≈ 0.32 eV for room temperature operation, corresponding to an aluminium mole fraction of $\sim 25\%$ in order to reduce the total leakage current to something below 100 A cm^{-2}. In practise, diode lasers do not operate at room temperature because of the problems associated with thermal dissipation. An operating temperature of 100°C leads to the requirement that the total band offset be ~ 0.4 eV, corresponding to $\sim 32\%$ Al. These are of course only estimates. A thicker p-AlGaAs layer with higher doping will reduce the requirement, but for more complete confinement, i.e. lower leakage current, the requirement is increased. However, confinement of the electrons and holes is but one consideration in the design of the heterojunction, and lasers with an AlAs content as high as 65% have been used [24] in order to increase the refractive index difference at the heterojunction. Such high Al content alloys can be grown without difficulty because of the lattice match, and the large band offset is a particular advantage. In other materials systems, however, the introduction of strain and non-radiative defects limit the composition.

The diffusive leakage of carriers over the barriers is one of the principle causes of a high temperature sensitivity in the threshold current of lasers. Carrier leakage over the barriers is both diffusive and activated. There is a net diffusive motion of carriers toward the barrier but only those with sufficient energy can cross. Far from the Fermi level the distribution of carriers will be given by the Boltzmann distribution, $\exp(-E/kT)$, so if the barrier is lowered by the action of the forward bias, or if the temperature is increased, the density of carriers able to cross the barrier is increased exponentially. For low-doped barriers, however, there may be a significant field inside the cladding material, and the diffusive component can be augmented by a drift component [25]. These considerations lie behind the empirically observed relationship

$$J_{th} = J_0 \exp\left(\frac{T}{T_0}\right). \tag{21}$$

There is no particular physical significance to the terms J_0 and T_0. They are simply parameters that characterise the laser structure such that the lower the sensitivity to temperature results from a higher value of T_0.

5.4.1 Recombination at interfaces

In the active region of the laser, carriers injected at one heterojunction either diffuse across to the other or recombine radiatively along the way. Those that reach the second junction will suffer one of three fates; traversal of the heterojunction as part of the leakage current, reflection, or interfacial recombination, usually assumed to be non-radiative. Interfacial, or surface, recombination is characterised by a recombination velocity S, such that the total recombination current is defined as,

$$J_r = q \cdot n \cdot S. \tag{22}$$

If the injected current is J_i, the total fraction of electrons available for radiative recombination is

$$\gamma = \frac{J_i - J_r}{J_i} = 1 - \frac{J_r}{J_i}. \tag{23}$$

Considering only injection at one heterojunction and recombination at the other, the density of electrons at the second heterojunction is

$$n(d) = n_i \exp\left(\frac{d}{L_D}\right) \tag{24}$$

where n_i is the density of injected electrons and d is the width of the active region. Equations (12), (22), and (23) give

$$\gamma = 1 - \frac{SL}{D} 1 + \exp\left(\frac{-d}{L}\right). \tag{25}$$

If $d \ll L$, which will be the case for active regions less then ~ 0.5 μm then

$$\gamma = 1 - \frac{SL}{D} \tag{26}$$

for one interface. Under these circumstances the electron concentration will be uniform and the second interface can be taken into account simply by doubling the recombination term, i.e.

$$\gamma = 1 - \frac{2SL}{D}. \tag{27}$$

Using typical values for L and D in GaAs results in the condition that S must be less than $\sim 10^3$ cm s^{-1}. There is no such thing as a typical surface recombination velocity; two nominally similar heterojunctions can have vastly different recombination properties because of the densities of traps that might exist. However, Nelson and Sobers [26] have reported values for AlGaAs/GaAs junctions of ~ 450 cm s^{-1}, well within the limit for efficient laser action in thin double heterostructure lasers. Van Opdorp and t'Hooft [27] have made a systematic study of 30 devices made by LPE and VPE by measuring the recombination lifetime in the high injection regime $p \approx n$, which is important because these conditions relate much more closely to the operation of a laser than the low injection conditions. They concluded that methods based on luminescence decay at lower values of the carrier density, as performed by Nelson and Sobers, may significantly overestimate the surface recombination velocity and concluded that for their best LPE devices $S \leq 270$ cm s^{-1}, while for VPE devices $S \leq 350$ cm s^{-1}. These values lie significantly below the required threshold and can be contrasted with $S \approx 10^6$ cm s^{-1} for a bare GaAs surface [1, p 75]. For the Al-free heterojunction, $Ga_{0.5}In_{0.5}P$-GaAs, [28], recombination velocities as low as 2 cm s^{-1} have been measured for low-doped GaAs, rising to ~ 200 cm s^{-1} for heavily doped GaAs.

5.5 Optical properties of heterojunctions; transverse mode control and optical confinement

The GaAs-AlGaAs heterojunction is the most widely investigated so it is a natural choice with which to illustrate the principles. The energy dependence of the refractive index and extinction coefficient have been given in figures 4.7 and 4.8. The band gap transition, which appears in the refractive index as a small kink in an otherwise smooth variation in a region of normal dispersion, and is shown in greater detail in figure 5.20 for low concentrations of aluminium ($\leq 15\%$). Also shown by way of contrast is the index for heavily doped GaAs. Figure 5.21 shows the variation of refractive index with mole fraction of aluminium at a photon energy of 1.38 eV. The variation is not quite linear, but very nearly so. Recalling the discussion in chapter 4, the formation

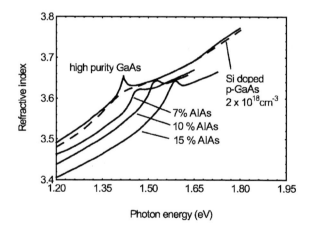

Figure 5.20. Normal dispersion around the band gap in low Al content AlGaAs. Heavily doped GaAs is shown as the dashed line, and has a much smaller resonance and a slightly lower index.

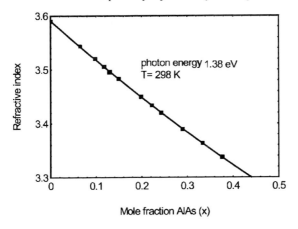

Figure 5.21. Refractive index as a function of Al mole fraction at 1.38 eV. (After Kressel and Butler [1, p 206].)

of a heterojunction leads to much larger changes in refractive index than occur in the homojunction laser. In consequence the confinement of the optical field is that much greater.

A typical near field pattern of the fundamental mode of a symmetric waveguide is illustrated in figure 5.22 for a wide active region laser (also called, "large optical cavity" – LOC) with a refractive index change of 0.1, corresponding from figure 5.21 to an Al content of ~12%. This is a transverse mode and not to be confused with a lateral mode, which describes the variation in light intensity at the output facet in the plane of the heterojunction. The intensity of the optical field at the core-cladding interface is ~5% of the maximum intensity, corresponding to a very strong confinement. Nearly all the photons generated within the laser are utilised for stimulated emission. Integration under the intensity distribution shows that the confinement is ~96%.

The near field pattern closely resembles the distribution of light within the waveguide. However, the far field pattern is more complicated. In a lengthy treatment which will not be reproduced here, Kressel and Butler [1, p 191] have shown that the emitting facet can be treated as a classical antenna aperture and that the radiation pattern is proportional to the Fourier transform of the aperture electric field.

$$I(\theta) = |g(\theta)|^2 \left| G\left(\frac{\sin\theta}{\lambda}\right) \right|^2 \tag{28}$$

where $g(\theta)$ is a so-called "obliquity factor" but in fact is closely approximated by $\cos(\theta)$, and $G(\sin\theta/\lambda)$ is the Fourier transform of the near field distribution. Numerical calculations are usually required to solve for $I(\theta)$. In general, large

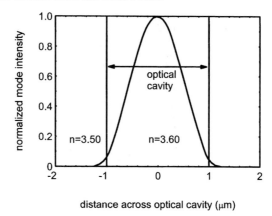

Figure 5.22. The fundamental transverse mode in a LOC DH laser. (After Kressel and Butler, p 206.)

cavities have side lobes associated with each of the transverse modes, so for the fundamental mode in figure 5.22, the far field pattern is as demonstrated in figure 5.23. There is a considerable angular deviation of the radiation pattern, which is essentially related to the aperture width. A narrower aperture (smaller cavity width) has a lower threshold current but will lead to a larger angular spread and conversely a narrower angular spread requires a higher threshold current. The radiation pattern of an antenna aperture generally gives rise to two sidelobes, and in fact the dominant mode, of number m, can be estimated from the far field pattern, by $m = 2 + 1$, where 1 is the number of minor lobes between the two major lobes. Thus in the first order transverse mode and far field pattern illustrated in figures 5.24 and 5.25, there are two major lobes only with no

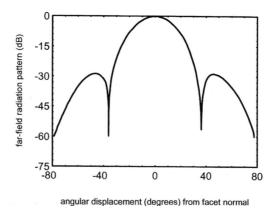

Figure 5.23. The far-field pattern of the fundamental mode in figure 5.22.

Optical properties of heterojunctions 111

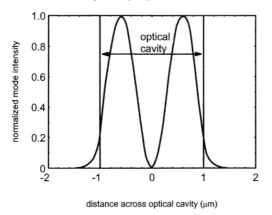

Figure 5.24. The first order ($m=2$) transverse mode of a similar waveguide in a DH laser.

minor lobes in between. For large optical cavity lasers the output beam pattern can be very complicated.

There are a lot factors all pointing towards the use of a narrow active region. The charge uniformity within the active region is improved, and the far field pattern, whilst more divergent, is simpler. The physics of gain and threshold also point in the same direction. Kressel and Butler have conveniently calculated the modal threshold gain for a similar LOC laser corresponding to a refractive index change of 0.06, and whilst it is not identical to the laser corresponding to figures 5.22–5.25 it is sufficiently similar to allow meaningful comparison. For a large cavity, the fundamental mode is not the first to be excited. Figure 5.26 shows the threshold modal gain for the first three modes

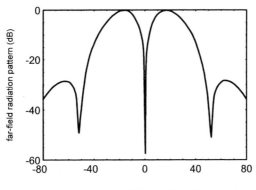

Figure 5.25. Far field pattern of the first order transverse mode.

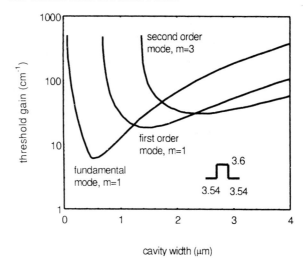

Figure 5.26. Modal threshold gain as a function of cavity width. (After Kressel and Butler.)

and for a cavity width $> 1.2~\mu\text{m}$ the first order ($m=2$) mode has a lower threshold gain than the fundamental. For cavity widths $> 1.7~\mu\text{m}$ the second order ($m=3$) mode has a lower threshold gain. Other modes follow in sequence. At $\sim 2~\mu\text{m}$ the order of modal excitation would be $m=2$, $m=3$, $m=1$, presuming that once a mode were excited others could also be excited. However, these are active waveguides and if the electron density is depleted by one mode others will not achieve threshold. In general, a wide cavity will operate at the highest mode, and where the cavity width corresponds to a crossover between modes, e.g. at $\sim 1.2~\mu\text{m}$ or $\sim 2.3~\mu\text{m}$ above, other factors such as the facet reflectivity may tip the balance in favour of one or the other.

The facet reflectivity has not been considered in figure 5.26 because the gain illustrated is actually the material gain G_{th}. Let the threshold condition be written as,

$$\Gamma(g_{th} - \alpha_{fc}) = (1 - \Gamma)\alpha_c + \alpha_{end} \tag{29}$$

where α_c is the absorption coefficient of the cladding, assumed to be identical on either side in a symmetrical waveguide, Γ is the optical confinement, and α_{fc} is the absorption coefficient of the free carriers in the inverted medium, and g_{th} is the threshold gain coefficient, and α_{end} is the effective absorption coefficient of the end of the cavity, which is simply the mirror loss.

$$\alpha_{end} = \frac{1}{2L} \ln\left(\frac{1}{R_1 R_2}\right). \tag{30}$$

The cladding losses are identified from the fractional powers in the respective regions of the guide, $a_{1,2,3}$, giving a total attenuation coefficient for the mode

$$\alpha = a_1\alpha_1 + a_2\alpha_2 + a_3\alpha_3. \tag{31}$$

For a symmetrical structure

$$\alpha = (a_1 + a_3)\alpha_1 + a_2\alpha_2 \tag{32}$$

and, in terms of the confinement,

$$\alpha = (1 - \Gamma)\alpha_1 + \Gamma\alpha_2. \tag{33}$$

Instead of the round trip losses, consider instead the losses occurring on reflection from one interface only and on traversing the cavity once only, i.e. $\alpha_{fc} = 0$.

$$\alpha = \frac{1}{L}\ln\left(\frac{1}{R}\right). \tag{34}$$

For a very large cavity length the facet reflection becomes less important and in the limit of an infinite cavity length $\alpha = 0$. For a non-zero value of α_1, $\alpha_2 < 0$. At threshold,

$$-G_{th} = \alpha_2 \tag{35}$$

so

$$(1 - \Gamma)\alpha_1 = \Gamma G_{th}. \tag{36}$$

Thus defined, G_{th} represents the recombination region gain coefficient at threshold for no free carrier absorption and no end loss. It is thus a property of the waveguide structure and the material, and therefore allows direct comparison between modes without having to consider the variation in facet reflectivity that inevitably occurs between modes.

Choosing a cavity width corresponding to a low threshold gain for the fundamental mode also ensures a high threshold gain for the first order mode so only the one mode will propagate. The far-field radiation pattern will also be the simplest possible, simpler even than indicated in figure 5.23. Botez and Ettenberg [29] have shown that for thin active regions the far-field pattern approximates to a Gaussian distribution. Defining the normalised thickness D of the active region to be

$$D = \frac{2\pi}{\lambda} \cdot d \cdot \sqrt{n_1^2 - n_2^2} \tag{37}$$

where n_1 is the refractive index of the core (actual thickness d) and n_2 is the refractive index of the cladding in a symmetrical junction, the Botez–Ettenberg approximation applies to $1.8 < D < 6$. By way of comparison the cavity illustrated in figure 5.22 has $D \sim 12$.

Within this approximation the field distribution in the waveguide is quite accurately described by a Gaussian function with the angular beam width perpendicular to the plane of the junction, $\theta_\perp$, given by

$$\theta_\perp = 2 \tan^{-1}\left(\frac{0.59 \cdot \lambda}{\pi w_0}\right) \tag{38}$$

where

$$w_0 = d\left(0.31 + \frac{3.15}{D^{3/2}} + \frac{2}{D^6}\right). \tag{39}$$

For $D < 1.5$ the electric field in the guide approximates to a double exponential as the field extends out into the cladding layers. Then,

$$\theta_\perp = \frac{4.09\left(\dfrac{d}{\lambda}\right)(n_1^2 - n_2^2)}{1 + 3.39\kappa\left(\dfrac{d}{\lambda}\right)^2(n_1^2 - n_2^2)} = \frac{0.65 D \sqrt{n_1^2 - n_2^2}}{1 + 0.086\kappa D^2} \tag{40}$$

with

$$\kappa = \frac{2.52\sqrt{n_1^2 - n_2^2}}{\tan^{-1}(0.36\sqrt{n_1^2 - n_2^2})} - 5.17. \tag{41}$$

Furthermore, the confinement factor can be expressed as [30]

$$\Gamma \cong \frac{D^2}{2 + D^2}. \tag{42}$$

Equation (chapter 3, 33) showed that the threshold current is inversely proportional to the optical confinement. For narrow active regions, $\Gamma \ll 1$, and a high threshold current can be expected. The offset current, J_t, required to bring about transparency, depends directly on the width of the active region via the number of carriers. This leads to the sort of behaviour illustrated in figure 5.27, where the threshold current at low d is dominated by the loss of optical confinement, but at high d is dominated by the transparency condition. The precise details will vary from laser to laser, but the trends are general among double heterostructure lasers. The optimum of low threshold current and fundamental transverse mode of the active region occurs at the point at which loss of optical confinement starts to become significant. This is usually at an active layer thickness of around 100–200 nm and the approximate expressions for the confinement and beam properties apply.

Although the concept of the double heterostructure laser is quite old, it is still far from obsolete, despite the many advances in laser technology that have

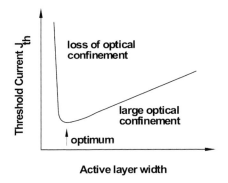

Figure 5.27. Illustration of the dependence of threshold current on active layer width.

taken place in the intervening years. For many applications and material systems the straight-forward double heterostructure design is still appropriate. For many years it has been the work-horse of optical communications systems, for example, and is still used in mid-infra-red applications. It is still the basis of many high power broad area devices, and though separate chapters will be devoted to some of these applications, a brief description of the materials and wavelength ranges of the lasers will be given here.

5.6 Materials and lasers

The wavelength ranges of some common III–V materials at 300 K is shown in figure 5.28. For the ternary compounds the end points represent either the band gaps of the binary components, e.g. GaAs, or the cross-over point from the

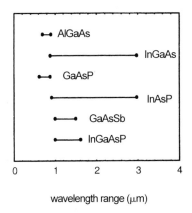

Figure 5.28. Wavelength range of some typical III–V laser materials. (After Casey and Panish.)

direct to indirect band gap. The data is taken from Casey and Panish and is by no means comprehensive. Developments in GaN, described in chapter 11, allow lasers out into the blue and developments in InAsSb lasers allow operation out to 3.4 μm [31] in double heterostructure systems. The particular feature of DH lasers which is relevant to figure 5.28 is the band gap and lattice parameter of the materials systems. The wavelength range of DH lasers is determined by the band gap of bulk-like material, for which a suitable lattice matched substrate must exist. This limitation does not apply to the use of very thin materials in quantum wells where so-called "strained layer" systems allow not only variations in composition outside the range of lattice matching but also emission at shorter wavelengths than would be expected from the bulk-like properties due to quantum confinement effects (see chapter 6). Here the emphasis is on lattice matched materials. The properties of the AlGaAs-GaAs system have already been described at length and will not be repeated. AlGaAs devices are currently grown by a number of methods, as described by Depuis in a personal historical perspective which not only describes the first growth of device quality AlGaAs by MOCVD but also includes a description of the MOCVD growth of other materials [32].

5.6.1 InP systems: InGaAs, InGaAsP, AlGaInP

InGaAs lattice matched to InP does not feature heavily in DH laser technology as the band gap corresponds to emission at ~1.7 μm, at the extreme of the range of wavelengths useful in lightwave technology. Historically the wavelengths 1.31 and 1.55 μm have been very important for long haul, high bit-rate communications systems because of the low loss and low dispersion achievable with suitable fibre designs. The emergence of erbium doped fibre amplifiers (EDFAs) with broad band gain makes this particular wavelength interesting again, but the current technological focus is on the design of monolithic tunable lasers rather than emitters at a single wavelength. Other applications might exist for 1.7 μm lasers in chemical spectroscopy but the focus is on longer wavelengths in the region of 2–5 μm. Lattice matched InGaAs is more commonly found in quantum well lasers.

This aside, room temperature operation of lattice matched InGaAs DH lasers was reported in 1976 (see [33] and references 17–19 therein). Operating at 1.67 μm, lasers fabricated by LPE had relatively high thresholds compared with lasers fabricated by MBE. Since then, growth of double heterostructures by MOCVD [34] resulted in material with a minority carrier lifetime (at low level injection) of $\approx$ 18 μs, clearly very good quality material. Chemical Beam Epitaxy (CBE) was used by Uchida *et al.* [35] to produce a structure with an active region 2 μm thick. The lattice mismatch was less than 5×10^{-4} and the output powers were comparable to LPE grown devices. Uchida's motivation was the desire to use CBE as a growth tool for quantum well devices in the surface emitting configuration, and so prepared DH lasers by way of a demonstration. Not surprisingly, the 2 μm device had a high threshold current density but normalised to the active layer thickness it was reasonable at 7.9 kA/cm^2 . μm. Normalising the threshold currents in this way is a technique that allows comparisons between different devices provided the threshold current is proportional to active layer thickness, which is true for $d > 0.2$ μm.

Nee and Green [36] have measured the optical properties of $In_{1-x}Ga_xAs$ lattice matched to InP ($x = 0.47$) by measuring the reflectivity and transmittivity of 800 nm thick undoped epitaxial layers on Fe-doped InP. The wavelength derivative of the spectra were measured simultaneously using phase sensitive detection techniques, and the spectra fitted to theoretical line-shape curves over the range 0.6 eV to 3.1 eV (figure 5.29). The fundamental band-gap is at 0.75 eV and over the near infra-red region the refractive index is close to 3.4 while that of InP [37] is ~3.1.

Martinelli and Zamerovski [38] studied long wavelength $In_{0.82}Ga_{0.18}As$ lasers emitting at 2.52 μm. In order to overcome a 2% lattice mismatch to the InP substrate a layer of graded InGaAs 20 μm thick was grown in which the In content ranged from 0.53 to 0.82. Graded layers in this manner have been used for many years in the manufacture of light emitting diodes (LED's) where a maximum gradient in the lattice of 1% per μm is used. The rate of change of crystal structure with thickness in this device is 0.1% per μm, ensuring a low density of dislocations over the gradation. The device characteristics were as follows: a cladding layer of $InSb_{0.1}As_{0.14}P_{0.76}$ provided a conduction band offset of 0.11 eV, a valence band offset of 0.3 eV, and with an active layer thickness in the range 0.5–0.7 μm, the optical confinement factor was 0.5 or more. The threshold current exhibited a strong temperature dependence with two characteristic T_0 temperatures; 41 K for $T < 180$ K and 29 K for $T > 180$ K. At

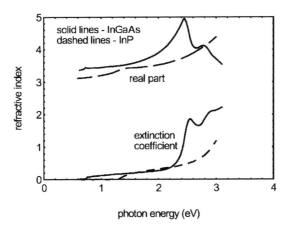

Figure 5.29. The refractive index of lattice matched InGaAs (solid line) and InP (dashed line).

80 K, however, the threshold current was 0.41 kA cm^{-2} averaged over 40 μ wide emitters with 200 μm long cavities.

By contrast, InGaAsP has been investigated extensively. The growth of this quaternary by LPE was a major advance in the development of this material system and because lattice matching can be achieved over the entire wavelength range 0.92 μm to 1.67 μm, lightwave communications technology received a considerable boost. Nahory and Pollack [39] reported detailed studies of the thickness dependence of the threshold current in devices grown by two-phase solution LPE at 635°C. In a two-phase solution the P is provided by solid single crystal platelets of InP floating on a melt consisting of accurately weighed In and undoped polycrystalline GaAs and InAs. As described earlier, phosphorus is very volatile and there are difficulties controlling the composition of the melt, but this method is automatic. The InP is dissolved at high temperatures but upon cooling, when supersaturation sets in, the InP acts as a seed for the deposition of InGaAsP, thus relieving the supersaturation. As an aside, it was not in fact until 1990, some 12 years later, that Thijs *et al.* [40] claimed the first report of an atomically abrupt InGaAsP-InP interfaces grown by OMVPE. Control of the phosphorus content was the essential difficulty and LPE remained the technique of choice for a long time. More information on MOCVD of InGaAsP can be found in a paper by Holstein [41].

The devices made by Nahory and Pollack, contained an active layer of $In_{0.75}Ga_{0.25}As_{0.54}P_{0.46}$ (nominally undoped, $n \sim 2 \times 10^{16}$ cm^{-3}) corresponding nominally to emission at 1.23 μm, but experimental wavelengths ranged from 1.21 to 1.24 μm. The thickness of the active layer ranged from 0.14 μm to 1 μm, the former being the smallest achievable with this method of growth. The devices were modelled by rearranging the basic condition for the threshold current to give

$$J_{th} = d\left[a + \frac{b}{\Gamma}\right] \tag{43}$$

where d and Γ are respectively the active layer thickness and optical confinement, and a and b are adjustable parameters, determined as $a = 1.91$ kA cm^{-3} and $b = 3.0$ kA cm^{-3}. This results in the curve shown in figure 5.30. The optical confinement was calculated assuming a refractive index for the quaternary active layer interpolated from refractive index of the four constituent binaries,

$$n_Q(\lambda) \approx 3.4 + 0.256y - 0.095y^2 \tag{44}$$

where y is the As fraction. The minimum threshold current occurs at ~ 0.2 μm active layer thickness and is ~ 1.5 kA cm^{-2}.

Nelson [42] reported the near equilibrium growth of InGaAsP emitting at 1.35 μm. Using a very similar device structure (typically 4 μm Sn-doped n-InP,

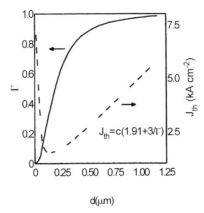

Figure 5.30. Optical confinement and threshold current for 1.23 μm emitting InGaAsP devices grown by LPE.

undoped 0.2 μm InGaAsP, and 2.5 μm Zn-doped p-InP, with a lattice mismatch for the active region of less than 2×10^{-4}) median threshold currents of 920 A cm^{-2} were obtained, the lowest being 670 A cm^{-2} at an active thickness of 0.1 μm. Arai and Suematsu [33] reported devices prepared by LPE with varying compositions capable of emitting over the range of 1.11–1.67 μm, the last representing lattice-matched InGaAs. Again the two-phase technique was used to produce solid solutions with the compositions shown in table 5.1. For the lasers emitting at 1.1 μm, the threshold current increased by 50% to about 8 kA cm^{-2} μm^{-1} compared with ~5 kA cm^{-2} μm^{-1} for longer wavelength emitters. Loss of electrical confinement occurs at $\lambda = 1.1$ μm as the difference in band gap between the active and cladding layers is only 0.23 eV.

Nelson and Dutta [25] reported compositions of $In_{0.74}Ga_{0.26}As_{0.6}P_{0.4}$ and $In_{0.6}Ga_{0.4}As_{0.9}P_{0.1}$ lattice matched to InP emitting at 1.3 μm and 1.55 μm

Table 5.1. Wavelength, As fraction (x) and Ga fraction (y) of two-phase solution LPE grown InGaAsP [33].

λ (μm)	x	y
1.11	0.343	0.157
1.25	0.547	0.252
1.35	0.676	0.312
1.45	0.792	0.368
1.55	0.899	0.419
1.65	0.997	0.466

respectively. For the 1.3 μm device, the surface recombination velocity was reckoned to be ≤ 1000 cm s^{-1}, and the recombination current contributing less than 5% to the threshold current at 300 K. The calculated diffusive leakage current was less than 2% of the threshold current. For devices 380 μm wide and 250 μm long and with active thickness 0.15 μm ± 0.05, the threshold current was 840 A cm^{-2} with an above threshold quantum efficiency of 0.22 mW mA^{-1} per facet.

InGaAsP lasers also exhibit low values of T_0, and for low-doped p-cladding layers field assisted carrier leakage may well be responsible. For higher doping, alternative mechanisms for the temperature sensitivity have been investigated. Non-radiative Auger recombination, in which an electron and hole recombine and transfer their energy to another carrier, is recognised as the main mechanism. As three carriers are involved the rate can be proportional to n^3, but it is often proportional to n^2, particularly in degenerate semiconductors. There are three main Auger mechanisms to consider, CCCH, CHHS, and CHHL, illustrated in figure 5.31 and defined as follows:

- CCCH involves a recombination transition from the conduction band to the heavy hole band at a different momentum, and subsequent transferral of the photon energy to a low lying conduction electron which is promoted to a higher energy within the conduction band. This last process necessarily requires a change in momentum, hence momentum is not conserved in the initial recombination event.
- CHHL and CHHS involve momentum conserving recombination from the conduction band to the heavy hole band. A light hole is promoted to the heavy hole valence band for CHHL and a hole in the split-off band is promoted to the heavy hole band for CHHS.

Figure 5.32 shows the temperature dependence of the recombination lifetime for the three processes calculated by Dutta and Nelson in direct band gap

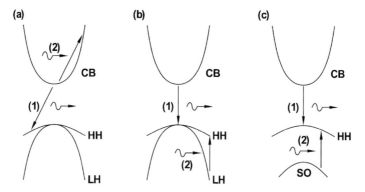

Figure 5.31. Auger process (a) CCCH, (b) CHHL, and (c) CHHS involving recombination and transfer of energy to other carriers.

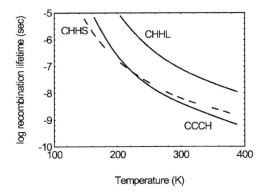

Figure 5.32. Temperature dependence of the three Auger recombination lifetimes in InGaAsAsP. (After Dutta and Nelson.)

material corresponding to 1.3 μm emission. The lifetime rapidly decreases with temperature and hence the efficiency of the stimulated emission process is drastically reduced, resulting in a low value of T_0.

The optical confinement was not described in detail by Nelson and Dutta. There have been numerous calculations of the refractive index as a function of the As fraction at specific wavelengths, some of which were compiled by Amiotti and Landgren for comparison with their own ellipsometric measurements [43]. Lattice matched films of $In_{1-x}Ga_xAs_yP_{1-y}$ grown on (100) InP by MOCVD were measured at 1.3 μm, 1.55 μm, and 1.7 μm. The data for 1.3 μm is shown in figure 5.33 by way of example, but the trends are similar for the other two wavelengths. Implicit in these curves is the application of Vegard's

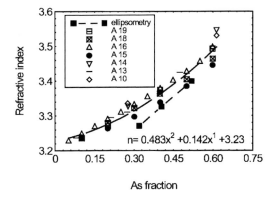

Figure 5.33. Refractive index vs. As fraction for lattice matched InGaAsP layers at 1.3 μm, where the legend A 10, for example, corresponds to reference 10 in Amiotti's paper. (After Amiotti *et al.* [43].)

law to quaternary InGaAsP [44] in which the lattice parameter $a(x, y)$ can be expressed in terms of the lattice parameters of each of the binary constituents

$$a(x, y) = xya_{GaAs} + x(1 - y)a_{GaP} + (1 - x)ya_{InAs} + (1 - x)(1 - y)a_{InP} \quad (45)$$

which becomes,

$$a(x, y) = 0.1894y - 0.4184x + 0.0130xy + 5.8696 \text{Å}. \quad (46)$$

The lattice parameter $a(x, y)$ is known from the condition of lattice-matching so specification of y allows specification of x and the complete composition. All the data from A10 to A19 has been calculated theoretically, which illustrates the difficulty of characterising the optical properties of epitaxial layers for design purposes. Whilst there is good agreement among the theoretical calculations, there are also clear differences, and the ellipsometric data does not agree with any of the calculated data. The differences may be small but comparable to the refractive index difference within the waveguide. The solid line corresponding to the quadratic equation is my average over all the data.

At visible wavelengths InGaAsP is used in a variety of optical data storage and processing applications. Lattice matching to the AlGaAs system provides much better material than matching to InP. Although AlGaAs itself, and even GaAsP, can be used at wavelengths around 650 nm InGaAsP has a higher quantum efficiency [45]. Figure 5.34 shows the quaternary system in greater detail. Lattice matching to InP is illustrated, as is lattice matching to the AlGaAs system. In principle wavelengths out to ≈ 630 nm are possible but the difference in band gap between the InGaAsP and AlGaAs would then be so small as to render the devices inefficient. For this reason interest is centred on $\lambda \approx 670$ nm.

Chong and Kishino [46] report a remarkable reduction in threshold current from over 5 kA cm^{-2} to 1.7 kA cm^{-2} just by increasing the Al content of the

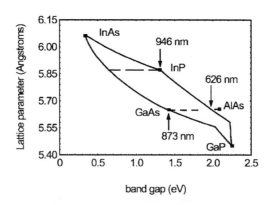

Figure 5.34. Close up of the GaInAsP quaternary lattice parameter vs. band gap.

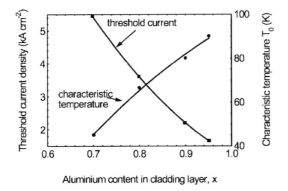

Figure 5.35. Threshold current reduction in visible GaInAsP laser as a function of Al content in the cladding layers. (After Chong and Kishino.)

cladding layers from 0.7 to 0.9 (figure 5.35). The composition of the active region was $Ga_{0.52}In_{0.48}As_{0.03}P_{0.97}$ emitting at 670 nm with an active layer thickness of 0.2 μm. Associated with the improvement in threshold current was an increase in the characteristic temperature, which Chong and Kishino took to be a consequence of the increased band offset from ~ 0.2 eV to ~ 0.3 eV. In an analysis of the leakage currents in this sytem, Lan et al. [45] reanalysed the data of Chong and Kishino and presented a different scheme for the band offsets based on the transitivity rule, which is nothing more than a method for estimating band offsets using Vegard's law. Whereas Chong and Kishino took as their basis band offsets of 0.22 eV for the GaAs-$In_{0.5}Ga_{0.5}P$ conduction band and 0.24 eV for the valence band, Lan et al. took 0.137 eV for the conduction band and 0.339 eV for the valence band and showed that the valence band of the AlGaAs matches the valence band of the $Ga_{0.52}In_{0.48}As_{0.03}P_{0.97}$ at an Al concentration of 65%, thereby explaining the high threshold currents reported by Chong and Kishino at an Al content of 70%. Lan's scheme seems to fit not only the data of Chong and Kishino but also a number of other measurements presented by Lan and would seem to be better. Nonetheless, the difference in opinion highlights again the fundamental difficulty in choosing between apparently conflicting data published in the open literature and it is important to keep an open mind.

AlGaInP is a direct competitor of InGaAsP for some wavelengths (~ 670 nm) but is better suited to red lasers emitting at $\sim 620-640$ nm. Kobayashi et al. [47] report the first use of an Al containing active layer in a DH laser. $Ga_{0.5}In_{0.5}P$ active layers with cladding layers of $(Al_{0.4}Ga_{0.6})_{0.5}In_{0.5}P$ operating CW at room temperature and emitting at 689.7 nm were reported by the same authors earlier in the same year [48]. The addition of Al to the active layer is essential to shorten the wavelength, and Kobayashi et al. managed to add 10% aluminium by MOCVD to produce $(Al_{0.1}Ga_{0.9})In_{0.5}P$ emitting

661.7 nm at room temperature. This was shorter than the shortest wavelength achieved in InGaAsP at that time. The threshold current density was 6.7 kA cm^{-2}.

Achieving shorter wavelengths in this system is not straightforward. The work by Hino *et al.* [49] on yellow emitting lasers at low temperatures demonstrates very well all the difficulties posed by this material system. The addition of Al to the active layer reduces the luminescence intensity through the creation of non-radiative recombination pathways, and a tenfold reduction in luminescence output with the inclusion of 15% Al was reported. At the same time, activation with zinc doping appears to be less effective so the cladding layers, which necessarily require even higher concentrations of Al than the active layer, become resistive [49, 50]. An additional problem arises from the sensitivity of the Al containing materials to environmental oxygen and water, which was overcome [49] through the use of $Ga_{0.5}In_{0.5}P$ capping layers. The developments in this system during the 1980's up until 1990 using MOCVD are shown in table 5.2 [49, 51–56].

The optical properties of the AlGaInP alloy system have been measured by various authors. Moser *et al.* [57] have measured alloy compositions $(Al_xGa_{1-x})In_{0.5}P$ for x = 0, 0.33, and 0.66 using a combination of transmission and ellipsometry for photon energies ranging from 0.01 to 2.2 (see figure 5.36). Kato *et al.* [58] measured the complex refractive index from 0.5 eV to 5.5 eV over the whole composition range, but only the end compositions are shown here (figure 5.37). Estimates of the variation of refractive index with Al fraction x for the important wavelengths 583 nm, 640 nm, and 660 nm are shown in figure 5.38 based on the data from Kato *et al.* As with InGaAsP, developments in this material system are now centred on quantum wells rather than strictly DH lasers. Indeed, with an active layer thickness of 0.08 μm for some of the devices in table 5.2, the tendency toward the thin active regions that define quantum well systems was already in evidence in 1990. These developments in quantum well lasers will be dealt with in chapter 6.

Table 5.2. The performance of MOCVD grown AlGaInP DH laser diodes in the 1980s.

Date	Operation		Wavelength (nm)	Active x $(Al_xGa_{1-x})_{0.5}In_{0.5}P$	Cladding x $(Al_xGa_{1-x})_{0.5}In_{0.5}P$	Active d (μm)	J_{th} kA/cm^2
1984	pulsed	77 K	579	0.32	0.74	0.1	5.6
1985	CW	RT	661	0.1	0.5	0.1	6.7
1986	CW	0°C	621	0.15	0.6	0.1	4.8
1986	CW	77 K	584	0.3	0.7	0.2	1.9
1987	CW	RT	640	0.15	0.6	0.1	4.2
1990	CW	RT	636	0.15	0.7	0.08	5.1
1990	CW	RT	638	0.15	0.7	0.08	5.0

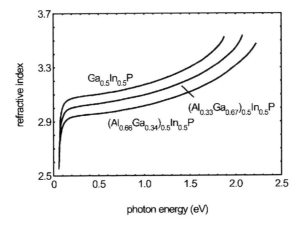

Figure 5.36. Refractive index of AlGaInP as determined by Moser *et al.* [57].

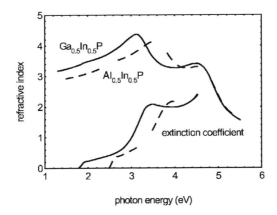

Figure 5.37. Complex refractive index of GaInP and AlInP (after Kato *et al.*). The variation in band gap is evident from the onset of absorption.

5.6.2 InAs-InSb lasers

InGaSbAs injection lasers emitting between 1.8 μm and 2.4 μm have been around for many years. The first report of uncooled operation occurred in 1987 [59] with a threshold current density as low as 5.4 kA cm^{-2} at $\lambda = 2$ μm, and 7.6 kA cm^{-2} at $\lambda = 2.4$ μm. These devices were made by LPE and utilised confining layers of GaAlAsSb. Longer wavelength DH lasers based on InAsPSb grown by hydride vapour phase epitaxy [60] emitting at 2.52 μm at 190 K have also been demonstrated, but the most significant developments have occurred but lately in high power DH devices emitting out to 5 μm [61].

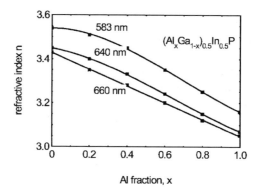

Figure 5.38. Refractive index against Al fraction based on the data of Kato *et al.* [58].

High power devices constitute a class of lasers in their own right and will be described in chapter 10.

5.7 Lateral mode control

The final topic of this chapter concerns the lateral modes, not to be confused with transverse modes, which are controlled by the thickness and refractive index of the active region and can lead to complex far-field patterns perpendicular to the junction. Similarly, complex far-field behaviour in the plane of the junction is also possible due to the lateral modes. In other laser systems, for example HeNe gas lasers operating cw, or solid state lasers such as Nd:YAG, spatial modes arise from misalignment of the resonator reflectors. Confusingly, however, these are referred to as transverse modes in conventional laser technology. An example of such transverse modes is shown in figure 5.39 from a HeNe laser. The term "misalignment" can imply unwanted, but it is perfectly possible of course that transverse modes in the output beam are

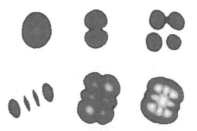

Figure 5.39. Transverse (lateral) mode structure of a HeNe laser showing regularly spaced bright spots in the output. The yellow colour is an artefact of the intensity.

desired. Nonetheless, the term "misaligned" will be used. Perfectly plane reflectors slightly off parallel will not lead to the transverse modes shown in figure 5.39 but if the reflectors are concave, as is commonly the case, misalignment can lead to closed round trip paths. That is, a beam traced round the cavity might make several trips striking the reflectors at several different points before returning to the original position. The output from the partially reflecting output coupler will therefore consist of regularly spaced bright spots, the number of which depends on the extent of the misalignment.

In a diode laser lateral modes do not arise from misalignment of the reflectors – these are always parallel because of the fabrication method – but from filamentation of the current. Filaments are spatially localised regions exhibiting high current flow and hence high brightness due to localised stimulated emission. In a broad area laser the near field pattern, which is essentially a representation of the lateral variation of light intensity within the active layer, can be complicated and exhibit chaotic behaviour in the time domain. Filaments are not necessarily static but can move across the active region, disappearing and "igniting" apparently at random. The subject of filamentation will be dealt with in more detail in relation to high power laser diodes, in which it is a common feature. It is sufficient to say here that historically the appearance of filamentation in the output led to severe problems in telecommunications where the core diameter of a single mode fibre was of the order of 10 μm across and coupling the output of broad area lasers to fibres was made difficult by the random location of the output along the facet. The answer lay in the use of a stripe contact in which the current is confined to a narrow region extending the length of the cavity (figure 5.40). The stripe is wide enough to ensure that the full width of the laser over which the current flows is illuminated. Sometimes the stripe is defined by an insulator, sometimes by implantation of hydrogen to render the semiconductor semi-insulating or diffusion of dopants to alter the conductivity type. Whatever mechanism is employed, it has become common practise to incorporate the stripe geometry into the routine fabrication of DH lasers, and indeed all the AlGaInP lasers, and most of the InGaAsP lasers, described within this chapter have used this configuration. This represents an extra complexity in the design and fabrication

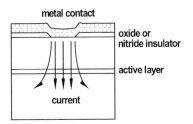

Figure 5.40. Schematic stripe contact in which the current is confined to flow in a confined region.

of the lasers but does not alter the physics of the optical and electrical confinement or the details of the threshold current.

5.8 Summary

This chapter has described the electrical and optical properties of heterojunctions in a variety of material systems. As well as the basic properties such as band bending and recombination, specific properties such as band offsets and refractive index differences are presented in detail.

For many of the DH lasers mentioned in this chapter technology has moved on. Whilst the lasers may still be fabricated and found commercially the current focus in research and on developments in fabrication revolves around other laser structures such as single or multiple quantum well devices. Even for these devices, however, the growth, refractive index profiles, carrier confinement, and optical confinement described here are all relevant to these devices also, and in the case of the mid-IR antimonide lasers, the double heterostructure is still the most important laser geometry.

5.9 References

[1] Kressel H and Butler J K 1979 *Semiconductors and Semi-metals* vol 14 (New York: Academic Press), Eds: R K Willardson and A C Beer
[2] *Properties of Gallium Arsenide, EMIS Datareviews Series No 2* 1986 (London: INSPEC)
[3] Singh J 1993 *Physics of Semiconductors and Their Heterostructures* (New York: McGraw-Hill)
[4] Lide D R (ed.) 1913–1995 *CRC Handbook of Chemistry and Physics* 75th edition (Boca Raton: CRC Press)
[5] Grovenor C R M 1989 *Microelectronic Materials* (Bristol: Adam Hilger) p 140
[6] *Properties of gallium arsenide, EMIS Datareviews* 1986 (London: INSPEC)
[7] Casey H C Jr and Panish M B 1978 *Heterostructure Lasers Part B: materials and operating characteristics* (Academic Press)
[8] Gebauer J, Lausmann M, Redmann F, Krause-Rehberg R, Leipner H S, Weber E R and Ebert Ph 2003 *Phys. Rev. B* **67** 235207
[9] Thomas H, Morgan D V, Thomas B, Aubrey J E and Morgan G B (eds) 1986 Gallium Arsenide for Devices and Integrated Circuits *Proc of the 1986 UWIST GaAs School* (IEE)
[10] Nelson R J 1979 *Appl. Phys. Lett.* **35** 654
[11] Olsen G H, Neuse C J and Ettenberg M 1979 *Appl. Phys. Lett.* **34** 262
[12] Razeghi M, Hersee S, Hirtz P, Blondeau R, de Cremoux B and Duchemin J P 1983 *Electron. Lett.* **19** 336
[13] Tsang W T, Reinhardt F K and Ditzenberger J A 1982 *Appl. Phys. Lett.* **41** 1094

[14] Anderson R L 1962 *Solid State Electronics* **5** 341
[15] Yu E T, McCaldin J O and McGill T C 1992 *Solid State Physics; Advances In Research And Applications* (H Ehrenreich & D Turnbull, eds, vol 46 pp 1–146) (Boston: Academic Press)
[16] Furtado M T, Loural M S S, Sachs A C and Shieh P J 1989 *Superlattices and Microstructures* **5** 507–510
[17] Guillot C, Dugay M, Barbarin F, Souliere V, Abraham P and Monteil Y 1997 *J. Elect. Mats.* **26** L6–L8
[18] Guillot C, Achard J, Barbarin F and Dugay M 1997 *J. Elect. Mats.* **28** 975–979
[19] Hall D J, Hosea T J C and Button C C 1998 *Semiconductor Science and Technology* **13** 302–309
[20] Abraham P, Perez M, Benyattou T, Guillot G, Sacilotti M and Letartre X 1995 *Semicond. Sci. Technol.* **10** 1585
[21] Casey H C Jr and Panish M B (1978) *Heterostructure Lasers Part A, Fundamental Principles* (Academic Press)
[22] Thompson G H B 1980 *Physics of Semiconductor Laser Devices* (Chichester: John Wiley & Sons) p 145
[23] Thompson G H B 1980 *Physics of Semiconductor Laser Devices* (Chichester: John Wiley & Sons) p 148
[24] Thompson G H B 1980 *Physics of Semiconductor Laser Devices* (Chichester: John Wiley & Sons) p 157
[25] Nelson R J and Dutta N K 1985 *Semiconductors and Semimetals* vol 22 part C p 14 (Academic Press) Edited by W T Tsang
[26] Nelson R J and Sobers R G 1978 *App. Phys. Let.* **32** 761
[27] van Opdorp C and t'Hooft G W 1981 *J. Appl. Phys.* **52** 3827–3839
[28] Ahrenkiel R K, Olson J M, Dunlavy D J, Keyes B M and Kibbler A E 1990 *J. Vac. Sci. & Technol.* **A8** 3002
[29] Botez D and Ettenberg M 1978 *IEEE J. Quant. Electr.* **QE-14** 827
[30] Botez D 1978 *IEEE J. Quant. Electr.* **QE-14** 230
[31] Wu D, Lane B, Mohseni H, Diaz J and Razeghi M 1999 *Appl. Phys. Lett* **74** 1194–1196
[32] Depuis R D 2000 *IEEE J. Select. Topics in Quant. Electr.* **6** 1040–1050
[33] Arai S and Suematsu Y 1980 *IEEE J. Quant. Electr.* **QE-16** 197
[34] Gallant M and Zemel A 1988 *Appl. Phys. Lett.* **52** 1686–1688
[35] Uchida T, Uchica T K, Mise K, Yokouchi N, Koyama F and Iga K 1990 *Jpn. J. Appl. Phys. Pt 1.* **29** 1771–1772
[36] Nee T W and Green A K 1990 *J. Appl. Phys.* **68** 5314
[37] Herzinger C M, Snyder P G, Johs B and Woollam J A 1995 *J. Appl. Phys.* **77**(4) 1716
[38] Martinelli R U and Zamerovski T J 1990 *Appl. Phys. Lett.* **56** 125
[39] Nahory R E and Pollack M A 1978 *Electronics Letters* **14** 726
[40] Thijs P J A, Montie E A, van Kesteren H W and t'Hooft G W 1990 *Appl. Phys. Lett.* **53** 971–973
[41] Holstein W L 1996 *J. Cryst. Growth* **167** 525–533
[42] Nelson J 1979 *Appl. Phys. Lett.* **35** 654
[43] Amiotti M and Landgren G 1993 *J. Appl. Phys.* **73** 2965
[44] Nahory R E, Pollack M A, Johnston W D Jr and Barns R L 1978 *Appl. Phys. Lett.* **33** 659

[45] Lan S, Chan Y C, Yu W J, Cui D L, Yang C Q and Liu H D 1996 *J. Appl. Phys.* **80** 6355–6359
[46] Chong T H and Kishino K 1989 *Electronics Letters* **25** 761–762
[47] Kobayashi K, Kawata S, Gomyo A, Hino I and Suzuki T 1985 *Electronics Letters* **21** 1162
[48] Kobayashi K, Kawata S, Gomyo A, Hino I and Suzuki T 1985 *Electronics Letters* **21** 931–932
[49] Hino I, Kawata S, Gomyo A, Kobayashi K and Suzuki T 1986 *Appl. Phys. Lett.* **48** 557–558
[50] Blood P 1999 *Mat. Sci. Eng.* **B66** 174–180
[51] Ikeda M, Honda M, Mori Y, Kaneko K and Watanabe N 1984 *Appl. Phys. Lett.* **45** 964–966
[52] Kobayashi K, Kawata S, Gomyo A, Hino I and Suzuki T 1985 *Electronics Letters* **21** 1162–1163
[53] Kawata S, Kobayashi K, Gomyo A, Hino I and Suzuki T 1986 *Electronics Letters* **22** 1265–1266
[54] Kawata S, Fujii H, Kobayashi K, Gomyo A, Hino I and Suzuki T 1987 *Electronics Letters* **23** 1327–1328
[55] Itaya K, Ishikawa M and Uematsu Y 1990 *Electronics Letters* **26** 839–840
[56] Ishikawa M, Shiozawa H, Tsuburai Y and Uematsu Y 1990 *Electronics Letters* **26** 211–213
[57] Moser M, Winterhoff R, Geng C, Queisser I, Scholz F and Dornen A 1994 *Appl. Phys. Lett.* **64** 235–237
[58] Kato H, Adachi S, Nakanishi H and Ohtsuhka K 1994 *Jpn. J. Appl. Phys.* **33** 186–192
[59] Drakin A E, Eliseev P G, Sverdlov B N, Bochkarev A E, Dolginov L M and Druzhinina L V 1987 *IEEE J. Quant. Electr.* **QE–23** 1089
[60] Martinelli R U and Zamerowski T J 1990 *Applied Physics Letters* **56** 125
[61] See for example Lane B, Tong S, Diaz J, Wu Z and Razeghi M 2000 *Mat. Sci. Eng.* **B74** 52–55

Problems

1. Calculate the band gap of $Al_{0.4}Ga_{0.6}As$ at room temperature, and assuming 40% of the band offset lies in the conduction band, calculate the position of the electron Fermi level relative to the bottom of the conduction band in the active region of a $GaAs$-$Al_{0.4}Ga_{0.6}As$ DH laser with an electron concentration of: (a) 10^{18} cm^{-3}; (b) 2×10^{18} cm^{-3}; (c) 5×10^{18} cm^{-3}; (d) 10^{19} cm^{-3}.
2. Plot n against F_{n2} on the same graph as p against $(E_v - E_f)$ to find the transparency density in GaAs.
3. Calculate the density of electrons injected over the barrier for the laser in (1). (Assume $N_c = 4.7 \times 10^{18}$ cm^{-3} and $kT = 0.0258$ eV.
4. Assuming charge neutrality in the active region calculate the hole density in this region of the DH laser.
5. Hence calculate the electron leakage current assuming a layer 1 μm wide with $D_n \approx 90$ cm^2 s^{-1} and $N_c = 4.7 \times 10^{18}$ cm^{-3} for the four cases above.

Problems   131

6. Consider the InGaAsP-InP laser described by Nahory and Pollack (equation 43 and ref. [39]). Figure 5.33 shows the median refractive index to be 3.44 at 1.3 μm, but it could be higher by at least 0.01. Assuming a refractive index for InP of 3.18 calculate the optimum thickness of the active region for both of these values. Hence show that $D < 1.5$ in both cases and calculate the optical confinement. Hint: use the Botez approximation for the confinement to express the threshold current as a function of the active region thickness d_a, and differentiate to find the minimum.

Chapter 6

Quantum well lasers

The quantum well laser has much in common with the double heterostructure laser. Both have active regions constructed from thin epitaxial layers of narrow band gap semiconductor sandwiched between wide band gap layers. The current flow is perpendicular to this layer and the optical emission is parallel to it. Both devices are therefore edge-emitting. Many of the materials are also common between the two. The differences between the two stem from the thickness of the active layer, which is 150–200 nm thick in the DH laser, but no more than 15 or so nm thick, and often much less, in the quantum well laser. This difference is sufficient to warrant a whole new approach to the emitting material, with a set of energy levels different from the bulk material of the DH laser and dependent on layer width. This affects the density of states and introduces a polarisation dependence into the stimulated optical transition that does not exist in bulk material. In addition, with such a thin layer there is the question of how effectively, and efficiently, are carriers captured and retained by the active region, as well as the question of the effectiveness of the optical waveguiding. These issues are discussed in this chapter, but we start with a description of exactly what is meant by the term "quantum well".

6.1 Classical and quantum potential wells

A "quantum well" is a potential well in which quantum effects occur. A particle confined to such a well cannot be treated using the laws of classical mechanics as developed by Newton. Such laws apply to the problems of our experience where we can define forces, calculate accelerations, and work out trajectories. Even then, Newton's method is not always the most appropriate, particularly for potential well problems. Take for example, a ball that rolls freely on a surface with negligible frictional loss. We can easily define the forces acting on it given the angle of the surface, but if the angle changes with distance the forces change. So, if the ball is contained within a vessel, say a bowl with sloping sides, then in trying to work out its motion it is necessary to account for the variation in force as the ball moves. This is not an easy problem to formulate generally. However, a force acting over a distance does work, so in moving away from the centre of the bowl the ball must do work against gravity and increase its potential energy. If it is then left to move freely from a position away from the centre it will move towards the centre and recover the potential energy in the form of kinetic energy. If the problem is formulated in terms of the interchange between kinetic and potential energy it becomes quite simple. We can work out how high the ball will ascend by working out how much energy it has to start with and comparing this to the total change in potential energy. It is obvious from this that the potential is a minimum at the bottom of the bowl, hence the term "potential well".

In quantum potential wells pretty much the same considerations apply. A total energy approach, rather than a force-based approach, allows for sensible solutions. For example, one of the simplest quantum wells in nature is the hydrogen atom, consisting of a positively charged nucleus and a single electron. The force between the nucleus and the electron is given by Coulomb's law and the potential energy arising from this force decreases with increasing distance, r, away from the nucleus as $1/r$. In principle any atomic system can be treated the same way but the presence of multiple charges distributed in space – the orbiting electrons – complicates matters enormously. With the hydrogen atom there are only two charges in the whole system so it is much simpler. Moreover, the masses of the two charges are very different, which further simplifies the problem. Were the mass of the electron and the nucleus equal, the mutual effect of each particle on the other would cause them both to accelerate towards each other and to orbit round a common centre of mass lying equidistant between the two. However, a nucleus is approximately 2000 times heavier than an electron so the effect of the electron on the nucleus is insignificant compared with the effect of the nucleus on the electron. The electron is therefore accelerated toward a nucleus which is effectively at rest relative to it.

So far there is nothing about this system that makes it appear different from, say, a gravitational system in which a planet orbits a star. Of course, electrostatic forces are much stronger than gravitational forces, so the masses are very much different and the distances involved are very much smaller. The scale of the problem is different but otherwise we would expect, by analogy with gravitational systems, to be able to calculate an orbital path for the electron. However, an electron is a charged particle, and according to classical electrodynamics an accelerating charge radiates electromagnetic energy. An electron orbiting a nucleus is, by definition, accelerating and according to classical physics, therefore, the atom will collapse as the energy is radiated away. It does not, though. Atoms are stable and when they do radiate energy it is at well defined wavelengths. Early in the development of quantum mechanics it was postulated that electrons have stable orbits and only exchange energy with their environment when changing orbits, but there was initially no explanation put forward as to why this should be so. This was the problem that Schrödinger set out to solve, and for which he developed his wave mechanics.

The wave equation exploits the conservation of energy, as described in chapter 4. The sum of kinetic and potential energies in a mechanical system is constant such that a body moving in a potential field exchanges kinetic energy for potential energy. You might argue that although the ball rolling in the bowl will ascend the sides of the bowl gaining potential energy as it does so the gain in potential energy will not be the same as the loss in kinetic energy because energy is dissipated through friction. This is true. As the ball ascends the sides it slows down, stops, and begins to move the other way. In an ideal system it moves from one side to the other *ad infinitum*, but of course we know in practice the ball eventually stops moving completely. The system is described as dissipative. The energy is transferred via friction into small scale random motions of the atoms of the bowl, and as such is not recoverable by the ball. The energy is lost rather than destroyed, so in this sense it is conserved, but it is not conserved by the ball, which is the body of interest to us. In the hydrogen atom, however, there are no other interactions and therefore energy is not dissipated.

This method of treating mechanical systems by conserving energy was developed by Hamilton, an Irish mathematician of the nineteenth century. The sum of kinetic and potential energies is known, therefore, as the Hamiltonian. It should be emphasised that Hamilton's mechanics does not fundamentally differ from Newton's in so far as the underlying concepts of position, velocity, momentum, etc. are all the same. It is a different approach that is better suited to problems of constraint (the particle is constrained to move within the well) and Schrödinger adapted the mechanics of Hamilton to describe quantum systems, developing wave mechanics in the process. Schrödinger explained the quantisation of atomic orbitals by treating the electron as a wave and assuming that the orbital path must equal a whole number of wavelengths, so the electron constructively interferes with itself. Schrödinger's wave equation was developed in chapter 4 and the treatment will not be repeated here. A reminder about some of the important points will be useful, however.

Wave mechanics uses the idea of a wavefunction to describe the basic properties of matter, but it should be emphasised that the wavefunction is a mathematical construction that has no particular physical significance. Whatever the reality of an electron may be, there are circumstances where it behaves either as a wave or as a particle. The physical reality lies in the square of the wavefunction, which, by analogy with a classical wave, represents intensity. In quantum terms the position dependent intensity represents the probability of finding the particle at a particular position, but it should be noted that there are other formulations of quantum mechanics that do not use wavefunctions, and it is not necessary to describe matter as a wave in order to solve quantum mechanical problems. Nonetheless, wave mechanics is one of the most common and easily understood formulations and is well suited to quantum well problems.

A plane wave propagating in the x direction, has the form

$$\psi = \psi_0 \exp[j(\omega t - kx)] \tag{1}$$

where ψ_0 is an amplitude and k is the wavevector. For convenience the time-dependent term (ωt) is usually omitted. The wavevector is given in terms of the energy of the particle relative to the potential, i.e.

$$k = \frac{1}{\hbar}\sqrt{2m(E-V)}. \tag{2}$$

If the total energy $E < V$ the wavevector is imaginary. From equation (2) this will lead to a real exponential decay in the wavefunction. Classically this would be interpreted as attenuation. The amplitude of the wave would decrease the further it propagates, but for a single particle this concept is meaningless. Either the particle exists or it does not; no physical meaning can be attached to the idea that it fades away to nothing as it propagates. We return, then, to the notion of probability and to the idea that the intensity of the wavefunction represents the probability of finding the particle. If the wavefunction decays to zero it indicates a decreasing probability of finding the particle in this region of space.

The wavefunction of equation (1) is simple to deal with; it is harmonic, has a fixed amplitude, and represents a travelling wave. Not all particles can be expressed by functions of this sort, though, especially if the particle is confined to some region of space. The travelling wave solution clearly will not apply, and in general the wavefunctions are more complicated. An arbitrary wavefunction can be constructed from other, known, wavefunctions according to equation 65 of chapter 4.

$$\psi(x) = \sum_n a_n \psi_{0n} \exp(jk_n x) = \sum_n a_n \psi_n(x) \tag{3}$$

where n is an integer and a_n is a coefficient. The wavefunctions $\psi_n(x)$ are said to form the basis set and (x) is said to be expanded in the basis set.

Equation (4) can be used to explain the very important property of orthonormality. The probability of finding the particle at any position in space has already been defined as the intensity (chapter 4, equation (41)). Integrating over all space,

$$\int \sum_m a_m^* \psi_m^*(x) \sum_n a_n \psi_n(x)\, dx = 1 \tag{4}$$

where the index of the complex conjugate has been changed to distinguish it from the wavefunction. This leads to a series of terms of the form

$$\int a_m^* \psi_m^*(x) a_n \psi_n(x) \, dx \qquad (5)$$

which leads to the interesting question of what happens if $m \neq n$? The integral has no physical meaning. Whereas we can appreciate that the integral over all space of one wavefunction multiplied by its complex conjugate represents the certainty of finding the particle a similar integral of the product of the wavefunction and the complex conjugate of an entirely different wavefunction can have no such meaning. Therefore

$$\int a_m^* \psi_m^*(x) a_n \psi_n(x) \, dx = a^*{}_m a_n \int \psi_m^*(x) \psi_n(x) \, dx = \begin{matrix} 0, \, m \neq n \\ a_m^* a_n, \, m = n. \end{matrix} \qquad (6)$$

This is ortho-normality. For $m \neq n$ the states are orthogonal and for $m = n$ the integral over all space is normalised to unity. Therefore

$$\int \psi^*(x) \psi(x) \, dx = 1 = \int \sum_n |a_n|^2 \psi_n^*(x) \psi_n(x) \, dx \qquad (7)$$

which is what we would expect intuitively. If the wavefunction is constructed from a basis set the probability of finding the particle is given by the weighted sum of the probabilities of each of the wavefunctions in the basis set. This is an immensely useful property of real quantum systems as often we do not know the form of the wavefunction. It has to be constructed from other known functions and the property of ortho-normality allows many terms to be eliminated and simplifies much of the mathematics.

To recap, there is a quantum particle moving in a potential. Its wavefunction may be harmonic or it may be a superposition of states. Its wavevector can be calculated from equation (2). Essentially this is all we need to know in order to treat a range of problems. If the particle is travelling in a region of space where the potential changes abruptly, known as a potential step (figure 6.1), the wavevector must change. By analogy with an optical wave incident on a refractive index step, some reflection will occur. Again, the reflection has to be interpreted in terms of probability. It makes no physical sense to suppose that the particle is divided into two with one part propagating onward and the other part being reflected. Instead, the reflection represents a probability that the entire particle is reflected at the boundary.

A quantum particle incident on a potential step such that the wavefunction changes from real to imaginary penetrates into the potential to some extent, and there is a finite probability of finding the particle inside the potential step. There is no classical analogy to this phenomenon. A classical particle incident

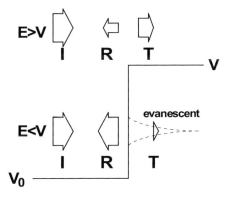

Figure 6.1. An electron incident on a barrier being reflected. For E < V the amplitude decays to zero inside the barrier.

on such a potential step would be entirely reflected. In quantum systems the particle is also reflected if the potential step is thick, but if the potential extends over a limited distance, such that it is said to form a barrier, there is a probability that the particle will appear on the other side of the barrier. This is known as quantum mechanical tunnelling (figure 6.2).

The behaviour of an electron inside a quantum well can now be understood. The simplest quantum well is a square well, of width 2L, centred on $x=0$ such that $V=0$ for $|x|<L$ and $V=\infty$ for $|x|>L$. The electron inside the well has a finite energy greater than V such that the propagation constant is real,

$$k = \frac{1}{\hbar}\sqrt{2mE} \qquad (8)$$

but within the barriers of the well the propagation constant is not only imaginary but infinite. The penetration into the barriers is therefore zero and can be neglected in this problem. This doesn't tell us the mathematical form of the wavefunction but the solution to Schrödinger's equation is particularly simple in this problem. Putting $V=0$ leads to

$$-\frac{\hbar^2}{2m}\frac{\partial^2 \psi}{\partial x^2} = E\psi \qquad (9)$$

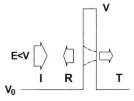

Figure 6.2. Partial reflection becomes tunnelling for a thin barrier.

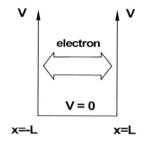

Figure 6.3. An electron in a well is confined by reflection at both barriers.

which is the equation of a simple harmonic oscillator with a known solution

$$\psi = A\cos(kx) + B\sin(kx). \tag{10}$$

The boundary conditions are similarly simple; because the wavefunction does not penetrate into the barrier the wavefunction goes to zero at $x = \pm L$. That is,

$$A\cos(-kL) + B\sin(-kL) = 0 = A\cos(kL) - B\sin(kL) \tag{11}$$

and

$$A\cos(kL) + B\sin(kL) = 0 \tag{12}$$

from which it is easy to show that

$$A\cos(kL) = B\sin(kL) = 0. \tag{13}$$

The trivial solutions occur when A or B are zero, but these are mutually exclusive. If $A=0$ then $B \neq 0$, by definition, otherwise the total wavefunction is zero and the particle doesn't exist. A similar argument applies to the condition $B=0$. Taking the odd functions first, i.e. $A=0$, then

$$kL = n\pi \tag{14}$$

where n is an integer, so that the sine function is zero to satisfy (16). Therefore,

$$E = \frac{\hbar^2 k^2}{2m} = \frac{\hbar^2 \pi^2 n^2}{2L^2 m}. \tag{15}$$

This wavefunction has to be normalised so that equation (6) applies, leading to

$$B = \frac{1}{\sqrt{L}}. \tag{16}$$

Similarly, for $B=0$, the quantisation condition becomes

$$kL = (n - \tfrac{1}{2})\pi \tag{17}$$

and

$$E = \frac{\hbar^2 k^2}{2m} = \frac{\hbar^2 \pi^2 (n-\frac{1}{2})^2}{2L^2 m}. \tag{18}$$

Normalisation again leads to

$$A = \frac{1}{\sqrt{L}}. \tag{19}$$

Therefore the general solution to this particular problem is

$$\psi = \frac{1}{\sqrt{L}} \sum_{n=1}^{\infty} \left[a_n \sin\left(\frac{n\pi x}{L}\right) + b_n \cos\left(\frac{(n-\frac{1}{2})\pi x}{L}\right) \right] \tag{20}$$

with

$$\sum_{n=1}^{\infty} |a_n|^2 + |b_n|^2 = 1. \tag{21}$$

Equation (21) expresses the fact that the particle may be in any one of a number of states so the coefficients a_n and b_n represent weighting factors that partition the probability among the states.

Several things stand out immediately from this solution.

- the wavefunctions are harmonic;
- each harmonic is associated with a higher energy;
- the energy is therefore quantised;
- the lowest energy state ($n=1$, even function) lies above the zero potential level by an amount

$$E_1 = \frac{\hbar^2 \pi^2}{8L^2 m} \tag{22}$$

called the zero point energy

- the energy depends inversely on the electron mass
- the zero point energy increases as the well width is reduced
- the separation between the states increases as the well width is reduced.

6.2 Semiconductor quantum wells

Real quantum systems do not have an infinite potential outside the well. Even natural wells, such as the hydrogen atom, which is in fact approximated very well by the infinite square well despite its $1/r$ dependence, do not have an infinite potential. The ionisation potential for the latter is 13.6 volts, and this is

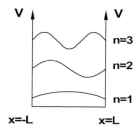

Figure 6.4. Wavefunctions in a square potential well.

effectively the depth of the well. All real wells therefore have a finite potential, but not all real wells necessarily have a square potential profile. The semiconductor growth technology described in relation to the fabrication of double heterostructure lasers provides the ideal method for manufacturing artificial finite potential square wells simply by reducing the thickness of the active region until it is no more than a few nanometres across. The spatial variation in the conduction and valence bands defines the quantum well system.

Of course, semiconductors are themselves quantum systems, irrespective of the dimensions of the layers. Energy bands are quantum constructs, but in conventional semiconductor physics this fact is often overlooked as the effective mass approximation allows the electron to be treated as a classical particle. This approximation no longer works in a quantum well system, and it is necessary to go back to the quantum basics in order to develop the system. Electrons in the conduction (c) or valence band (v) are described by wavefunctions of the form

$$\psi_{(c,v)k}(r) = B u_{(c,v)k}(r) \exp(jkr) \qquad (23)$$

where k is the wavevector and r is a 3-dimensional position vector. B is a normalisation constant that satisfies equation (6). This is essentially the wavefunction for a plane wave but with a coefficient $u_{(c,v)k}(r)$ that has the same periodicity as the lattice. It describes an electron that extends throughout the volume of the crystal but with a spatially modulated probability density corresponding to $|u^*_{(c,v)k}(r) \cdot u_{(c,v)k}(r)|$. Equation (23) is known as a Bloch function and $u_{(c,v)k}(r)$ is a unit cell function (see appendix III), and is formally the solution to an electron propagating in a periodic potential (figure 6.5). This, rather than a constant potential, is the most realistic representation of the potential inside a crystal lattice where the presence of spatially distributed charges on the atoms has to be taken into account in the motion of the electron. Band structure calculations based on equation (23) lead to well defined parameters such as the effective mass and the band gap.

The concept of effective mass is very useful. The conduction band and the valence band are regarded as being parabolic with k, which will always be true

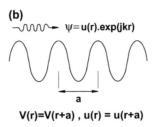

Figure 6.5. A plane wave in a constant potential (2) and its modification to a Bloch function in a periodic potential (b).

for a limited range of energies, as can be seen by expanding the energy in a Taylor series about the wavevector. To second order,

$$E(k_0 + \delta k) = E(k_0) + \delta k \frac{dE(k_0)}{dk} + \frac{1}{2}(\delta k)^2 \frac{d^2 E(k_0)}{dk^2} + \ldots \quad (24)$$

At the centre of the Brillouin zone ($k=0$) or at the edge ($k = \pm \pi/a$), in fact wherever a minimum in the conduction band occurs,

$$\frac{dE}{dk} = 0 \quad (25)$$

and hence

$$E(k_0 + \delta k) = E(k_0) + \frac{1}{2}(\delta k)^2 \frac{d^2 E(k_0)}{dk^2}. \quad (26)$$

Assuming $k_0 = 0$ so $\delta k = k$, and writing the electron energy

$$E(k) = E(0) + \frac{(\hbar k)^2}{2m^*} \quad (27)$$

it is apparent that the effective mass is inversely related to d^2k/dE^2. In short, provided the variation of E with k is approximately parabolic the Taylor expansion is valid to second order, an effective mass can be defined without difficulty. In essence such a mass is defined because the semiconductor is treated as a one particle system, the particle in question being either an electron or a hole. In reality the semiconductor is a many-body system and each electron interacts with every other electron via Coulomb and exchange forces, the combined effect of which is to cause the electron to behave as if it has a mass different from the free electron mass. The effective mass is therefore an approximation that allows us to ignore all the other electrons or holes.

The bulk effective mass is defined by the band structure and is derived naturally from the detailed quantum mechanics. The difficulty in a quantum well is that not only that the material properties change abruptly at the interfaces so that an electron traversing the interface must also change its properties, but also that quantisation of the electron wavevector occurs so that the behaviour in the x, y, and z directions become very different. It is not clear, therefore, what effective mass is appropriate in a quantum well, but it is clear that the electron or hole must have an effective mass. Not only does it determine the quantisation energy (equations 15 and 18) but as the electrons and holes are essentially free to move in the x-y plane of the well there will also be dispersion in $E(k)$ (equation 27). Appendix III gives some insight into the origins of the conduction and valence bands and the derivation of the effective mass. For the remainder of this chapter it will be taken that an effective mass can be defined inside the well in both the plane of the well (called the in-plane effective mass) and in the growth direction of the quantum well (quantisation effective mass), though these may not be the same. The in-plane effective mass determines the threshold condition, and the quantisation effective mass determines the energy levels and therefore the lasing wavelength.

The effective mass approximation effectively smooths out the potential fluctuations inside the semiconductor. The presence of the lattice is ignored and the electron is regarded as a particle propagating freely through the solid at a specific energy determined by its wavevector. In effect the conduction band is assumed to have a constant energy throughout real space rather than the rapid variations on the scale of the lattice. The only fluctuations of note are therefore the real variations introduced by a change in the material properties, such as the presence of an atomically abrupt interface between two materials. This leads to well-defined conduction and valence band offsets, as described in chapter 5. A single heterojunction therefore provides a potential step, and two heterojunctions in close proximity form the quantum well system comprising the well in both the conduction and the valence bands.

In this sense a quantum well laser is a natural extension of the DH laser. An electron injected into the active region of the latter enters a potential box defined by the heterojunctions, but the width of the active region is such that this is considered a bulk, rather than a quantum, system. Therefore the electron resides at the bottom of the conduction band and not at the zero point energy. However, if the dimensions of the low band-gap active region are reduced from a few hundred nanometres down to just a few, usually less than 20 nm or so, the double heterostructure becomes a quantum well. The nature of the semiconductors used in heterojunctions means that the band offsets are commonly no more than a fraction of an electron volt, perhaps 0.1 or 0.2 eV, depending on the materials used. With such a small potential step the wavevector within the barrier, where $E<V$, is finite but imaginary. The penetration depth of the wavefunction into the barrier, given by the inverse of the wavevector, is non-zero and may in fact be quite large. Therefore it is necessary in a finite well to

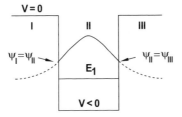

Figure 6.6. Matching the wavefunctions at the edge of the well.

match the wavefunctions at the boundaries of the well (figure 6.6) in order to solve Schrödinger's equation. This is an additional complication over the infinite well that makes the solution of the equations much more difficult to understand mathematically and to visualise.

Before going into detail, it is instructive to employ intuitive arguments to understand the influence of the well width and the potential height on the energy levels. Suppose for example, a GaAs quantum well is grown within $Al_xGa_{1-x}As$ barrier layers. The conduction band offset depends only on the composition, x, of the $Al_xGa_{1-x}As$, and once the structure is made both the width and the depth of the well are fixed. Imagine, however, that it is somehow possible to reduce the dimensions of the well smoothly until the well disappears, as in figure (6.7a), and the material becomes the same as the bulk AlGaAs. We would expect a smooth transition in the wavefunction of the electron from a confined state of the quantum well to an extended state at the bottom of the conduction band in the $Al_xGa_{1-x}As$. As the well width decreases the zero point energy increases until at the point of well closure the energy of the electron coincides with the top of the well, corresponding to the potential of the bulk material. Obviously higher energy states will also rise in energy until they too no longer correspond to bound states of the well.

As the zero point energy increases, the penetration of the wavefunction into the barrier also increases as the difference |E-V| decreases. The electron is said to be less bound. At the point when E = V the wavevector will of course

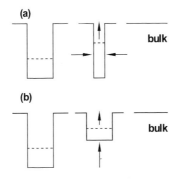

Figure 6.7. Changing a quantum well to bulk material.

be zero, and the penetration of the wave into the barrier will effectively be infinite. This corresponds to the extended state wavefunction of an electron in the conduction band, which, through Heisenberg's uncertainty principle, is regarded as occupying the whole of the crystal. Similarly, if the width of the well were to be kept constant but the height reduced, the energy of the first quantum state of the well would rise but the magnitude of the zero point energy would actually decrease relative to the bottom of the well. Thus, the state will eventually become indistinguishable from a conduction band state of the bulk barrier material. Similar arguments apply to the penetration of the wavefunction into the barrier. For finite quantum wells, then, both the width of the well and the depth of the well are important in determining the energy of the electrons and holes within the well. The arguments given above have of course been developed for electrons in a conduction band well but essentially equivalent arguments apply to holes within the valence band. However, the valence band of typical III–V materials is considerably more complex than the conduction band, and there are additional factors to be considered, such as the existence of three bands, two of which are degenerate in the bulk. The electronic structure of the valence band is described in more detail in appendix III, where the effect of the bulk band structure on the quantised states is also described.

The steady evolution of the quantised state to a bulk state as the well is extinguished also has implications for the effective mass. In GaAs, for example, the conduction band effective mass is 0.0665 but in $Al_{0.3}Ga_{0.7}As$ it is 0.095 [1]. Suppose in the analogy above that we start not with a quantum well but with a double heterostructure in AlGaAs-GaAs and reduce the width of the active region so that the system changes from bulk GaAs to a quantum well and then to bulk AlGaAs as the well is extinguished. It is clear that the effective mass must change from that of GaAs to that of AlGaAs. Moreover, in the quantum well the mass becomes anisotropic as the in-plane effective mass is distinguished from the quantisation effective mass in the growth direction. The quantisation effective mass is often taken to be the effective mass of the well material. However, for the in-plane effective mass, which describes the propagation of the particle within the plane of the well, it would be reasonable to suppose as the state rises in energy and the wavefunction penetrates further into the wide band gap material it assumes more of the characteristics of this material. Indeed, Bastard [2] gives the in-plane effective mass of the nth quantised state in terms of the well barrier masses as

$$\frac{1}{m_n^*} = \frac{1}{m_W^*}[1 - P_B(E_n)] + \frac{1}{m_B^*}P_B(E_n) \tag{28}$$

where

$$P_B(E_n) = 2\int_L^\infty \chi_n^2(z)\,dz \tag{29}$$

is the integrated probability of finding the electron in the barrier. Here χ_n is the envelope function for the nth state. Envelope functions are derived from the property of the Hamiltonian that the x, y, and z, dependences can be separated out. The physical justification for this is that the x and y directions lie in the plane of the well and are distinguished from the z-direction. Therefore

$$\nabla^2 = \frac{\partial^2}{\partial x^2} + \frac{\partial^2}{\partial y^2} + \frac{\partial^2}{\partial z^2} = \nabla^2_{xy} + \frac{\partial^2}{\partial z^2} \tag{30}$$

so if the wavefunction is expressed as a z-dependent amplitude with an x-y dependent oscillatory term,

$$\psi(r) \propto \chi(z) \exp(jk_{xy}r_{xy}) = \chi(z) \cdot \psi_{xy}(r_{xy}) \tag{31}$$

where the index of the state, n has been dropped for convenience and $\psi_{xy}(r_{xy})$ is a Bloch function. Then

$$-\frac{\hbar^2}{2m^*} \nabla^2_{xy} \psi_{xy}(r_{xy}) = E\psi_{xy}(r_{xy}) \tag{32}$$

and

$$-\frac{\hbar^2}{2m^*} \frac{\partial^2(z)}{\partial z^2} + V(z)\chi(z) = E\chi(z). \tag{33}$$

The solution to equation (32) is

$$E_{xy} = \frac{\hbar^2 k^2_{xy}}{2m^*} \tag{34}$$

and for equation (33) the solution is the quantised energy of equation (22) subject to the finiteness of the well potential. The envelope functions are therefore seen to be the wavefunctions described in equations (11) to (13) and depicted in figure 6.6.

Returning to the effective mass in equation (28), you may query why the barrier and well effective masses should be combined in this way. Even though the electron propagates partly within the barrier and partly within the well, it does not appear to be propagating in the conduction band of the barrier, but at some energy substantially below it. Hence it is not clear that barrier effective mass is appropriate or that the weighted average of the well and barrier effective masses is strictly valid. However, envelope functions are formally constructed mathematically from admixtures of conduction band states from both the barrier and the well materials, so in this sense it is not inappropriate to use the conduction band properties of the electron.

An alternative view has it that the non-parabolicity of the bands determines the effective mass. It has been shown [1] that the in-plane effective mass of a conduction band electron is increased in an infinite quantum well by

$$m_{plane}(E) = m[1 + (2\alpha + \beta)E] \tag{35}$$

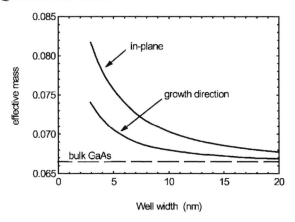

Figure 6.8. The effective mass of an electron in a GaAs quantum well as a function of well width. (After Ekenberg [1].)

where E is the energy of the electron, and α and β are constants, equal to 0.64 eV^{-1} and 0.70 eV^{-1} respectively for GaAs. The quantisation effective mass, which is applicable only to the calculation of quantisation energies, is

$$m_{quant}(E) = \frac{m}{2\alpha E}[1 - (1 - 4\alpha E)^{1/2}]. \tag{36}$$

In a finite well the expressions are more complicated, but essentially similar results hold, as shown in figure 6.8 for GaAs. Measurements of the in-plane effective mass via electrical measurements have been compared with values reported in the open literature and with the non-parabolicity theory [3]. Although the experimental data is scattered there is broad agreement, and moreover, the quantisation effective mass doesn't change by very much with well width and therefore supports the idea that the quantisation effective mass can be approximated by the bulk effective mass of the well material. Similar trends have been reported for InGaAs quantum wells confined in GaAs barriers [4]. Additionally, the effects of band non-parabolicity were shown to be similar to the effects of wavefunction penetration into the barrier (equation (28)) for

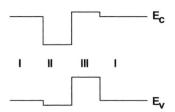

Figure 6.9. Illustration of a type-II quantum well where I represents the cladding layers and II and III the electron and hole confining layers respectively.

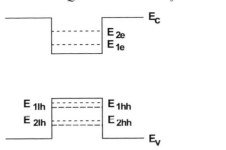

Figure 6.10. A type-I quantum well with the first and second confined electron and hole states. The light hole states lie deeper in the valence band.

the in-plane effective mass in the conduction band. The in-plane mass of the heavy hole was measured as a function of well-width and indium content, and again the mass increased as the well width decreased.

So far attention has been focused on the conduction band, but of course a heterojunction will also imply some sort of offset for the valence band. Not all heterojunctions are well suited to the formation of quantum well systems, though. The division of the offset between the conduction and valence bands may well mean that the valence band is not confined, in which case a second heterojunction is needed (figure 6.9). In such a system, called a type-II quantum well, the confined electrons are spatially separated from the confined holes so the probability of recombination is generally much reduced. Much more common is the type-I quantum well in which the offsets in both the valence and conduction bands lead to confinement within the same layer. The quantum well systems described in this chapter are of this type.

The hole states are in general more complicated than the electron states. Appendix III contains a detailed discussion of the valence band structure, the hole effective masses, and the effects of confinement. Here it will simply be recognised that there are three valence bands corresponding to the light and heavy holes, and the split-off band. The split-off band can be ignored for most purposes because the split-off energy Δ is often such that the split-off holes are not in an energy region of interest. For example, in GaAs $\Delta = 0.341$ eV [5], which for many quantum well structures will be more than the valence band offset. That still leaves the light and heavy hole states, each of which gives rise to a set of quantised levels in its own right (figure 6.10).

6.3 Quantised states in finite wells

Looking at the states themselves, for the conduction band the envelope functions are essentially odd or even with respect to the centre of the well. For even functions,

$$\chi_n(z) = A \cos(kz) \tag{37}$$

for $|z| < L$, and in the barrier

$$\chi_n(z) = B \exp[-\kappa(z - L)] \tag{38}$$

for $z > L$, and

$$\chi_n(z) = B \exp[\kappa(z + L)] \tag{39}$$

for $z < L$. κ is the propagation constant inside the barrier. Similar arguments apply to the odd functions. The barrier wavefunctions are identical to (38) and (39), but within the well

$$\chi_n(z) = A \sin(kz). \tag{40}$$

The energy of the electron within the well is given by

$$\varepsilon_n = \frac{\hbar^2 k^2}{2m_W^*} - V_0 \tag{41}$$

where $V_0 < 0$ is the potential confining the electron, but the energy inside the barrier ($V = 0$) is

$$\varepsilon_n = -\frac{\hbar^2 \kappa^2}{2m_B^*} \tag{42}$$

for bound states where the energy is less than zero. m_W^* is the effective mass inside the well and m_B^* is the effective mass inside the barrier. Both the wavefunction and

$$\frac{1}{m^*} \frac{\partial \chi}{\partial z} \tag{43}$$

must be continuous at the boundary $z = \pm L$. Continuity of the wavefunction is obvious as the square of the wavefunction is related to the probability of finding the electron and a discontinuity would represent a physical impossibility. The second continuity condition is different from the standard condition quoted in quantum mechanical texts because the electron has an effective mass. A single electron moving along x gives rise to a current

$$j = e \cdot \frac{dx}{dt} = \frac{ep}{m^*} \tag{44}$$

where e is the electronic charge and p is the momentum, which must also be continuous across the boundary. The momentum operator is proportional to the differential of the wavefunction, and as the effective mass changes also at the boundary the continuity condition must also consider this.

For even functions, these continuity conditions yield

$$A \cos\left(\frac{kL}{2}\right) = B$$

$$\frac{Ak}{m_W^*} \sin\left(\frac{kL}{2}\right) = \frac{\kappa B}{m_B^*} \tag{45}$$

from which, by straightforward division,

$$\frac{k}{m_W^*}\tan\left(\frac{kL}{2}\right) = \frac{\kappa}{m_B^*}. \quad (46)$$

Similarly, for the odd functions,

$$\frac{k}{m_W^*}\cotan\left(\frac{kL}{2}\right) = -\frac{\kappa}{m_B^*}. \quad (47)$$

There are no simple solutions to these equations. They have to be solved numerically or graphically, which is a big difference between the idealised infinite potential well and the finite well. Some further manipulation is required. Setting

$$k_0^2 = \frac{2V_0 m_W^*}{\hbar^2} \quad (48)$$

allows κ to be expressed in terms of k by setting the energy of the levels within and without the well equal to each other. Therefore

$$\kappa^2 = \frac{m_B^*}{m_W^*}(k_0^2 - k^2) \quad (49)$$

and hence, for the even wavefunctions,

$$\tan^2\left(\frac{kL}{2}\right) = \frac{m_W^*}{m_B^*}\left(\frac{k_0^2}{k^2} - 1\right). \quad (50)$$

Figure 6.11 shows the graphical form of this equation using $kL/2$ as the variate and making the assumption that the ratio of effective masses is unity, and taking values of $(k_0L/2)^2$ of 1,2,3 and 10, by way of illustration. Note that only the ratios of k_0 to k are important in (50), hence the transformation to $k_0L/2$ to match the left-hand side. It is evident that there is always at least one energy level given by the intersection of the LHS and RHS terms. In fact, by making the assumption that the effective masses in the well and the barrier are equal, which will not always be the case but simplifies the problem, it is possible to transform equation (50) further using the identity

$$\tan^2(x) = \frac{1}{\cos^2(x)} - 1 \quad (51)$$

to give

$$\cos\left(\frac{kL}{2}\right) = \frac{k}{k_0}. \quad (52)$$

Similar manipulations are possible for the odd wavefunctions, giving

$$sin\left(\frac{kL}{2}\right) = \frac{k}{k_0}. \tag{53}$$

Equations (52) and (53) are simplified eigenvalue equations for the electron wave-vectors.

6.4 The density of states in two-dimensional systems

Having discussed in detail the nature of the quantised states, and the modifications brought about by a finite potential of the conduction band, it is necessary then to consider the density of such states as a prelude to the calculation of gain. The assumption of a parabolic in-plane dispersion makes the density of states calculation very simple, and is one of the reasons that such an approximation is preferred in quantum well laser structures. Treating the electron as a particle in a box of dimension L, the wavevector must be quantised in each direction according to

$$\frac{m\lambda}{2} = L \tag{54}$$

where m is an integer. This is a purely general statement and applies as much to a bulk semiconductor in which L is large as it does for a quantum well

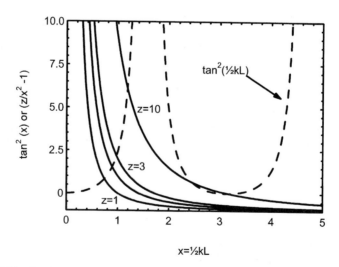

Figure 6.11. Graphical solution of the eigenvalue equation (50) for the energy levels in a finite well.

The density of states in two-dimensional systems

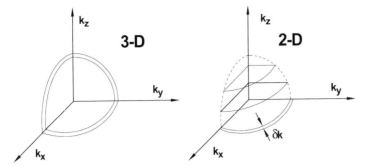

Figure 6.12. Constant energy surfaces in bulk and in 2-D where k_z is quantised.

sample for which L is small in one direction but large in the other two. Sticking with the bulk case for now, the wavevector is also quantised

$$m = \frac{kL}{\pi}. \tag{55}$$

The energy of the electron is

$$E(k) = \frac{\hbar^2 k^2}{2m^*} = \frac{\hbar^2}{2m^*}(k_x^2 + k_y^2 + k_z^2) \tag{56}$$

and in three dimensions this represents any point on the surface of a sphere of radius k (figure 6.12). Strictly, $k_{x,y,z} < 0$ are not of interest as the quantised states are standing waves and back propagation is implied, so in fact only the octant of the sphere corresponding to positive k is important. However, k is quantised, so in order to calculate the density of states it is necessary to calculate the number of quantised states, Z, that can fit into a particular volume. Then

$$Z = 2 \times \frac{1}{8} \cdot \frac{4}{3} \cdot \pi m^3 \tag{57}$$

where the coefficient 2 arises because there are 2 spin states (↑↓) per quantum state, and the coefficient $\frac{1}{8}$ accounts for the octant. This leads to a total number of states that depends on $E^{3/2}$, but the density of states between energies E and $E + dE$ is given by

$$g(E) \cdot dE = \left[\frac{dZ}{dE} \right] \cdot dE \tag{58}$$

which gives the well known $E^{1/2}$ dependence for bulk semiconductors given in chapter 4. In two dimensional systems, though, the wavevector in the growth direction is quantised so rather than considering the surface of a sphere, the surfaces of constant energy lie on the circumference of a circle defined by k_x

and k_y at a specific value of k_z. Again, only positive values of k are of interest, leading to an available fraction of 1/4 rather than 1/8 for the bulk states.

Hence the number of states Z is, for a particular value of k_z,

$$Z = 2 \times \frac{1}{4} \cdot \pi m^2 = \frac{1}{2\pi} k^2 L^2 \tag{59}$$

where L^2 is the cross-sectional area of the quantum well system. In terms of the energy of the electrons,

$$Z = \frac{L^2}{2\pi} \cdot \frac{\hbar^2 k^2}{2m^*} \cdot \frac{2m^*}{\hbar^2} = \frac{L^2 m^*}{\pi^2} \cdot E. \tag{60}$$

The density of states per unit area per unit energy is thus

$$g(E) \cdot dE = \frac{1}{L^2} \frac{dZ}{dE} \cdot dE = \frac{m^*}{\pi \hbar^2} \tag{61}$$

and is therefore constant within a particular value of k_z. If the Fermi energy is increased such that additional quantised values of k_z are allowed, the density of states increases in a step wise manner.

Throughout this chapter it has been emphasised that the quantisation properties of the semiconductor would change smoothly to the bulk properties if the dimensions of the active region were to be varied smoothly. The same is true of the density of states. Using equation (15) for the energy of the first quantised state, and substituting into the 3-D density of states it can be shown that

$$g_{3D}(E_1) \cdot dE = \frac{1}{L_z} \frac{m^*}{\pi \hbar^2} = \frac{g_{2D}}{L_z}. \tag{62}$$

In short, if the bulk density of states is normalised to the thickness of the quantum well it becomes equal to the 2-D density of states at the energy of the confined state. Therefore, if the width of the quantum well were to be increased smoothly the spacing between quantised levels would decrease and the density of states would become equal to the bulk density of states (figure 6.13).

This simple picture is somewhat of an idealisation. It is a pretty good approximation for the conduction band but, as already discussed, the valence

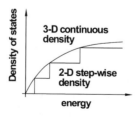

Figure 6.13. The density of states in a 2-D system.

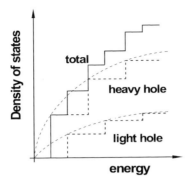

Figure 6.14. The total density of states in the valence band is the sum of the light and heavy hole contribution and can be quite complicated.

band is much more complicated. The density of states depends upon the effective mass (equation (61)) so both the light and heavy hole densities are different. If the states were parabolic we would expect step-wise increases in the density of states of each, the sum of which gives rise to the total density of states as illustrated in figure 6.14. In addition, any deviation from parabolicity, and the valence band states in quantum wells very often are not parabolic over a wide range of energy, will mean that the constant energy surfaces are not circles, leading to a density of states that is not constant in energy. As well as the complicated step structure the density of states may increase between steps and smooth out the profile somewhat. The net result is that the valence band density of states is often much higher than the conduction band density of states, with the result that the quasi-Fermi level for electrons penetrates the conduction band far more than the hole quasi-Fermi level penetrates the valence band.

6.5 Optical transitions in semiconductor quantum wells

The energies of the quantized states determine the emission wavelength, and the density of states contributes to the gain, as described in chapter 4. The strength of the optical transitions is determined by the matrix element, and quantum wells differ substantially from the bulk in this respect. Of primary interest is the process of stimulated emission, but many texts on quantum wells deal with absorption. Bear in mind that absorption and stimulated emission of radiation are essentially the same interaction in quantum mechanics. Which of the two processes occurs depend only on the initial state of the electron, with the matrix element for the two processes being identical.

In the description of bulk transitions given in chapter 4, the Hamiltonian for the interaction of electromagnetic radiation with electrons and holes was shown to be proportional to $A \cdot p$, where A is the vector potential of the electromagnetic field and p is the momentum operator. Clearly, A is related to $\exp(jk_0 \cdot r)$, where k_0 is the wavevector of the optical radiation and r is a position vector. In general the wavelength of light involved in bandgap absorption is long compared with the dimensions of the quantum well so we can regard the exponential as approximating to unity and only the amplitude of the optical field need therefore be considered.

For the sake of definiteness, assume that the incident light is plane polarised in a particular direction, say x. This means that the light can be incident either in the plane of the quantum well and propagating in the y-direction, or it can be incident perpendicular to the plane of the quantum well and propagating in the z-direction. In either case the interaction Hamiltonian will contain a term of the form $A_0 \cdot p_x$, because the electron momentum must be acting in the same direction as the optical electric field for an interaction to occur. Fermi's golden rule, also given in chapter 4, gives the probability of an interaction in terms of the initial (I) and final (f) states according to the matrix element

$$M_{fi} \propto \int_{volume} \psi_f^*(r) \cdot A_0 \cdot p_x \psi_i(r) \cdot d^3r \qquad (63)$$

where some of the coefficients have been left out for convenience. The exact nature of the initial state is important here. So far the nature of the Bloch functions has been glossed over, and whilst it is not the intention to describe their origin in great depth, some recognition of their properties is important. Specifically, as we are concerned with transitions from the light and heavy hole states to the conduction band, the Bloch functions relevant to these bands are needed. There are two Bloch functions per band corresponding to the two spin states, and the Bloch functions themselves are constructed from linear combinations of the basis functions u_x, u_y, and u_z [6]. The basis Bloch functions are related to the p-orbitals of the atomic wavefunctions and have similar symmetry properties. It is the directionality of the basis Bloch functions that is key to understanding the optical transitions.

Dealing with only one spin state (the opposite spin state differs only in the signs of some of the terms) the heavy hole Bloch function can be written as

$$u_{hh} = -\frac{1}{\sqrt{2}}(u_x + ju_y) \qquad (64)$$

and the light hole Bloch function is

$$u_{lh} = -\frac{1}{\sqrt{6}}(u_x + ju_y - 2u_z). \qquad (65)$$

Looking first at the heavy holes, the Bloch functions contain no z-dependence so the states can be split into their separate x-y and z dependences as before (equation (33)). Recognising that the momentum operator is a differential, the term in ψ_i of equation (63) becomes

$$p_x \cdot \psi_i = \chi(z) \cdot p_x \cdot \psi_i(xy) \tag{66}$$

where the envelope $\chi(z)$, being independent of x, can therefore be taken out of the integral. The integral over all space can then be separated into a product of integrals over x-y and z, i.e.

$$M_{fi} \propto \int_z \chi_f^*(z) \cdot \chi_i(z) \cdot dz \int_{xy} \psi_f^*(xy) \cdot A_0 \cdot p_x \psi_i(xy) \cdot d^2 r_{xy} \tag{67}$$

where r_{xy} is the position vector in the x-y plane. Concentrating first on the terms in x-y, expansion of the wavefunctions into their plane wave form using the complex exponential (equation (28)) leads to a product of the form

$$\exp(j[k_i - k_f] \cdot r_{xy}) \tag{68}$$

where $-k_f$ arises because of the complex conjugate. This is an oscillatory term so the integral over all space must average out to zero for any non-zero value of $(k_i - k_f)$. Therefore any states for which $k_i \neq k_f$ can be disregarded. In short, momentum is conserved in the x-y plane. If $k_i = k_f$, however, the complex exponential simply becomes unity and the matrix element is greatly simplified. After further manipulation it can be shown that

$$M_{fi} = -\frac{e \cdot A_0}{2m_0} \int_{xy} u_c^*(r) \cdot p_x u_{hh}(r) \cdot d^3 r \int_z \chi_c^*(z) \cdot \chi_{hh}(z) \cdot dz = -\frac{e \cdot A_0}{2m_0} p_{c-hh}$$

$$\times \int_z \chi_c^*(z) \cdot \chi_{hh}(z) \cdot dz \tag{69}$$

where the subscripts have been changed from I (initial) to hh (heavy hole) and from f (final) to c (conduction), and $p_{c\text{-}hh}$ is the momentum matrix element (the first integral) as defined in equation (11) of Appendix III. This integral is non-zero because, from equation (62), u_{hh} has a component in the x-direction. The integral over the envelopes χ is known as the overlap integral and is non-zero only if the states have the same parity. That is, a transition from the $n=1$ state of the valence band to the $n=1$ state of the conduction band ($hh1 \rightarrow e1$) is strongest, $hh2 \rightarrow e1$ is forbidden as the integral goes to zero, and $hh3 \rightarrow e1$ may occur if the bound states exist but will be weak. However, $hh2 \rightarrow e2$, $hh3 \rightarrow e3$, are allowed because of the overlap, provided, of course, that the states are bound. Light hole to conduction band transitions can also occur because the Bloch functions u_{lh} also contain terms varying in x that can interact with the momentum operator. However, the prefactor $1/\sqrt{6}$ means that the transition is weaker by a factor of three [2, p 236] as the transition probability is related to the square of the matrix element. In addition, the light hole state is at a higher energy in the valence band so the heavy hole transition is the more prominent.

156   *Quantum well lasers*

It is possible therefore to sum up the selection rules for normal incidence:

- transitions are allowed for plane polarisation in both the *x* and *y* directions (only *x* has been treated above but *y* is identical);
- momentum must be conserved in the *x-y* plane;
- the overlap integral restricts transitions among states of similar parity and in practice this usually means from the lowest lying heavy hole to the lowest lying conduction band.

For light incident in the plane of the well polarised in the *z* direction it is the heavy hole Bloch function rather than the envelope function that can be taken out of the integral as being independent of *z*.

$$p_z \cdot \psi_{hh} = \psi_{hh}(xy) \cdot p_z \cdot (z). \tag{70}$$

Therefore the matrix element contains a term of the form

$$M_{c\text{-}hh} \propto \int_{xy} \psi_c^*(xy) \cdot \psi_{hh}(xy) \cdot d^2 r_{xy} \int_z \chi_c^*(z) \cdot A_0 \cdot p_z \cdot \chi_{hh}(z) \cdot dz. \tag{71}$$

It is now the plane wave terms that appear as an overlap integral but these states are orthonormal and all the integrals go to zero. Therefore the transition probability goes to zero, so an electric vector in the growth direction cannot interact with heavy hole states (figure 6.15). For light hole states, though, the Bloch function contains a term in u_z. Therefore the Bloch functions cannot be extracted entirely out of the integral which then becomes non-zero. For light hole states transitions between the valence band and conduction band are therefore possible for light polarised perpendicular to the well. The implications for the operation of a laser are as follows:

- Light can propagate perpendicular to the plane of the well or in the plane of the well.
- For light propagating in the plane of the well, the laser is a conventional edge-emitting device.

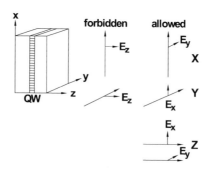

Figure 6.15. Polarisation selection rules for heavy holes.

Optical transitions in semiconductor quantum wells   157

- As with a DH laser the light must be guided by a refractive index difference and will propagate as either a TE mode or TM mode.
- The TE mode will dominate as the heavy hole transition is more probable than the light hole transition and is at a lower energy.
- The output from an edge-emitting quantum well laser is therefore predominantly polarised in the TE mode.
- Light propagating perpendicular to the well requires a different cavity to be constructed. Such lasers, called vertical cavity surface emitting lasers (VCSELs), are indeed possible but will be described in chapter 7.

Before leaving the topic of polarisation selection rules it is appropriate to mention the intra-band transitions that occur when the polarisation lies in the growth direction, i.e. from $n=1$ in the conduction band to $n=2$ in the conduction band and similarly for electrons. These transitions are not involved in the lasers discussed in this chapter, but they are exploited in the quantum cascade laser (chapter 12) and it makes sense to deal with the physics behind the transition selection rules at the same time as other transitions are described. In essence, the overlap integrals do not all go to zero as the confined states, being within the same band, are derived from the same basis set. Moreover, the momentum operator, being a differential, flips the parity of the initial state so transitions between sub-bands, as these states are called, of different parity are made possible.

6.5.1 Gain in quantum wells

Equations (86) and (87) of chapter 4 give the gain in a semiconductor when the light is polarised in a particular direction. A slight modification is required for heavy hole states in a quantum well. In bulk material the electron wavevector is not constrained but only those electrons with wavevector in the same direction as the electric vector of the incident radiation can undergo a stimulated transition. Hence the matrix element has to be averaged over all directions and is reduced by a factor of 1/3 (the factor of 1/6 appears because of an existing factor of $\frac{1}{2}$). For heavy holes in a quantum well, however, there are no electron wavevectors in the z-direction so the averaging has to be performed over only two directions. Therefore the gain is increased by a factor of 3/2 over the bulk. Furthermore, the matrix element decreases with the total energy of the electron, and becomes [7]

$$|M_{QW}|^2 = \frac{3}{4}|M_{avg}|^2 \left[1 + \frac{E_z}{E_e - E_{g,c}}\right] \quad (72)$$

where E_z is the quantised energy of the electron, E_e is the total energy of the electron

$$E_e = E_z + \frac{\hbar^2}{2m^*}(k_x^2 + k_y^2) \quad (73)$$

and $E_{g,c}$ is the conduction band edge. The matrix element enhancement will be different for light holes because the electron wavevectors will average over space differently according to the different selection rules.

The matrix element is also polarisation dependent. Quite general expressions exist for the matrix elements when the electron wavevector lies at an arbitrary angle to the electric vector [8], but the simplest and most convenient is a matrix element of the form

$$|M_T|^2 = \frac{3}{2}|M|^2(1 - e_z^2) \tag{74}$$

for heavy holes and

$$|M_T|^2 = \frac{1}{2}|M|^2(1 + 3e_z^2) \tag{75}$$

for light holes, where e_z is the unit vector along the direction of the vector potential A. For TE polarised modes $e_z = 0$ leading to a matrix element for heavy holes that is a factor of three greater than for light holes. For a TM mode where $e_z = 1$ the matrix element for the heavy hole transition is zero but the matrix element for the light hole transition is two. These forms take into account the enhancement due to averaging of the polarisations in the quantum well.

With these ideas it is now possible to look at the gain spectrum. We can define a maximum gain from equation (chapter 4, 89) when the difference between the quasi-Fermi levels is $+1$. The reduced density of states is simple to derive in a quantum well because the step increase in the idealised density of states for both electrons and holes differs only by the mass, which can be factorised out in the form of a reduced mass (equation chapter 4, 86) with separate expressions for the light and heavy holes. Typically for GaAs the maximum gain is $\sim 10^4$ cm^{-1}. A difference in Fermi occupations of $+1$ implies that the quasi-Fermi levels are well into the bands but the transition energy is well below the Fermi level separation. In fact, the minimum transition energy is the quantum well band gap, i.e. from the first quantised state of the conduction band to the first quantised state of the heavy hole band (see figure 6.16).

The gain will therefore rise sharply as the density of states increases in a step-wise manner, but as the photon energy increases the gain will decrease to zero at the Fermi level separation, where both Fermi functions are equal at $\frac{1}{2}$. At high photon energies, when the difference in Fermi functions is -1 the gain is a minimum and becomes equal to the bulk absorption. As the Fermi levels are driven further into the bands the maximum gain remains at the quantum well band edge but the zero gain is pushed higher in energy. In addition the density of carriers in the second quantised state increases, but because of the asymmetry in the densities of states between the conduction and heavy hole states this is likely to happen in the conduction band first. Figure 6.17 illustrates the effect on the carrier density.

Optical transitions in semiconductor quantum wells 159

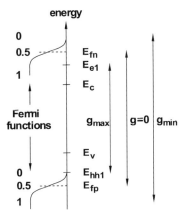

Figure 6.16. Identifiable points in the gain spectrum super-imposed on the energy level structure. The Fermi functions for the conduction and valence bands and the corresponding gain are shown.

Once the second quantised state becomes populated transitions from this state become possible, even if the hole quasi-Fermi level has not penetrated the second heavy hole state. This sounds counter-intuitive because the overlap integral only allows transitions from the same principal quantum number,

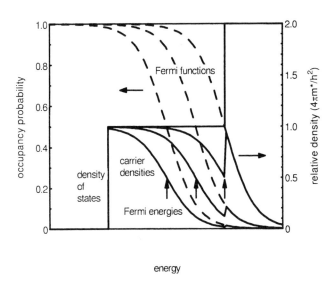

Figure 6.17. The Fermi functions (dashed lines) and carrier densities (solid lines) for three quasi-Fermi level positions (vertical arrows) between the first and second quantised states which is shown as the box structure.

and we would therefore expect both Fermi levels to have to enter the second level in order to ensure population inversion. However, the condition for positive gain is that the photon energy must be less than the Fermi level separation so a transition is possible between two $n=2$ levels even if the quasi-Fermi levels have not entered both states. Whether the laser output is dominated by these transitions or transitions from the $n=1$ states depends primarily on the gain at threshold, because the carrier density tends to be clamped at the threshold value and the Fermi level no longer moves with current. If the gain maximum corresponds to the $n=1$ level this comprises the output, but if the gain from the $n=2$ level is higher at threshold then this will dominate. An exception can occur in quantum well devices operated at high power, where junction heating causes both a reduction in band gap and an increase in the carrier density of the second level via the effect of the increased temperature on the Fermi function. If this is sufficient to tip the gain maximum in favour of the second level the output wavelength will switch during operation [9].

The expected gain at a moderate carrier density is illustrated in figure 6.18, along with the maximum gain. The maximum gain depends inversely on the photon energy and therefore decreases with increasing energy, which trend is accentuated by the dependence of the matrix element on electron energy, until the increase in the density of states causes the gain to rise. In practice the gain (solid lines) decreases much more rapidly with energy than the maximum gain because of its dependence on the Fermi occupation functions, i.e.

$$g(E) = g_{max}(E)(f_c - f_v). \tag{76}$$

This is very much an idealisation, for a number of reasons:

- First, at low levels of injection there will be a single triangular profile from the $n=1$ transitions only, but before the injection levels bring the second set of transitions into play the light hole states should also contribute. However, the polarisation selection rules dictate that for TE polarised radiation the light

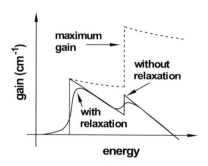

Figure 6.18. Schematic illustration of the gain from a quantum well laser.

hole transitions are a factor of three weaker so at most the light holes represent a small perturbation. For TM modes, of course, heavy hole transitions are forbidden and only light hole states are observed. Given the difference in energy between first quantised states of the light and heavy holes we would expect a similar profile as in figure 6.18 but shifted higher in energy.
- Second, the failure of the parabolic band model caused by band mixing in the valence band (see Appendix III) will mean that the simple step function approximation to the density of states is unrealistic.
- Third, the finite well height can lead to wavefunctions that penetrate quite deeply into the barriers so that the overlap integrals are non-zero for odd-to-even transitions, for example $n=2$ in the conduction band to $n=1$ in the valence band [10]. This is particularly true for non-zero values of the in-plane wavevector where the total energy of the electron or hole is such that the confinement is reduced, and these extra transitions should also contribute to the gain.
- Fourth, the idea of a sharp transition at a well-defined transition energy is not physically realistic and a finite line width for the transitions leads to a smoothing of the gain profile.

The last process is generally known as relaxation, and incorporates carrier-carrier and carrier-phonon scattering into a single timeconstant. In fact, carrier-carrier scattering processes tend to be sub-picosecond whilst carrier-phonon interactions are slightly longer at about a pico-second or more [11]. This leads to a line width for the transition which is limited fundamentally by Heisenberg's uncertainty principle, and is inversely proportional to the lifetime of the electron in the excited state, i.e. $\hbar/\tau$. The most common values lie in the range 0.06 ps to 0.1 ps for both bulk material and quantum wells [12] with the upper value being commonly used [13]. The line width is then ~ 17 meV. The line width represents a spread of energies for the transition, so for a pair of states at an energy E there is a chance that the photon will be emitted at an energy $\hbar\omega \neq E$. However, in order to use the concept a line width function – an expression for the probability of finding a transition within this range of energies – is needed. If this function is $L(E)$ the gain is

$$g(E) = \int_{\infty}^{\infty} g_{max}(E)(f_c - f_v)L(E) \, . \, dE. \qquad (77)$$

It is the effect of the line width that causes the rounding off of the sharp features in the gain profile, and even leads to a small but non-zero gain below the energy of the transition where the density of states is zero. The simplest line width function, and the one most commonly used, is the Lorentzian

$$L(E, \hbar\omega) = \frac{1}{\pi} \left[\frac{\hbar/\tau}{(\hbar/\tau)^2 + (E - \hbar\omega)^2} \right] \qquad (78)$$

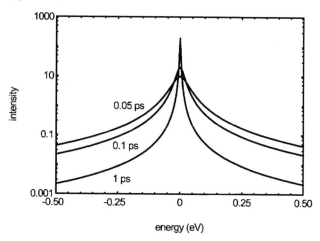

Figure 6.19. Lorenztian functions for three different time-constants from 0.05 ps to 1 ps.

which is quite narrow when plotted on a linear scale but when plotted on a logarithmic scale, as shown in figure 6.19 for three different time constants, there is a small probability of a transition extending out to a few hundred meV from the central transition energy $E = \hbar\omega$. The influence of these low energy transitions on the total gain is excessive and the Lorentzian function is actually not the best. It can in fact lead to a small sub-bandgap absorption [11], which is not physical. An asymmetric function with less emphasis on low energy transitions is better suited [13]. Alternatively an energy dependent timeconstant can be used in conjunction with the Lorentzian function to bias the line width in favour of the higher energy transitions.

Experimentally, gain functions have been measured on a number of different systems. In general the findings correspond well with the purely theoretical expectations given above. The gain in multiple quantum well (MQW) diodes in GaAs/AlGaAs shows a strong polarisation dependence with the TE polarisation switching on at a lower energy than the TM polarisation [12,14]. However, the strong asymmetry in the gain profile caused by the sharp rising edge at low energy with the tail at higher energies is not always observed. One of the most common methods of measuring the gain is due to Hakki and Paoli [15]. The gain is measured at sub-threshold currents where the carrier density is not clamped and significant variations in Fermi level position are possible. It is a fairly simple matter of optics to relate the field amplitude outside the laser cavity to the field amplitude inside, which of course depends on the gain. The field amplitude depends on whether constructive or destructive interference occurs within the Fabry-Perot cavity so by taking the ratio of neighbouring maxima and minima in the output the gain at a particular wavelength can be extracted. This is not the material gain defined by equation (77) but the modal gain G, where

$$G(\lambda) = \Gamma_p(\lambda) g(\lambda) \qquad (79)$$

where $\Gamma_p(\lambda)$ is the polarisation and wavelength dependent optical confinement. The optical confinement is not strongly dependent on the wavelength but it represents an extra distortion in the measured gain. In consequence the gain from each transition tends to resemble a smooth, rounded, symmetric peak which saturates as threshold is reached [10]. Fitting equation (77) to the experimentally measured gain functions allows the timeconstant to be determined.

Finally on the subject of gain, it remains to say something about the effect of high carrier densities. The foregoing treatment is essentially a one-particle treatment in which carriers are considered in isolation and the total effect is simply proportional to the number of carriers, with the exception of the carrier-carrier scattering represented by the Lorentzian lineshape function. This is clearly not a one-particle effect, and in fact, other many body effects can occur. Most important is the effect of a high carrier density on the bandgap. Typical threshold carrier densities are of the order of $\sim 10^{18}$ cm^{-3} and at such high densities electrostatic screening can interfere with the atomic potentials, causing the electron wavefunctions to overlap to a greater extent and reducing the bandgap. Balle [11] has considered analytical approximations to the gain equations in quantum well devices including the effect of bandgap re-normalisation, as the phenomenon is most commonly known, and shows that apart from a redshift the gain profile is to all purposes independent of the bandgap shrinkage. Mathematically, the renormalisation is considered to be directly related to the average separation of the carriers and is therefore proportional to the cube root of the carrier density [16].

6.6 Strained quantum wells

The concept of a strained layer has been met already in connection with double heterostructure devices. To recap, strain occurs in heterostructures when the lattice parameter of the substrate does not match the lattice parameter of the deposited layer, which therefore deforms in order for the two lattices to register with each other. However, strained layers must be thinner than the critical thickness, after which plastic deformation occurs via the creation of dislocations. These can act as non-radiative centres and in most materials will destroy the laser action. As the critical thickness may be no more than a few nanometers, strain has to be avoided at all costs in the DH laser, but quantum wells can be fabricated quite easily below the critical thickness. Such layers are therefore stable and can be used to construct diode lasers, even for operation at high power.

Strained layer lasers exhibit improved performance over lattice matched lasers, principally because of the effect of strain on the valence band, which is described in more detail in Appendix IV. In essence, strain in bulk material alters the band gap and lifts the degeneracy of the light and heavy hole bands. The effect of quantum confinement, which is to separate these states according to effective mass, is superimposed on this initial separation. The peak in the gain spectrum is shifted in wavelength to match the bandgap, so the well width has to be different from that in unstrained material of the same composition to give a specified output wavelength. In compressively strained materials the band gap is increased and the light hole band is pushed deeper relative to the heavy hole. Therefore the quantum confined heavy hole and light hole states are even further apart in energy. Furthermore, the heavy hole states have their in-plane effective mass reduced, thereby reducing the density of states. Tensile strain decreases the bandgap and causes the light hole to rise in energy relative to the heavy hole band. If the strain is large enough the light hole forms the band gap. In the plane of the well, however, the effective mass is increased.

The band mixing effects in compressively strained lasers are much reduced by the greater separation and the valence band therefore approximates much more to a parabolic dependence of energy on wavevector than in the unstrained case. This further reduces the density of states, especially in narrow wells. Figure 6.20 shows the valence band density of states calculated by Pacey *et al.* [17] for 1% compressively strained InGaAsP quantum wells designed to emit at 1.3 μm. For 2.5 nm wells two levels only contribute to the density of states, the topmost heavy hole and the topmost light hole states. For the 4.0 nm wide well the second heavy hole state contributes before the light hole state and for the 10.0 nm well the third heavy hole state contributes before the light hole state. The deviation from the simple step function is due to the non-parabolicity

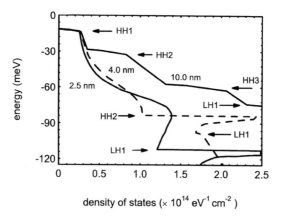

Figure 6.20. The valence band density of states in 1% compressively strained 1.3 μm InGaAsP quantum well lasers. (After Pacey *et al.*)

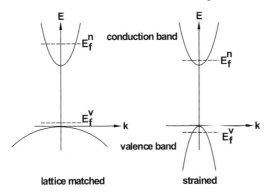

Figure 6.21. A Reduced density of states in the valence band leads to a greater penetration of the Fermi level into the valence band.

of the bands, but even so it is clear that for narrow wells the density remains low over some 30 meV. The condition for charge neutrality within the well determines the density of holes which, with a lower density of states caused by the two mechanisms above, can only be achieved by a deeper lying Fermi level (figure 6.21). Therefore, while the reduction in density of states would appear to reduce the gain somewhat, the effect is more than offset by a more balanced distribution of the Fermi levels between the two bands, which means that it is easier to achieve population inversion and the threshold current is lowered.

The redistribution of the Fermi levels also affects the differential gain. To recap, the differential gain is the rate of change of gain with carrier density, and it influences several aspects of diode laser performance. In telecommunications it affects the modulation rate as well as the output spectrum and in high power diode lasers it affects the formation of filaments, or lateral modes, observed in broad area structures. The maximum in gain appears at photon energies corresponding to the band edge in quantum well devices, the smoothing effects of the time constant notwithstanding, so if the quasi-Fermi level lies high in the conduction band but relatively low in the valence band, possibly not even entering the valence band in the extreme case of a very high density of valence band states, the change in carrier density at the band edges is relatively insensitive to changes in the Fermi level. The reduction in the density of valence band states in a compressively strained quantum well leads to a greater sensitivity to changes in carrier density.

For the reasons given above, early work on strained quantum wells concentrated on compressively strained systems, but in tensile systems the valence band edge can be changed from heavy hole to light hole. Figure 6.22 shows the threshold current and polarisation of tensile strained 11.5 nm wide GaAsP QW lasers as a function of GaP content and shows clearly the switch from TE to TM polarised output [18]. The rise in threshold current at large GaP

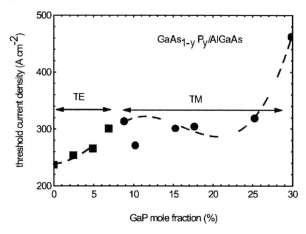

Figure 6.22. Threshold current in GaAsP QW lasers showing changes in the output polarisation with increasing GaP content. The lines are for guidance only. (After Tolliver et al. [18].)

content is a consequence of the smaller band offset and the loss of carriers from the well. In lattice-matched quantum wells the matrix element for the TM mode is larger than for the TE mode so the switch in polarisation should also result in a larger gain. However, there is an additional effect of strain; strain of whatever type can also enhance the relevant matrix elements through spin-orbit coupling, which occurs when the split-off band lies close in energy to the light hole band [19]. It is not necessary to consider the effect here, but interested readers are referred to the paper by Chang and Chuang [19] where analytical expressions for the spin-orbit effect are given in terms of the fundamental material parameters available in reviews such as that by Vurgaftman et al. [20].

Modelling of InGaAs wells on InGaAsP layers lattice matched to InP showed that tensile strained lasers emitting in the TM mode can exhibit not only a high gain but also a high differential gain [21]. The light-hole to split-off energy separation is 150 meV for InGaAs and 0 meV for InGaP [20] so spin-orbit coupling will be very strong for the latter, which is often used for visible red-emitting lasers. In InGaAsP wells in InGaAs barriers, the spin-orbit effect enhances the TE matrix elements for the light hole under compressive strain by a factor of three from 0.5 to $1.5\times$ the bulk momentum matrix element whilst suppressing the TM matrix element almost to zero. Under tensile strain, however, the inverse applies, with the TM matrix element being enhanced to $3\times$ the bulk whilst the TE matrix element is suppressed. By contrast, strain alone has hardly any effect on the matrix elements near the zone centre ($k_{x,y} \sim 0$). Figure 6.23 illustrates what might typically be expected from strained InGaAsP lasers [22].

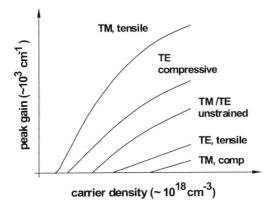

Figure 6.23. Illustrative effects of strain on the polarised gain as a function of carrier density. TE polarisation is enhanced under compressive strain and TM polarisation under tensile strain. The differential gain is given by the slope and is similarly enhanced by strain. (After Jones and O'Reilly [22].)

6.7 Optical and electrical confinement

This discussion of semiconductor quantum wells and lasers began with the assertion that a quantum well is to great extent a natural extension of the DH. Although the physics of the energy states and transition selection rules is very different, as far as the growth is concerned the techniques are very similar. The essential difference lies in the width of the active layer. Consideration of the optimum active layer dimensions of the DH laser reveals that anything thinner than 150–200 nm is likely to have a large threshold current due to the loss of optical confinement, and we would expect that to be true also of the quantum well. A TE mode will propagate in an infinitely thin symmetric waveguide, so we can still expect optical confinement of some sort, but it will be small, and because of it the threshold current density will be high. Using multiple quantum wells (MQW) increases the optical confinement because the presence of several layers of high refractive index leads to an effective refractive index across a thicker active region. One of the first reports of multiple quantum well lasers, published in 1979 [23], used 13.6 nm GaAs wells in 13 nm AlGaAs barriers and the combined effect led to a confinement of ~0.35. Incidentally, the authors of this particular work were keen emphasise the quality of the interfaces in this structure. The amount of work in the open literature on multiple quantum wells since that time is enormous and such structures are grown routinely in a wide range of materials. We take it for granted that the quality of the material is high but in 1979 the increase from 2 interfaces to 28 interfaces gave the authors reason to question whether the device would work at all. It did, and demonstrated the suitability of MBE for the growth of high quality opto-electronic materials and interfaces.

Although the optical confinement is enhanced in MQW devices, the effect is most pronounced when the wells and barriers are a similar thickness. If the wells are thinner than the barriers the effective index is low, and if the barriers are thinner than the wells the probability of tunnelling between wells must increase, changing the MQW structure to a superlattice (SL). In a superlattice the electronic communication between wells leads to the formation of minibands of states rather than isolated states at discrete energies (see chapter 12). Detailed modelling of MQW structures designed to emit at 830 nm, 130–0 nm, and 1550 nm [24] shows that for total thickness below 50 nm the width of the optical mode increases rapidly with decreasing thickness as the optical field extends into the cladding layer. Above 50 nm the optical confinement increases slowly with the total thickness but the effect saturates with the number of wells, and little is to be gained from having more than 5 wells. That is not to say that quantum well lasers having more than five wells are not useful. Increasing the number of wells is an obvious way to increase the active volume, and hence the power output, without increasing either the cavity length or the aperture. Obviously, the threshold current will increase with the additional wells.

Many multiquantum well devices reported in the literature are intended for high power use, but in addition to increasing the power output the modulation bandwidth is also increased with an increase in well number, but the bandwidth eventually saturates as transport effects begin to limit the response of the lasers [25,26]. That is to say, transport is generally considered to be a consequence of carrier capture, carrier escape, and diffusion across the barriers [27]. The finite time taken for carriers to diffuse from the contacts to the centre of the quantum well, overcoming the multiple barriers in the process, increases the turn-on time and limits the modulation frequency [28, 29]. If the barriers are high, and numerous, then even under steady state conditions there may be significantly more carriers at the edge of the MQW structure than at the centre, severely affecting the gain in the centre wells. Hole transport in particular seems to be the limiting factor [30,31]. Multiquantum well lasers therefore do not have large numbers of wells, and a common geometry is to place the wells in a separate confinement structure [32,33].

The separate confinement heterostructure (SCH) concept is illustrated in figure 6.24 for single quantum wells, though any number of wells can be incorporated provided the critical thickness is not exceeded. Additional optical confinement can be provided by an extra pair of heterojunctions (figure 6.24a) with the barrier layers for the carrier confinement lying within these. As the refractive index is generally inversely related to the bandgap the outer layers form the waveguide so that the optical field is confined by this junction, whilst the carriers are of course confined within the quantum well. The issues involved in this structure will not be immediately apparent but they number several.

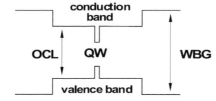

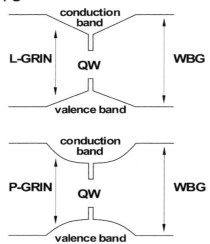

Figure 6.24. Separate confinement heterostructure configurations, including a step (a) and linearly (L-GRIN) and parabolically (P-GRIN) graded index structures (b).

- The materials have to be lattice matched across wide compositions so that dislocations are avoided at every interface. Dislocations generated at any interface in the structure will be propagated through the epitaxial growth so even the cladding layers have to be defect free.
- Low refractive index layers can be easily achieved in the AlGaAs-GaAs system, and with the use of quaternary layers in other materials systems, but if the band gap is too large contacting the materials becomes difficult.
- Given an upper limit on the bandgap of the cladding layers, the band gap of the optically confining layer (quantum well barrier) has to be chosen to lie between this and the quantum well band gap. A large offset at the optical interface will lead to strong optical confinement but possibly poor carrier confinement, and on the other hand, a high quantum well barrier will lead to good carrier confinement but poor optical confinement. Low optical confinement will result in a low modal gain, whilst low carrier confinement

will lead to population of the quantum well barrier, increasing the threshold current.

The SCH therefore represents an additional degree of complexity in the design and the structure of the laser which ensures that not only is the optical field confined but also so are the carriers, and to some extent these two are mutually incompatible.

Optimisation can be by experiment or by design. Experimentally, the simple step GaAs SCH laser has been investigated by Givens et al. [34] who built five lasers with different barrier compositions but with a fixed cladding composition of 85% Al. For a total barrier width of 200 nm (100 nm either side of the quantum well) and for a fixed cavity length of 815 μm, the optimum threshold current was found to be ~ 150 A cm^{-2} at a barrier composition of 23% Al. Contrast this with the threshold currents > 10^4 A cm^{-2} observed in the first homojunction lasers. Clearly, this approach to optimisation is limited because many of the parameters have to be fixed. An alternative is to model the laser, taking into account the waveguide structure, the gain, the facet reflectivity, the material absorption etc. For a discussion of optical modelling techniques the reader is referred to Czyszanowski et al. [35,36] who have constructed a comprehensive optical model of not only the SCH laser but also other optically confining structures in AlGaAs-GaAs. Finally, it is worth pointing out that the lattice matching in AlGaAs-GaAs tends to induce the belief that any thickness of AlGaAs is possible, but dislocations do occur, especially on other growth orientations, and their formation and avoidance is discussed by Gutiérrez et al. [37].

Two variations on the SCH structure are shown in figure 6.24(b), in which the band gap, and hence the refractive index, is graded (GRIN) either linearly or parabolicly. These structures can offer advantages over the step-SCH in so far as a poor carrier confinement effect can be offset somewhat by the smaller volume of material at the lower band gap. Thus, if the confinement in the quantum well is poor because of a low barrier the in-built electric field due to the band gradient will concentrate the carriers near the edge of the well. This will naturally contribute to a lower threshold current as carriers are not wasted filling up states within the barrier layer. Yamamoto et al. [38] have defined the optimum structure for low threshold currents in 1.55 μm GRIN-SCH lasers in strained InGaAsP, where the GRIN layer was formed by stepped compositional changes. Formation of a continuously graded structure again implies not only lattice matching over a wide range of composition, which is relatively straightforward in AlGaAs-GaAs, but also the control over the growth of a complicated quaternary system to get the right grading in the InGaAsP composition. Hence the use of steps in composition is common in InGaAsP [39]. Detailed modelling reveals that the optimum is a three-step GRIN layer, each step being 100 nm thick, with bandgap wavelengths of 1.0 μm, 1.05 μm and finally 1.1 μm at the edge of the well. The calculated carrier density outside the well is minimised in this structure, and fabrication of a 10 nm wide strained well produced a threshold current of 98 A cm^{-2}. Optimised multi-quantum well lasers can be very stable against temperature variations in the threshold current, with characteristics temperatures rising from ~ 150–200 K in a single well to something in excess of 300 K with multiple wells [31].

GRIN structures can also be exploited in carrier transport across the SCH layer. Transport across the step-SCH is by diffusion, and is characterised by an ambipolar diffusion coefficient that averages out the difference between the electrons and holes. The ambipolar diffusion coefficient in AlGaAs is 2.5 cm^2 s^{-1} and 3 cm^2 s^{-1} in AlGaInAs cladding layers in InGaAs lasers [30]. Transport across GRIN layers is assisted by the in-built field and can be much faster, as demonstrated directly by time-resolved photoluminescence [40]. Carriers injected into the structure optically by a fast pulse are transported to the well and captured. It takes time, therefore, for the photoluminescence from the well to rise to its maximum intensity before it decays through the steady radiative recombination of carriers. Alternatively, or sometimes in addition, the decay of the luminescence from the barrier can be monitored. An extensive investigation into optically monitored transport in quantum well systems, together with a review of other relevant literature, is provided by Marcinkevičius and co-workers [30,41,42].

It is important to distinguish between capture rate and capture efficiency. In time-resolved luminescence measurements a change in the intensity of the luminescence may be caused by carrier capture or carrier emission, but other processes are occurring and some analysis is required before capture rates can be extracted. The ultimate fate of all carriers injected into the system is to recombine so there must be equal numbers of electrons and holes injected from each contact. In the ideal device all of these carriers will end up in the quantum well, so the charge density will be zero. However, it takes time for the carriers to reach the well and to be scattered into it, so charge neutrality will not exist initially. The charge will be distributed throughout the OCL and the quantum well, and will change with time as the carrier densities change. Moreover, if carriers are re-emitted from the well, which can happen if the barrier is low, it is unlikely that equal numbers of electrons and holes will be emitted because the barriers will be different. Therefore charge neutrality within the well may be violated, but taken over the whole of the OCL and the quantum well together charge neutrality must exist. If large numbers of carriers exist outside the well, i.e. if the capture is inefficient, internal electric fields may be set up within the OCL layer as the charges redistribute themselves. These electric fields affect the transport and the capture rates. Full dynamic simulations involving time-dependent solution of Poisson's equation are possible, but sometimes a simpler approach is adopted.

Morin *et al.* [40] measured the time-resolved luminescence in GaAs-AlGaAs step-SCH and GRIN-SCH structures, as illustrated in figure 6.23(a) and (b). The evolution of the carrier density within the cladding layer is given by

$$\frac{dn_b(z)}{dt} = D_{amb}\left[\frac{d^2n_b}{dz^2} - \frac{v}{D_{amb}}\frac{dn_b}{dz} - \frac{n_b}{\tau_b} - f(z)\frac{n_b}{\tau_w}\right] \quad (80)$$

where z represents the direction of drift. The last two terms represent respectively the decay of the carriers in the barrier and the capture of carriers within the well, and $f(z)$ is a capture parameter set equal to unity if the carrier is within a certain distance of the well and zero if the carrier lies outside. The capture area is similar to the well width and simply represents the reality that a carrier well away from the well will not be captured whereas a carrier close to the well will. Quantum wells are extremely efficient at capturing carriers so this simple approximation is not far from the truth. The time constants τ_b and τ_w represent respectively the lifetime of carriers in the barrier and the capture time into the well. The first term in (80) is simply a diffusion term with the ambipolar diffusion coefficient D_{amb}, but the second represents the drift caused by the internal electric field, and is simply the divergence of the particle flux.

$$divJ = v \cdot \frac{dn}{dz} \quad (81)$$

where v, the velocity of the carriers, is expressible in terms of the band gap gradient, via

$$v = \mu E = \frac{q}{kT}D_{amb}\frac{dE_g}{dz}. \quad (82)$$

Using this model, in conjunction with experimental data, it is possible to show that capture times are of the order of 1ps, possibly varying from 0.8 ps at low temperature to 3 ps at room temperature, with capture/transport times in the step-SCH being considerably longer than in the GRIN-SCH. In fact, in both the L-GRIN and the P-GRIN structures carriers are swept to the edge of the well where there is a delay of a few ps before capture.

Equation (80) is known as a rate equation. When the carrier density is coupled to the photon density, rate equations can be used to model the modulation performance of laser structures. Such modelling will be described in detail in chapter 8, but here a single equation in the carrier density is used to extract timeconstants. It has to be emphasised that this is an indirect technique. Observed decays have to be modelled in order to extract the time constant for capture and transport. As well as time resolved photoluminescence, four wave mixing has also been used to give capture times in the range 1.5–1.8 ps and escape times of 8 ps in InGaAs wells [43]. By way of analysis, Monte Carlo simulations of capture yielded capture times into 8.5 nm wells of 0.56 ps for electrons and 0.44 ps for holes, rising to 1.7 ps and 1.1 ps respectively for 7.5 nm wells. Direct determination of the modulation response using electrical impedance measurements is another method which again depends on rate equation analysis [44] and this has revealed a carrier dependent capture time in multi-quantum well InGaAs-AlAs-GaAs devices [45] varying from 14 ps at low currents (1 mA) to 1 ps at high currents (50 mA). Similarly, the intrinsic capture time in InGaAs multiquantum wells, separated from any diffusion effects has been determined to be between 0.2 and 0.5 ps [46]. Effective capture times that include an element of diffusion are, of course, longer at 2–5 ps, but escape times were found to be very much longer, typically 2 ns. Sometimes the rise times are reported without any analysis. A characteristic rise time of 45 ps has been observed for as-grown 5 nm InGaAs wells in GaAs barriers falling to 20–25 ps for wells mixed by ion implantation [47].

Determination of the capture times in quantum wells is an ongoing subject of experimental and theoretical research. The capture times, as well as the transport across the OCL directly affects the modulation response of the lasers. The picture is far from clear. The measured data varies enormously, from fractions of a picosecond to over 60 ps, and theoretically the picture is just as uncertain. The earliest calculations, by Brum and Bastard [48], predicted that capture rates would oscillate with the well width, a prediction that has been reproduced in more than one theoretical work [49–55]. Others have expressed the phenomenon as an oscillatory mobility [56]. The principal cause is the wave-mechanical description of the electron. The electron wavevector changes at the well-barrier boundary and some reflection is inevitable (figure 6.25) at both interfaces. Phase coherence between the reflected and transmitted components leads to an oscillatory reflection and transmission which are of course out of phase with each other. The reflection is high when the transmission is low. If the well width were to be varied smoothly the transmission would rise and fall as the phases between the components making up the transmission align, and in the ideal case of normally incident mono-energetic electrons not scattered this oscillatory behaviour would carry on

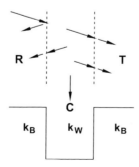

Figure 6.25. Reflection, transmission, and capture of an electron "wave" incident on a quantum well.

indefinitely as the well width increased. Two things occur to modify the picture. First, the phase coherence is reduced somewhat by the variation in incident angles among the electrons, which leads to corresponding variations in the normal component of the wavevector, so the maxima and minima are smoothed out somewhat as well width increases. Second the electrons are scattered as they traverse the well. Without scattering there can be no capture, because to be confined within the well the elcctrons must lose energy. Therefore the transmission resonances are damped with increasing well thickness, and the average transmission probability decreases with well thickness.

The oscillatory capture rate arises from the phase coherence, and is in phase with the transmission, because a high transmission is also associated with a high probability of finding the electron within the well, and of course the scattering probability is related not only to the probability of finding the electron within well boundaries but also the probability of finding a state into which the electron can be scattered. Increasing the well width causes virtual bound states – the states associated with the higher probability of finding the electron within the well – to descend in energy, eventually becoming bound states. Brum and Bastard assumed that electrons traversing the well emitted optical mode phonons and were captured by these newly bound states, after which intra-band scattering dominates the relaxation to the lowest level of the well. This process was not considered in detail. A matrix element was defined to describe the probability of scattering into the newly bound state, but, as you might expect from the above, the probability of this occurring depends on the state being within an optical mode phonon energy of the virtual state, and of course the energy separation will also vary with well width. Therefore there are two oscillatory components to the capture process; the reflectivity and the scattering.

The theoretical picture is still very much under development. The calculated capture times tend to lie in the range 10–100 ps, only becoming much lower at the capture resonances and for wide wells [50,52]. There are, in addition, other ways of calculating the scattering probability than defining a matrix element and using Fermi's Golden Rule, which can overestimate the scattering rates and become unphysical if the coherence length is shorter than the well width [51,54], but the details are not needed here. Experimentally measured capture times range the order of 1–2 ps to ~50 ps in InGaAs [57], InAs [58] and InGaAsP [59]. An oscillatory dependence of capture rate on well width is not observed experimentally unless both the carrier density and temperature are low, and Baraff has considered the question of electron-electron scattering to explain the observation. At high carrier densities the dephasing time for electrons is less than 10 fs, and ~60 fs for holcs, which suggests that for well widths >5 nm the electron phase coherence is destroyed and the oscillatory capture rates will not apply in lasers operated at high temperatures and carrier densities.

Capture times, then, would appear to be of the order of a picosecond. Many of the measurements of relatively long capture times have been made on low doped materials may not be relevant to laser structures where the carrier density is high. Certainly, the discrepancy between theory and experiment in relation to the oscillatory capture is almost certainly a consequence of the high carrier densities. However, it is important to distinguish between capture times and capture efficiency, by which is meant the proportions of carriers within the well compared to those in the optically confining layers. Capture times are intrinsically fast, but escape times are variable because the electrons and holes have to gain energy to escape. If they are tightly bound to the well escape times will be long, but if the quantised levels are relatively shallow re-emission from the well is likely to occur and escape times will be fast. Another way of looking at this is to consider the time averaged occupancy, as expressed by the Fermi function. In a shallow well under population inversion such that the Fermi level lies above the first quantised state, the band edge of the barrier layer will also, by definition, lie close to the Fermi level and a significant carrier density must exist in the barrier layers. In short, although quantum wells are intrinsically good at capturing carriers the steady state occupancy of the well relative to the barrier layers will depend on the details of the electrical confinement, on the relative densities of states between well and barrier, and on the position of the Fermi level relative to the band edge.

Such a situation can arise if the band offsets are not evenly distributed, causing one well to be much shallower than the other (figure 6.26) [60]. A large density of barrier states adversely affects the threshold current and the noise properties of the lasers because recombination in the barrier layers gives rise to emission which is well away from the gain maximum and therefore represents a waste of injected carriers. However, such emission can also be used to evaluate the capture times into the wells, as demonstrated in step-SCH strained GaInAs/InP lasers [61], and neatly demonstrates the difference between capture times and capture efficiency. Spontaneous emission profiles were used to measure the relative efficiency of compressively strained, lattice matched,

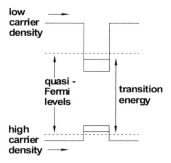

Figure 6.26. A low barrier density of electrons and high barrier density of holes caused by a difference in band offsets.

and tensile strained lasers to assess the suitability for high power operation. The spectrum at low currents consists of emission from the well only, but at higher currents an emission from the OCL can be seen. At threshold the intensity of the well emission saturates, as would be expected from the clamping of the carrier density, but the OCL emission continues to increase in intensity. The fraction of integrated intensities, $I_{OCL}/(I_{WELL} + I_{OCL})$, emitted from the OCL is a measure of the efficiency of the capture process, and also of the laser. Fitting theoretical models to plots of this against the injection current yields 0.1 ps for hole capture, rising in some cases to 0.25 ps, with an electron capture time estimated at 1 ps, in agreement with the preceding observations.

Although the capture times are very fast in all the structures examined the compressively strained laser has a higher proportion of its current going into the OCL emission than the unstrained and tensile-strained lasers. Lattice-matched and tensile-strained lasers would therefore be suitable for high power operation but the compressively strained single quantum well laser would not. The difference between the structure, of course, lies in the different indium concentration in the well. Not only are the band offsets altered, but so also is the width of the well in order to achieve emission at a set wavelength. The tensile strained laser had an In content of 33% and a well width of 12 nm, but the compressively strained device had 70% In and a well width of only 3 nm. Not only are the band offsets changed considerably, but intuitively one would expect a wider well to be more efficient at capturing the carriers. State filling, as this phenomenon is often called, affects not only high power operation, but also the modulation response through the differential gain. With a large number of carriers going into the OCL layer the gain in the well is relatively insensitive to changes in carrier density so the differential gain is reduced [62].

One solution to state filling and carrier overflow from the wells is to use superlattices in the OCL [63]. A superlattice resembles a multiquantum well structure but the barriers are generally thinner and of the same order as the well width so that tunnelling between wells can occur. The non-zero probability of finding an electron in more than one well effectively leads to an overlap of electron wave-functions and, as with the formation of energy bands described in chapter two, this leads to the formation of energy bands in the superlattice. Therefore, the electrical confinement is increased, with the effective barrier height being increased by over a factor of two in some cases [64,65] (see figure 6.27). The advantage of the superlattice approach is that the use of complicated quaternaries, especially graded quaternaries, is avoided. The effective index of the OCL is essentially determined by the ratio of well/barrier width, and if this is graded the refractive index can also be graded without changing the material composition. The disadvantages are that the devices are more complicated and the effective refractive index of the OCL is reduced compared with a step-SCH. However, the increased confinement can lead to higher powers [66], as well as a reduced sensitivity of the threshold current to the effects of temperature [67,68].

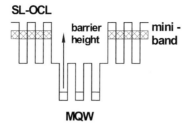

Figure 6.27. Increased electron confinement through the use of a superlattice barrier layer.

6.8 Optimised laser structures

The preceding discussion on optical and electrical confinement suggests a bewildering array of possibilities for practical laser structures. It is useful, therefore, to conclude this chapter with a brief look at a couple of optimised laser structures that have been described in the literature to illustrate how the opposing requirements can be balanced.

It is not always necessary to use the techniques described above if the materials allow for effective confinement. Such is the case in the recently developed strained layer SCH SQW InGaAs/GaAs laser emitting at 980 nm [69]. High power lasers emitting at 980 nm are used to pump Er-doped fibre amplifiers and therefore need to be highly reliable. Such lasers will be incorporated into under-sea systems, which are expensive to lay, and therefore must last the lifetime of the cable. Most practical lasers emitting at this wavelength are based on AlGaAs/InGaAs in order to take advantage of the lattice matching properties of AlGaAs to GaAs substrates, as well as its low refractive index and high thermal conductivity. A single pseudomorphic (strained) InGaAs well provides the gain. The development of GaAs/InGaAs lasers has been described as an obvious step [69].

In order to design the laser full simulation of the above threshold operation was undertaken, with Poisson's equation for the electric field as a function of charge density, the continuity equations for electron and hole densities, the current as a function of quasi-Fermi levels, the gain, and the wave propagation in the optical guide all being solved simultaneously and self-consistently. The principles of all these simulations have been described within these pages. A couple of simplifying assumptions were made, however. First the effective mass approximation was used as this simplifies considerably the gain calculation with changes in both well width and composition, and second, the matrix elements were based on Fermi's Golden Rule, which strictly applies to a one-electron system and does not take into account the many-body effects leading to band-gap renormalisation. Nonetheless, such simulations provide a good starting point for determining the effects of design changes without the

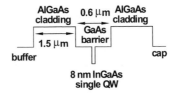

Figure 6.28. Optimised SCH-SQW InGaAs-GaAs 980 nm laser.

need to grow and fabricate the lasers, and therefore play an important part in the optimisation of laser structures. There is never perfect agreement between the simulators and the measured properties, with the practical threshold currents generally being slightly higher than the predicted threshold because of the inevitable imperfections in the material, such as well-width fluctuations. Such simulations therefore provide a lower limit on the performance.

The optimised structure is shown in figure 6.28 and consists of an n-type (5×10^{17} cm^{-3} Si doping) Al$_{0.3}$Ga$_{0.7}$As cladding layer 1.5 μm thick on a GaAs buffer, 0.3 μm of undoped GaAs, an 8 nm undoped In$_{0.21}$Ga$_{0.79}$As well, 0.3 μm of undoped GaAs, 1.5 μm of Be doped (5×10^{17} cm^{-3}) Al$_{0.3}$Ga$_{0.7}$As, and a Be doped GaAs (3×10^{19} cm^{-3}) cap to provide the top contact. The measured threshold current was 280 A cm^{-2} and at 50 mW output, there was no appreciable sign of degradation after 1000 hours. This is due in part to the choice of 0.3 μm barrier layer. An alternative structure consisting of 0.1 μm barrier with 70% Al cladding showed a much reduced threshold current with a significantly increased optical confinement in the well, but for high power operation emission from a larger, rather than smaller, aperture is better for reducing the chances of catastrophic optical damage (see chapter 10). The alternative structure would be suitable for high speed modulation at low powers, but transmission of optical signals at 980 nm is not common.

The second laser structure of interest is red edge-emitting lasers at 630 nm, 650 nm and 670 nm for use at high power output [70]. The basic structure consists of a single GaInP quantum well with an AlGaInP waveguide region and AlInP or GaInP cladding layers. Some structures included a graded index layer between the cladding and the waveguide (figure 6.29). If the Al content of the waveguide layer were to be increased not only would the carrier confinement in the quantum well increase but the refractive index difference

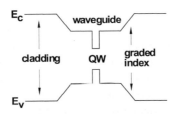

Figure 6.29. Red lasers with a graded optical confinement layer [70].

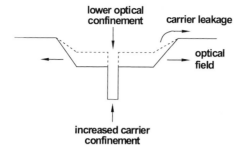

Figure 6.30. Increased leakage from a higher Al content in the optically confining layer of figure 6.29.

between the cladding layer and the waveguide would also decrease. Hence the optical confinement would decrease as the optical field spreads into the cladding layer, which has both a beneficial effect and a deleterious effect for high power operation. Beneficially, the far field pattern is improved because of the larger effective emitting area of the aperture, but detrimentally, the optical intensity in the quantum well is reduced and hence so is the rate of stimulated emission, all other things being equal. Whilst the laser output characteristics may be improved an increased threshold current density may be expected through the combined action of reduced carrier confinement and lower optical field (figure 6.30).

The layer structure was therefore modified using a combination of modelling and experiment. First a waveguide mode solver was used to determine the optimum waveguide modes, after which a full laser simulator package was used to fine tune the structure. The laser simulator took into account carrier transport, carrier confinement, and optical confinement, and was able to predict the far field radiation patterns as well as the threshold currents. There was general agreement between the simulations and the experimentally measured performance, but not sufficiently so that the simulations could be relied on alone to predict the optimum structure. The final step, therefore, was to produce some lasers by solid source MBE around the optimum designs to determine the best performance. The result was the insertion of two additional layers; one to increase the confinement in the quantum well and one to increase the optical spreading into the cladding layers whilst maintaining good carrier confinement in the waveguide.

This type of structure decouples the carrier confinement from the optical confinement. Carrier confinement in the well is determined by the barrier layer which is thin enough to have a negligible optical effect, despite having a reduced refractive index within the optical structure brought about by the larger band gap of the barrier layers. Carrier confinement within the waveguide is determined now by the barrier between the waveguide and the confining layers but the thickness of the confining layer is such that it has an optical effect.

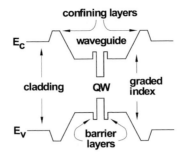

Figure 6.31. The result of optimisation is to introduce extra layers to decouple the optical and electrical confinement.

The confining layer has a lower refractive index than either of the adjacent layers and therefore has no optical confining properties. In consequence the optical field is diverted in both directions; into the waveguide structure, and into the cladding layer. The increased optical field inside the waveguide increases the optical confinement whilst the increased penetration into the cladding increases the effective area of the aperture and improves the far-field pattern. The precise thickness and composition of each layer depended on the emission wavelength, which could be controlled either by the quantum well composition or the width. The reader is referred to [70] for detailed descriptions of the structures and performances compared with similar lasers in the literature. In general, high power outputs exceeding 1W were achieved from these broad area structures.

6.9 Summary

Quantum well lasers are fundamentally different from bulk structures by virtue of the quantum confinement, which affects not only the electron and hole energy levels but also gives rise to polarisation dependent matrix elements. The transition energy can be changed not only through the composition but also through the well width. However, changes in composition also affect the band offsets and the strain state of the layer, and detailed calculations are required to determine the effect on the energy levels. Simplifications such as the infinite well approximation and the effective mass approximation can be used to give an estimate of transition energies and gain, but for accuracy solution of the $k \cdot p$ Hamiltonian, including the effects of strain, is required. Strain particularly affects the valence band structure, which is in any case complicated enough because of the interaction and overlap of the three valence bands; the light and heavy hole bands and the split-off band. Compressive strain leads to lower threshold currents and higher differential gains, while tensile strain can change the polarisation of the output from TE to TM by raising the light hole state above the heavy hole state.

In addition to these fundamentals, real structures can be characterised by poor optical or electrical confinement. To an extent these are mutually contradictory, and a range of ideas exist within the published literature on the effects and how to mitigate them. In particular, slow transport can affect the modulation rate, and carrier leakage and band filling effects can affect the high power operation. If the material properties of the heterojunctions do not allow for effective confinement, superlattices can be grown which can improve electrical confinement by increasing the barrier height by a factor of two or more. Other structures can be grown, however, and an example of a high power red emitting device has been given in which additional layers decouple the optical and electrical confinements.

6.10 References

[1] Ekenberg U 1989 *Phys. Rev. B* **40** 7714–7726
[2] Bastard G 1988 *Wave Mechanics Applied to Semiconductor Heterostructures*, Les Editions de Physique, France
[3] Celik H, Cankurtaran M, Bayrakli A, Tiras E and Balkan N 1997 *Semicond. Sci. Technol.* **12** 389–395
[4] Lee K-S and Lee E H 1996 *ETRI Journal* **17** 13–24
[5] Aspnes D E, Olsen C G and Lynch D W 1975 *Phys. Rev. B* **12** 2527–2538
[6] Coldren L A and Corzine S W 1995 *Diode Lasers and Photonic Integrated Circuits* (John Wiley & Sons) p 493
[7] Asada M and Suematsu Y 1984 *IEEE J. Quant. Electr.* **QE-20** 745–753 and Yan R H, Corzine S W, Coldren L A and Suemune I 1990 *IEEE J. Quant. Electr.* **QE-26** 213–216
[8] Yamanishi M and Suemune I 1984 *Jap. J. Appl. Phys.* **23** L35–L36 and Chen P A, Chang C Y and Juang C 1994 *J. Appl. Phys.* **76** 85–91
[9] Dutta N K, Jaques J and Piccirilli A B 2002 *Electronics Letters* **38** 513–515
[10] Avrutin E A, Chebunina I E, Eliachevitch I A, Gurevich S A, Portnoi M E and Shtengel G E 1993 *Semicond. Sci. & Technol.* **8** 80–87
[11] Balle S 1998 *Phys. Rev. A* **57** 1304–1312
[12] Yamada M, Ogita S, Yamagishi M, Tabata K, Nakaya N, Asada M and Suematsu Y 1984 *Appl. Phys. Lett.* **45** 324–325
[13] Newell T C, Wright M W, Hou H and Lester L F 1999 *IEEE J. Sel. Topics. in Quant. Electr.* **5** 620–626
[14] Kobayashi H, Iwamura H, Saku T and Otsuka K 1983 *Electronics Letters* **19** 166–167
[15] Hakki B W and Paoli T L 1973 *J. Appl. Phys.* **44** 4113–4119
[16] Kleinman D A and Miller R C 1985 *Phys. Rev. B.* **32** 2266–2272
[17] Pacey C, Silver M, Adams A R and O'Reilly E P 1997 *Int. J. Optoelectr.* **11** 253–262
[18] Tolliver T R, Anderson N G, Agahi F and Lau K M 2000 *J. Appl. Phys.* **88** 5400–5409
[19] Chang C-S and Chuang S L 1995 *Appl. Phys. Lett.* **66** 795–797

[20] Vurgaftman I, Meyer J R and Ram-Mohan L R 2001 *J. Appl. Phys.* **89** 5815–5875
[21] Chang C-S and Chuang S L 1995 *IEEE J. Sel. Topics. in Quant. Electr.* **1** 218–229
[22] Jones G and O'Reilly E P 1993 *IEEE J. Quant. Electr.* **QE-29** 1344–1354
[23] Tsang W T, Weisbuch C, Miller R C and Dingle R 1979 *Appl. Phys. Lett.* **35** 673–675
[24] Kraus J and Diemel P P 1993 *J. Lightwave Technol.* **11** 1802–1805
[25] Keating T, Jin X, Chuand S L and Hess K 1999 *IEEE J. Quant. Electr.* **35** 1526–1533
[26] Hochholzer M and Harth W 1995 *IEE Proc. – Optoelectron.* **145** 232–236
[27] Irikawa M, Ishikawa T, Fukushima T, Shimizu H, Kasukawa A and Iga K 2000 *Japn. J. Appl. Phys.* **39** 1730–1737
[28] Lu M F, Juang C, Jou M J and Lee B J 1995 *IEE Proc. – Optoelectron.* **145** 237–240
[29] Liang K, Pan Q and Green R J 1999 *IEEE J. Quant. Electr.* **35** 955–960
[30] Hillmer H and Marcinkevičius S 1998 *Appl. Phys. B* **66** 1–17
[31] Dion M, Li Z M, Ross D, Chatenoud F, Williams R L and Dick S 1995 *IEEE J. Sel. Topics in Quant. Electr.* **1** 230–233
[32] Nagarajan R, Kamiya T and Kurobe A 1989 *IEEE J. Quant. Electr.* **QE-25** 1161–1170
[33] Phillips A F, Sweeney S J, Adams A R and Thijs P A J 1999 *IEEE J. Sel. Topics in Quant. Electr.* **5** 401–412
[34] Givens M E, Miller L M and Coleman J J 1992 *J. Appl. Phys.* **71** 4583–4588
[35] Czyszanowski T, Wasiak M and Nakwaski W 2001 *Optica Applicata,* **XXXI** 313–323
[36] Czyszanowski T, Wasiak M and Nakwaski W 2001 *Optica Applicata,* **XXXI** 325–336
[37] Gutiérrez M, Herrera M, Gonzàlez D, Aragón G, Sánchez J J, Izpura I, Hopkinson M and García R 2002 *Microelectronics Journal* **33** 553–557
[38] Yamamoto N, Yokoyama K, Yamanaka T and Yamamoto M 1997 *IEEE J. Quant. Electr.* **33** 1141–1148
[39] Namegaya T, Katsumi R, Iwai N, Namiki S, Kasukawa A, Hiratani Y and Kikuta T 1991 *IEEE J. Quant. Electr.* **QE-29** 1924–1931
[40] Morin S, Deveaud B, Clerot F, Fujiwara K and Mitsunaga K 1991 *IEEE J. Quant. Electr.* **QE-27** 1669–1675
[41] Marcinkevičius S, Fröjdh K, Hillmer H, Lösch R and Olin U 1998 *Mat. Sci & Eng.* **B51** 30–33
[42] Fröjdh K, Marcinkevičius S, Olin U, Silfvenius C, Stålnacke B and Landgren G 1996 *Appl. Phys. Lett.* **69** 3695–3697
[43] Paiella R, Hunziker G and Vahala K J 1999 *Semicond. Sci. & Technol.* **14** R17–R25
[44] Giudice G E, Kuksenkov D V and Temkin H 2001 *Appl. Phys. Lett.* **78** 4109–4111
[45] Klotzkin D, Zhang X, Bhattacharya P, Caneau C and Bhat R 1997 *IEEE Phot. Technol. Lett.* **9** 578–580
[46] Esquivias I, Weisser S, Romero B, Ralston J D and Rosenzweig J 1996 *IEEE Phot. Technol. Lett.* **8** 1294–1296

[47] Dao L V, Johnson M B, Gal M, Fu L, Tan H H and Jagadish C 1999 *Appl. Phys. Lett.* **73** 3408–3410
[48] Brum J A and Bastard G 1986 *Phys. Rev. B* **33** 1420–1423
[49] Abou-Khalil M, Goano M, Reid B, Champagne A and Maciejko R 1997 *J. Appl. Phys.* **81** 6438–6441
[50] Kàlna K, Moško M and Peeters F M 1996 *Appl. Phys. Lett.* **68** 117–119
[51] Register L F and Hess K 1997 *Appl. Phys. Lett.* **71** 1222–1225
[52] Moško M and Kàlna K 1999 *Semicond. Sci. & Technol.* **14** 790–796
[53] Crow G C and Abram R A 1999 *Semicond. Sci. & Technol.* **14** 1–11
[54] Levetas, S A and Godfrey M J 1999 *Phys. Rev. B* **59** 10202–10207
[55] Baraff G A 1998 *Phys. Rev. B* **58** 13799–13810
[56] Daniels M E, Bishop P J and Ridley B K 1997 *Semicond. Sci. & Technol.* **12** 375–379
[57] Wang J, Greisinger U A and Schweizer H 1996 *Appl. Phys. Lett.* **69** 1585–1587
[58] Brübach J, Silov A Y, Haverkort J E M, van der Vleuten W and Wolter J H 2000 *Phys. Rev. B* **61** 16833–16840
[59] Güçlü A D, Rejeb C, Maciejko R, Morris D and Champagne A 1999 *J. Appl. Phys.* **86** 3391–3397
[60] Kurakake H 1998 *J. Appl. Phys.* **84** 5643–5646
[61] Hirayama H and Asada M 1994 *Optical and Quantum Electronics* **26** S719–S729
[62] Zhao B, Chen T R and Yariv A 1992 *Appl. Phys. Lett.* **60** 1930–1932
[63] Ginty A, Lambkin J D, Considine L and Kelly W M 1993 *Electronics Lett.* **29** 684–685
[64] Usami M, Matsushima Y and Takahashi Y 1993 *Electronics Lett.* **29** 684–685
[65] Usami M, Matsushima Y and Takahashi Y 1995 *IEEE J. Sel. Topics. Quant. Electr.* **1** 244–249
[66] Takayama T, Imafuji O, Hashimoto T, Yuri M, Yoshikawa A and Itoh K 1996 *Jpn. J. Appl. Phys.* **35** L493–L495
[67] Pan J-W, Chau K-G, Chyi J-I, Tu Y-K and Liaw J-W 1998 *Appl. Phys. Lett.* **72** 2090–2092
[68] Chyi J-I, Chen M-H, Pan J-W, Shih T-T 1993 *Electronics Lett.* **35** 1255–1257
[69] Bugajski M, Mroziewicz B, Regiński K, Muszalski J, Kubica J, Zbroszczyk M, Sajewicz P, Piwonski T, Jachymek A, Rutkowski R, Ochalski T, Wójcik A, Kowalczyk E, Malag A, Kozlowska A, Dobrzanski L and Jagoda A 2001 *Optoelectronics Review* **9** 35–47
[70] Dumitrescu M, Toivonen M, Savolainen P, Orsila S and Pessa M 1999 *Optical and Quantum Electronics* **31** 1009–1030

Problems

1. Plot the first three wavefunctions ($A = 0$) for an electron in an infinite quantum well extending from -2.5 nm $\leq z \leq 2.5$ nm and find the positions within the well where the probability of finding the electron is maximised and minimised. Give the value of $|\psi|^2$ at these points.
2. Calculate the energy levels for the three levels above assuming both a free electron mass and an effective mass of 0.068.

184   *Quantum well lasers*

3. Assuming a 5 nm wide quantum well of GaAs in $Al_{0.4}Ga_{0.6}As$ with 40% of the band offset in the conduction band, plot the eigenvalue equation to determine the first quantisation energy. Assume for simplicity that the effective mass in both the well and the barrier is the same at 0.068.
4. Perform the same calculation for the heavy hole ($m^* = 0.45$) and hence estimate the transition energy and wavelength for a quantum well laser based on this system. Contrast this with the emission wavelength for a bulk GaAs laser.
5. Estimate the penetration depth of both the electron and the hole inside the barrier.

Chapter 7

The vertical cavity surface emitting laser

The DH and quantum well lasers considered in the last two chapters are edge-emitting devices. Light is emitted in the plane of the wafer, so dicing is an essential part of the fabrication of the cavity. Manufacture of such lasers is therefore a relatively high cost operation compared with other semiconductor fabrication technologies. It is not possible to perform the sort of wafer scale quality and reliability checks as part of the manufacturing process that are normally employed in semiconductor fabrication, nor is it easy to integrate the diode lasers with other opto-electronic components in a single monolithic fabrication process. The functionality of high speed integrated circuits that can generate and process signals at several tens of GHz could be increased by integrating laser emitters, but edge-emitting devices have to be bonded to the circuit. Not only does this provide a potential mechanism for failure – any external connections that have to be bonded or soldered are potential weak spots – but the number of emitters that can be incorporated is clearly limited. Surface emitting lasers that do not have to be diced therefore have the potential not only to be integrated with other functional devices but arrays can be made and parallel optical processing becomes possible. Moreover, in the vertical cavity configuration the emitting aperture is often circular leading to a relatively well conditioned symmetric output beam of low divergence. By contrast a typical edge-emitting device is essentially a line source and the output beam is also highly astygmatic, with the divergence in one direction far exceeding the divergence in the other.

The vertical cavity surface emitting laser (VCSEL) was first proposed in 1977 by Kenichi Iga of Japan [1], who worked pretty much alone on the device until about 1988 when, significantly, room temperature operation was reported. A number of other groups began to work on the devices and eventually in about 1999 the technology reached the stage where a number of devices were being transferred from the laboratory to production [2]. The most common wavelength at the moment is 850 nm, which is used as a high speed optical source for local area networks (LANs). What makes this wavelength so attractive is the use of the well established GaAs/AlGaAs material system. Devices in the visible (650–670 nm) and in the near infra-red (1310 nm and

1550 nm) have been extensively investigated but as yet have not reached the production stage. More recently still, blue-green emitters based on GaN have also been investigated.

VCSEL technology is still very much under development but aspects of the technology are sufficiently mature to warrant a fairly extensive discussion. However, some aspects of this device, such as the definition of the cavity length and the solution of the electromagnetic equations for the waveguide modes, are considerably more complicated than their counterparts in edge-emitting structures. Whilst some of this can be handled analytically other parts have to be treated using numerical methods, which makes a full description difficult. By their nature, numerical solutions are specific to a set of circumstances, and extracting from them ideas and concepts that can be applied more generally is not easy. Moreover, the computational methods employed to extract the numerical solutions are topics in themselves and are beyond the scope of this book. For this reason, specific examples of numerical calculations will be described where appropriate and the reader is referred to more extensive texts for the details. This chapter proceeds with a description of the cavity and some of the issues involved, followed by a consideration of the wave guiding mechanisms before finishing with a description of developments in both long wavelength VCSELs suitable for optical communications and visible VCSELs.

The vertical cavity

The geometry of the VCSEL is very different from that of the conventional edge-emitting device (figure 7.1). It is immediately obvious that the VCSEL requires mirrors and contacts to be fabricated in such a way that epitaxial growth of the active layer material on top is possible. Needless to say, the

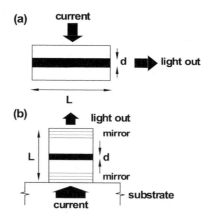

Figure 7.1. Edge-emitting structure (a) compared with VCSEL structure (b).

fabrication of such mirrors is not straightforward. The difficulties involved are one reason why it has taken so long for the VCSEL to reach production from the first proposal, but such mirrors have now been perfected for operation at 850 nm. They consist of multiple layers of thin films. Partial reflections off each interface add together to produce a high reflectivity (>95%). However, in the edge-emitting device the reflection occurs at a particular, well-defined point in space (the mirrors are said to be "hard") whereas in the VCSEL the optical field penetrates some distance into the multi-layer stack. The definition of the cavity length is therefore somewhat ambiguous. Other differences between the two can also be identified. In the edge-emitting device the current flow is perpendicular to the optical cavity so population inversion and gain exists over the whole of the cavity. In the VCSEL, the active region forms but a small part of the cavity. This, together with the ambiguity over the cavity length, affects the threshold condition as well as the output characteristics of the VCSEL.

In chapter 3, the idea that the threshold gain is proportional to the threshold current was developed (equation chapter 3, 23) so it is only necessary to identify the factors that affect the threshold gain to determine the important influences on threshold current. The cavity can be divided into three sections (figure 7.2). Suppose for the sake of convenience the active region is situated exactly in the middle of a cavity of effective length L ("effective" in recognition of the ambiguity). Starting at the left mirror the intensity of a beam propagating to the right will first be attenuated by an amount $\exp[-\alpha(L-d)/2]$, then amplified by $\exp(\gamma d)$, attenuated again by $\exp[-\alpha(L-d)/2]$, and finally attenuated on reflection by R_2. There will be a slight reflection on crossing from the lossy region to the active region because of the inevitable refractive index mismatch, but this loss can be subsumed into the loss coefficient α. On traversing back across the cavity to complete one round trip the amplitude must be similarly attenuated and amplified before a final attenuation of R_1 at the right hand mirror. At threshold the net effect of this round-trip must be to restore the intensity to its original value. Strictly, the optical confinement factor Γ should be included in each of these arguments but it is neglected here because in the VCSEL the optical confinement is defined differently according to how much of the active region occupies the cavity. Moreover, the lateral distribution of light is not so easily defined as in the DH laser or the quantum well laser with a narrow optically confining layer where there is a definite wave-guide defined

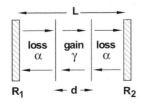

Figure 7.2. The three-section model of the VCSEL cavity.

by the geometry of the active region and surrounding layers. In the VCSEL optical propagation will be occurring within regions where a well defined wave-guide might not exist, but even if it does exist it is likely to be several micrometres in diameter and the lateral optical confinement can be expected to be high. Taking together all the terms in sequence, the threshold gain γ_{th} is therefore defined by the condition that

$$\exp[-2\alpha(L-d)]\exp(2\gamma_{th}d) = \frac{1}{R_1 R_2} \qquad (1)$$

which leads to a threshold gain

$$\gamma_{th} = \frac{1}{2d}\ln\left(\frac{1}{R_1 R_2}\right) + \frac{\alpha}{d}(L-d) \qquad (2)$$

where the second term is simply the material loss weighted by the length of the loss region. The threshold current

$$J_{th} = \frac{\gamma_{th}}{A} + J_t = \frac{1}{2Ad}\ln\left(\frac{1}{R_1 R_2}\right) + \frac{\alpha}{Ad}(L-d) + J_t \qquad (3)$$

where, as before, A is a constant. Equations (4) and (5) tell us all we need to know: the mirrors should be highly reflecting so that the product of their reflectivites is as close to unity as possible to offset the effect of a small d relative to L. The logarithmic term is then very small and the intrinsic material gain, γ, can be correspondingly smaller. The magnitude of d is not so obvious. A large d implies a small loss but the transparency current will be higher. Ideally, d should be small, especially if the loss coefficient is small and the second term becomes relatively unimportant. In a double heterostructure system, for example, where the cavity is composed mainly of a wide band gap material relative to the active region, the intrinsic material loss can be expected to be almost zero.

The question resolves essentially into the choice between a double heterostructure or a quantum well system, a dichotomy which is nicely illustrated by a couple of papers published within a few pages of each other in the 1991 issue of the IEEE Journal of Quantum Electronics. Iga's group were reporting the results of a VCSEL device based on a $Al_{0.3}Ga_{0.7}As$ – GaAs – $Al_{0.3}Ga_{0.7}As$ double heterostructure device with a 2.5 μm thick active region defined into a circular cavity 7 μm in diameter [3], whereas Geels et al. [4] were reporting on the use of InGaAs quantum wells as active layers for emission at 980 nm. These two devices provide a good starting point for the discussion on practical VCSEL devices.

The Iga device (figure 7.3) was constructed upside down by depositing the DH layers on a GaAs substrate, including a low – Al layer ($Al_{0.1}Ga_{0.9}As$) to

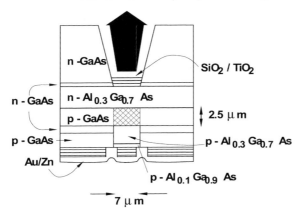

Figure 7.3. Iga's bottom-emitting VCSEL [3].

assist with current injection from the contact. Etching through the top AlGaAs layers and regrowing current-blocking GaAs layers on to the etched surfaces defined the cavity width by means of current restriction. The cavity was formed by depositing quarter-wavelength pairs of SiO_2/TiO_2 layers onto the top $Al_{0.1}Ga_{0.9}As$ layer and the bottom $Al_{0.3}Ga_{0.7}As$ layer after etching through the substrate and was 5.5 μm long. These layers act as a dielectric mirror. Unlike the edge-emitting device, there are no cleaved facets. It would be possible to use a growth surface as a similar reflector, but with the refractive index of a typical semiconductor being in the region of 3–3.5 in the near infra-red, the reflectivity of the surface exposed to air will be around 25%–30%. This works in the edge-emitting device because the long cavity length allows sufficient amplification at even a low material gain to overcome the losses, but it is nowhere near high enough for the short cavity of the VCSEL. In the device illustrated in figure 7.3 one of the mirrors could, in principle, be metallic as it also coincides with the contact but a simple metal overlayer rarely provides a good contact to a semiconductor without some sort of heat treatment which can disrupt the interface. In Au/Zn alloys on *p*-GaAs heating will cause an in-diffusion of Zn to form a heavily doped *p*-type layer with a corresponding out-diffusion of Ga into the Au. Such diffusion is not uniform and the non-uniform interface has a low reflectivity. However, Ag on GaAs can be used in this way for LEDs [5] though the technology is not employed in lasers.

Even if a metal could be used for the bottom contact, it is inappropriate for the top contact, which must be partially transmitting. A thin metal layer can be both reflecting and transmitting but it is also absorbing, which would not only increase the threshold gain required but would also lead to heating and possible damage under operation at even moderate output powers. The dielectric mirror provides a non-absorbing, partially reflecting structure. The principle of operation is illustrated in figure 7.4 and is very simple. Light incident on any

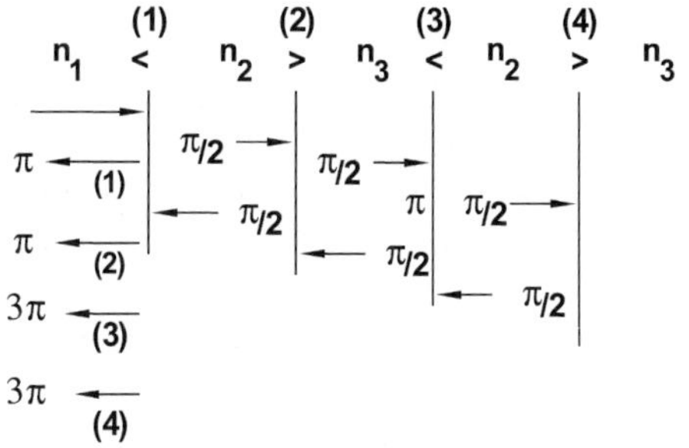

190   The vertical cavity surface emitting laser

| | (1) | | (2) | | (3) | | (4) | |
| | n_1 | < | n_2 | > | n_3 | < | n_2 | > n_3 |

$\pi \leftarrow \overline{(1)}$ $\pi/2 \rightarrow$

$\leftarrow \pi/2$ $\pi/2 \rightarrow$

$\pi \leftarrow \overline{(2)}$ $\pi \pi/2 \rightarrow$

$\leftarrow \pi/2$

$3\pi \leftarrow \overline{(3)}$ $\leftarrow \pi/2$

$3\pi \leftarrow \overline{(4)}$

Figure 7.4. In-phase reflections from quarter-wavelength pairs.

interface where a refractive index difference exists is partially reflected according to the ratios of the difference and sum of refractive indices. For normal incidence, and for simplicity we can assume that the light is propagating normally to the interface, the well-known Fresnel amplitude reflection and transmission coefficients are, for a wave travelling from medium of refractive index n_1 to n_2,

$$r = \frac{n_1 - n_2}{n_1 + n_2} \tag{4}$$

and

$$t = \frac{2n_2}{n_1 + n_2}. \tag{5}$$

Some reflection therefore occurs at each interface, but the key to the quarter-wavelength stack lies in the phase of each reflection. If $n_1 > n_2$ r is positive, but if $n_1 < n_2$ r is negative, which simply corresponds to a phase change of radians, or equivalently one half-wavelength. In crossing a quarter-wavelength layer the light undergoes a phase change of $\pi/2$, so if the total phase of a ray is traced out from the point of entry to the point of reflection and back, then each pair of interfaces reflect in phase. Other interfaces are shifted by multiple of 2π. No matter from what interface it is reflected the wave will arrive back at the front of the stack shifted in phase from other reflected waves by a whole number of wavelengths. In short the waves constructively interfere and the amplitudes are added. It is not immediately obvious from the above, but if the stack terminates on air or a low refractive index material an odd number of layers is required to achieve a high reflectivity. Dielectric mirrors are therefore constructed with a final half-pair or single layer.

The number of pairs of thin films required to achieve a high reflectivity depends on the refractive index difference between the constituent materials. Silicon dioxide and titanium dioxide are common materials used in dielectric mirrors for a wide variety of applications because they are transparent over a very wide range of wavelengths, hard-wearing, thermally stable, do not interdiffuse, are easily deposited by sputtering, and have a large refractive index difference. In fact, rutilated quartz (titanium oxide is also known as rutile) is a naturally occurring state of quartz caused by phase segregation from TiO_2-SiO_2 mixtures which are stable at high temperatures and pressures but which separate out on cooling, leading to a quartz matrix with rutile inclusions. Titanium dioxide has different phases with slightly different refractive indices, each of which has a high dispersion. The refractive index varies from 2.5 to nearly 2.8, depending on the wavelength, so each interface in the dielectric stack has an amplitude reflection coefficient at ~0.25 or 25%. This means that only five pairs of films are needed to produce the high reflectivity output mirror for the VCSEL, with a power reflectivity of >95% and a transmittivity of ≤5%. For the other mirrors, the reflectivity should ideally be as near 100% as possible, which requires additional pairs of films. Iga's structure actually helps in this regard, because the contact is deposited over the mirror structure, and will provide a highly reflecting surface for any light not turned back within the length of the dielectric stack. The metal makes contact with the AlGaAs in the form of an anulus etched into the mirror stack (figure 7.3). It is imperative in this type of structure that injection into the active region is uniform so a very low resistivity layer above the active region adjacent to the contact is required.

It is quite clear from figure 7.3 that no discernible waveguiding structure exists over the whole of the cavity length. In fact, the bottom AlGaAs layers surrounded by GaAs form an anti-guiding structure because the surrounding refractive index is higher. Within the active region itself, which forms a little under half of the total cavity, gain guiding occurs in a similar manner to the *p–n* homojunction device. In the top AlGaAs layer next to the output coupling mirror there is neither gain nor a refractive index difference so there is no mechanism to confine the optical field. Analysis of the optical propagation is therefore quite complex, but the key to the structure is the gain guiding in the active region. Although the guide is short, there is a clear guiding mechanism and well defined modes of a cylindrical waveguide must exist. Whatever effects may occur in the rest of the cavity, these modes, or at least something very similar, must exist throughout the whole cavity so that radiation reflected from the mirrors can couple into the waveguide modes of the active region.

Contrast this structure with that published by Geels at the same time. As illustrated in figure 7.5, the structure is fully integrated with the quarter-wave stacks (sometimes also called distributed Bragg reflectors, or DBRs) were grown into the structure as epitaxial AlAs/GaAs quarter-wavelengths pairs. The active region consisted of two InGaAs quantum wells (980 nm) with $Al_{0.2}Ga_{0.8}As$ spacer layers on either side grown between the DBRs. These DBRs are intended to act not only as mirrors but also as contact layers, so the growth and design of such structures is made more complicated and demanding than Iga's structure. For a start, the refractive index difference (2.95 to 3.52)

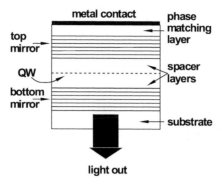

Figure 7.5. Schematic of the bottom-emitting 980 nm VCSEL of Geels *et al.*

between these mirror materials is not so large as with TiO_2-SiO_2 so the Fresnel coefficient is correspondingly smaller (0.0881) and hence more pairs are needed. In this case 18.5 pairs were grown adjacent to the substrate through which the radiation is emitted. In common with Iga's device, this is a bottom-emitting laser but unlike the device in figure 7.3 it is not necessary to etch through the substrate as GaAs is transparent at 980 nm. Strangely, the method used to define the cavity width was not described so it is not clear what size aperture was used. Instead the authors concentrated on modelling the reflectivity of the mirrors and attempting to measure the intrinsic loss within the cavity, and indeed, this was one of the first papers in which detailed modelling of the cavity and the threshold condition was allied to an experimental investigation.

It is apparent that the growth of a laser such as this is inherently more difficult than the laser in figure 7.3. The mirrors are an integral part of the laser and have to be grown within the same sequence as the active layers. With a total mirror thickness approaching 3 μm, it is imperative that lattice matching be maintained in order to ensure that the active region on top is grown in good quality material. This effectively restricts the choice of material to the GaAs/AlAs system. Moreover, control of the growth rate has to be maintained to an accuracy greater than 1% so that the peak wavelength of the dielectric stack coincides with the intended peak wavelength. One of the features of dielectric stacks with small refractive index differences is that the bandwidth decreases with the number of stacks. Figure 7.6 shows the effect of increasing the number of AlAs/GaAs pairs in a free-standing stack (bounded on either side by air) assuming a constant refractive index for each layer over the calculated wavelength range. Such a structure is not physically realistic, of course, and the reflectivity of a real stack has to be calculated taking into account not only the wavelength dispersion but also the the effect of the layers on either side. The layers within the cavity of a VCSEL are themselves quite thin and therefore exhibit phase effects in their own right, which modulates the reflectivity of the dielectric stack and can reduce the bandwidth further.

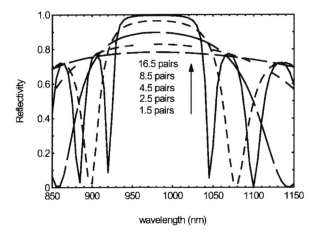

Figure 7.6. Reflection bands from AlAs/GaAs quarter-wavelength mirrors. Note the narrowness of the reflection range for 16.5 pairs compared with 4.5 or less.

Spectra such as that shown in figure 7.6 can be calculated by a standard transfer matrix formulation for the reflectivity in which the total amplitude is calculated from the bottom of the stack upwards, taking into account the phase change traversing each layer. Transmission matrices provide an exact method of calculating the mirror properties for any wavelength and refractive index, but the details are complicated and will not be given here. Instead, the reader is referred to [6], and here the simple formula provided by Geels *et al.* for such calculations, which is based on the transmission matrix approach, will be used. Each layer is exactly a quarter wavelength thick and the phase change on traversing each layer is exactly $\pi/2$, so at the peak wavelength where all the reflections are in phase the total reflectivity can be calculated simply by adding a series of terms corresponding to the reflections from each interface. For small amplitude reflection coefficients, r_i, at each interface, given by equation (6), the effective amplitude reflectivity of a quarter-wave dielectric stack is

$$r_{\mathit{eff}} = \tanh\left[\sum_{i=1}^{N} \tanh^{-1}(r_i)\right] \qquad (6)$$

where, for $i=1$ $r=0.064$ in [4] rather than 0.0881 because the spacer layer is $Al_{0.2}Ga_{0.8}As$.

Strictly, equation (8) applies to abrupt interfaces, but the requirement that the quarter-wave stacks also act as contacts means that abrupt interfaces are not the most appropriate. Although the total offset will be the same, a graded interface will present a smaller barrier to current flow under the influence of an electric field (figure 7.7). Recall that the carrier densities will be equal within the active region in order to ensure charge neutrality. By definition the electric

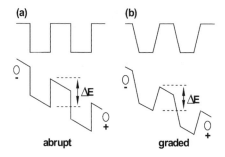

Figure 7.7. Field lowering of the barrier in a graded DBR.

field must also be zero and most of the voltage will be dropped across the contacts, even if they are highly doped. For graded optical interfaces it is necessary to use coupled mode theory in which two counter propagating waves are described in terms of a coupling coefficient that couples energy from one wave into the other. This theory is described in greater detail in chapter 9. For an abrupt interface the coupling coefficient is

$$\kappa_{abr} = \frac{2\Delta n}{\lambda_0} \tag{7}$$

and for a graded interface where the grading occurs over a distance $2w$ the coupling coefficient is

$$\kappa_{gr} = \kappa_{abr} \cdot \frac{\sin\left(\frac{2\pi w}{T}\right)}{\frac{2\pi w}{T}} \tag{8}$$

where T is the period of the grating. Geels *et al.* showed that for an interface graded over 18 nm

$$\kappa_{gr} = 0.98 \kappa_{abr} \tag{9}$$

leading to a reduction of the amplitude reflectivity of the matrix from 0.9974 to 0.9970 with corresponding reductions in the power reflectivity from 0.9950 to 0.9941. These are very small reductions but in the context of the threshold of a VCSEL where the cavity is short, any reduction in the reflectivity manifests itself as an appreciable effective loss coefficient because the round trip length is so small. The way to overcome this is to add an extra pair of dielectrics, and for a 19.5 pair stack of graded layers the power reflectivity rises to 0.9958.

Finally, it remains to discuss the nature of the active region within these two lasers. The volume of the active region in Iga's device works out at about 96 μm^3, compared with ~0.8 μm^3 for a similar area in Geels' device. This is a dramatic difference and we can expect the transparency current contribution to the threshold current to be correspondingly different, all other factors being equal. In fact, compared with a conventional DH edge-emitting laser with a cavity length say 300 μm, an active region 0.2 μm thick, and a stripe 3 μm wide, Iga's device achieves only a factor of two reduction in the active volume. Given the additional complications in the fabrication, such as etching through the substrate and depositing external mirror layers, there is seemingly little benefit in a surface emitting laser based on this type of design compared with the conventional edge emitter. Even the possibility of integration seems remote given the fabrication process. Geels's device, on the other hand, can be much more easily integrated with other device technologies as all the growth is done *in situ*. In addition the volume of the active region is much reduced, leading potentially to very low threshold currents.

This doesn't necessarily answer the question as to whether a quantum well system is intrinsically better than a classical bulk laser based on the DH structure. After all, one could just as easily have a double heterostructure system 0.1 μm thick. In fact, anything over 30–40 nm thick will behave pretty much like a bulk semiconductor, so if it is simply a question of the volume of the active region there is plenty of scope within the DH structure to have a much smaller transparency current. As the previous chapter showed, the physics of the quantum well is fundamentally different from from the physics of a bulk semiconductor so it is not just a matter of volume. There are good reasons for supposing that quantum wells will out-perform the DH structure. In particular, two differences stand out:

- the degeneracy of the light and heavy hole states is lifted within a quantum well leading to a lower density of states around the valence band maximum and a more even distribution of the Fermi levels between the valence and conduction bands;
- the polarisation selection rules for heavy holes, which form the valence band maximum, lead to an enhancement of the matrix element over bulk material by 3/2.

A muliple quantum well VCSEL will therefore have a lower threshold current and a higher gain than a single layer of equivalent thickness. With such thin layers, however, the position of the layer within the cavity is crucial.

It is easy to forget that the Fabry–Perot cavity modes are standing waves with nodes and anti-nodes, because in a long cavity edge-emitting device it is simply not important. Although there will be periodic variations in light intensity along the cavity, with consequent periodic variations in the rate of stimulated emission and carrier density, carrier diffusion means that the carrier density will not simply increase indefinitely at the nodes where the light intensity is lowest, but will be moderated by diffusion to the anti-nodes. Furthermore there will be several thousand nodes and antinodes in such a cavity so the net effect along the cavity length will be an average of all these separate nodes. In a VCSEL, however, the cavity length is much smaller (in Iga's device it was about 5.5 μm, and in others it is a fifth of this at ~1 μm) so there will be a handful of nodes and antinodes within the standing wave pattern. The active region is located in space by the position of the quantum well, so if this is inadvertently placed at a node the gain will be effectively zero until thermal effects or the high carrier density affect the optical properties of the cavity and shift the node pattern slightly. Either the device will not lase or the threshold current will be very, very high.

Additionally, the cavity resonance has to be tuned to the gain maximum. Suppose, by way of example, a simplified cavity consisting of two metal mirrors bounding a quantum well system consisting of AlGaAs barriers approximately 0.5 μm thick, of refractive index 3.4, and InGaAs designed to emit at 980 nm. We'll assume that the light propagates paraxially (though it will be shown that this might not happen in practise) so the cavity modes can be assumed to be true Fabry–Perot modes. Let's also ignore the refractive index of the quantum well because its thickness is much less than that of the barriers. In essence, then, there is about 1 μm of wide band gap material between the mirrors. It is an easy matter to show that for a thickness of exactly 1μm the resonant frequency of the cavity is actually 971 nm, which might well fall within the gain spectrum but will not coincide with the gain maximum. The laser will work but the threshold current will be high. For emission at 980 nm the cavity has to be 1009 nm thick corresponding to a mode index of 7, or alternatively 865 nm or 1153 nm thick for mode indices of 6 and 8 respectively. Conversely, for a cavity at 1009 nm thick, the resonant wavelengths corresponding to mode indices of 6 and 8 respectively are 1143 nm and 857 nm.

The discrepancy between the cavity resonance and the gain maximum leads to the concept of the intrinsic voltage, which is the minimum voltage that can be applied to the device to achieve threshold. It is a more appropraiate measure of the threshold performance than the current density because VCSEL devices often have large series resistances associated with the DBRs, whereas in edge-emitting devices the relatively wide cross-section of the active region means that series resistance tends to be low and most of the voltage is dropped across the active region. An ideal diode under forward bias will exhibit an exponential dependence of the forward current on bias, but in the presence of a large series resistance the forward bias characteristics are almost ohmic as nearly all the additional voltage is dropped across the resistance rather than the diode (figure 7.8). In a VCSEL the principal source of series resistance are the mirrors, so if the current-voltage characteristic is ohmic above threshold the mirror resistance, R_m, can be determined. If the applied voltage at threshold is V_{th} and the voltage dropped across the mirrors is $I_{th} \cdot R_m$ the intrinsic voltage is then

$$V_i = V_{th} - I_{th} R_m. \tag{10}$$

The vertical cavity surface emitting laser 197

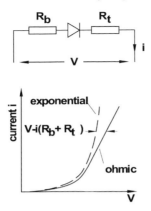

Figure 7.8. Forward bias characteristics of a VCSEL diode in the presence of the bottom and top mirror resistances R_b and R_t, respectively.

The intrinsic voltage increases with the cavity detuning and represents the physical reality that the carrier density must be higher to bring the gain up to the minimum required for threshold. In fact, the intrinsic voltage can be taken to be equivalent to the Fermi level separation, which must increase if the gain maximum does not coincide with the cavity resonance. This was demonstrated in nominally 980 nm VCSELs [7]. These near-IR lasers were designed to have the same gain maximum but the cavity resonances differed slightly because of growth non-uniformity across the wafer. Modelling the lasers assuming a Fermi level separation equal to the intrinsic voltage and using a gain expression derived from semiconductor Bloch equations (often referred to as SBEs) showed that a single threshold gain of 500 cm^{-1} explains all the experimental data (figure 7.9).

The details behind this type of modelling are beyond the scope of this book, but an appreciation of the ideas is useful. SBEs treat the electromagnetic field classically taking into account the polarisation but the active medium is treated quantum mechanically. Also included are expressions for the electron density (hence the ability to extract the Fermi level separation) as well as many body effects such as Coulomb enhancement of the recombination. This last is a many body effect. Free carriers in the conduction band are in fact attracted to free carriers in the valance band via Coulomb forces except at very high carrier densities, when dielectric screening renders the effect negligible. At the carrier densities typically used in a VCSEL these many body effects still occur. The consequence of Coulomb attraction is to reduce the separation between electrons and holes and enhance the dipole moment operator for the transition. These ideas are very important for VCSEL modelling. The conventional rate equation approach applicable to long cavity edge-emitting lasers is limited, partly because of the many-body effects, but also because the electromagnetic field within the VCSEL has a well-defined polarisation and position-dependent intensity.

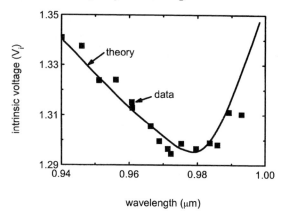

Figure 7.9. Measured intrinsic voltage (symbols) plotted against the laser wavelength as determined by the cavity resonance for near-IR lasers. The solid line represents a theoretical model assuming that the intrinsic voltage is equivalent to the Fermi level separation. (After Chow et al.)

The calculated mode spacings relative to the cavity resonance illustrate the need for growth to much greater than 1% accuracy, which is reinforced by the concept of the minimum voltage illustrated in figure 7.9. However, given accurate knowledge of the layer thicknesses, it is still not clear what the effective cavity length is. Not only does the refractive index change from the barrier layers of the quantum well system to the mirrors, but the reflection does not occur at a fixed point in space. The calculation of the mode properties in a cavity with distributed reflectors is not as simple as the previous calculation suggests, though it is not a bad approximation for the fundamental mode. Fortunately coupled mode theory comes to our aid again by allowing the distributed reflector to be replaced by an effective metal mirror. The details need not concern us here but can be found in Coldren and Corzine [6, p 93] and [8]. For the central wavelength of the dielectric stack, the effective length of the mirror is given by

$$L_{eff} = \frac{1}{2\kappa} \tanh(\kappa L_g) \qquad (11)$$

where κ is the coupling coefficient (equations 9 and 10) and L_g is the grating length where

$$\kappa L_g = 2mr \qquad (12)$$

with m equal to the number of dielectric pairs and r being the amplitude reflection at each interface. L_{eff} corresponds to the distance a metal mirror

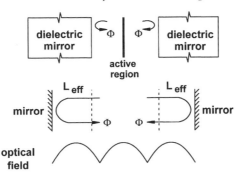

Figure 7.10. The effective mirror model of the VCSEL cavity. The hand mirrors are positioned to maintain the phase Φ.

would have to be placed in a medium of the same refractive index as the cavity material in order to produce the correct phase at the boundary between the cavity and the dielectric mirror. The model of the cavity is therefore as shown in figure 7.10, with the active region placed at an anti-node of the standing wave. If more than one quantum well is used, it is not possible to place all the wells at an anti-node, so Geels *et al.* modelled the effect of the number of quantum wells on the threshold current using a self-consistent model of the quantum well gain, including the energy levels, the density of states, and the strain. They concluded that the optimum number of wells is three. Any more than this and the standing wave enhancement is reduced, and most VCSELs today have three wells in the active region.

7.1 Fabry–Perot and waveguide modes

In any laser structure where waveguiding occurs the Fabry–Perot modes are modified somewhat. A mode propagating down a guide does so with an effective wavevector different from the free-space (or free-material) wavevector by virtue of the guiding. This complication is not overly important in long cavity structures such as a *p–n* homojunction or a DH laser where any number of modes can exist within the gain region. The detailed mode structure is not usually of much interest except where it is necessary to control the modes, for example to minimise dispersion in optical fibre systems. Moverover, the waveguide structure in these devices can often be approximated to a slab waveguide for which the propagation is consistent with any polarisation selection rules that might apply. In a bulk active region, for example, both TE or TM modes will experience gain. In a quantum well laser, on the other hand, the selection rules dictate that only the TE mode will experience gain, but in either case the mode properties allow gain. In a VCSEL the situation is somewhat different. A vertical cavity is usually symmetric, with either a circular or square cross-section. The modes, by which is meant the lateral intensity profile, are more complicated, and it is not clear that such modes do correspond to waveguide modes.

In the Iga structure of figure 7.3 we could reasonably expect guiding over a significant portion of the cavity but in the structure of figure 7.5 no physical guiding mechanism is apparent. However, these are not passive optical structures, and do not bear comparison with such as optical fibres, in which propagation over any distance requires a well-defined guided mode. Propagation of an arbitrary intensity distribution is always possible over short distances in a fibre but without guiding the intensity will decay with distance propagated as radiation is lost to the environment. In a VCSEL the propagation distance is very small but feedback occurs at the mirror and any attenuation will be overcome by the gain at threshold so intensity profiles that might not be stable in a well-defined guide might well be stable in a VCSEL. Moreover, the gain profile will not be uniform across the cavity, and one would expect in the first instance that gain would be highest in the centre of the cavity, diminishing to zero at the edge of the active region. One might reasonably expect the intensity profile to match the gain profile.

Other effects might also occur within the cavity. The gain region itself may well act as a self-focusing element if the refractive index does indeed vary across the cavity. A normally incident ray at the edge of the active region will experience a slightly smaller phase retardation than a ray at the centre, which will of course have the effect of steering the beam toward the centre of the cavity. In addition, heat generated by the current in the contact regions will diffuse laterally and might also lead to a thermally generated refractive index profile across the cavity. These effects are likely to be small but they might be sufficient to induce an effective waveguide with an effective mode structure. The word "effective" is used because the structure acts as if there is a well-defined guide along its whole length.

The circular aperture is easiest to deal with. Ideally the refractive index profile will be as in figure 7.11, with a core index $n_2 > n_1$. For a large core diameter the mode will be strongly confined and for the lowest waveguide mode the intensity distribution will be almost Gaussian, with the propagation almost paraxial. There will be a component of the electric field in the direction

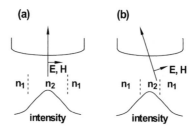

Figure 7.11. Non-paraxial propagation as the cavity radius decreases.

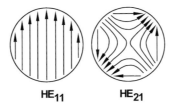

Figure 7.12. Linearly polarised modes of a circular guide.

of the guide but it will be small, and to all intents negligible. For a narrower guide the guiding is weaker and the component of the electric field in the direction of the guide increases, effectively tilting the direction of propagation. The mode family can be either TE, TM or hybrid, labelled either HE_{mn} or EH_{mn}. The subscripts m and n refer to the Bessel functions from which these particular solutions are derived, but this detail will not be described here, where only a description of the modes and the scalar wave equation is provided. The reader is referred to Snyder and Love [9] for a comprehensive description of waveguide theory.

The lowest waveguide mode is designated HE_{11} and is sometimes referred to as a linearly polarised (LP) mode. The electric field is perpendicular to the direction of propagation (figure 7.11a), uniformly aligned along one direction (x or y), as illustrated in figure 7.12, and may be described by a scalar equation of the form [ref 9, p 285]

$$E_{(x,y)} = F_0(r) \exp(j\beta z). \quad (13)$$

This mode pattern is consistent with the selection rules for heavy hole transitions in quantum wells and is therefore ideal for the VCSEL. The higher order mode HE_{21} is not strictly linearly polarised because the direction of polarisation depends upon the radial position within the guide.

The scalar nature of equation (3) allows the mode to be derived from a solution of the scalar wave equation using an effective index model of the VCSEL first derived by Hadley [10]. The essence of the effective index model is that an effective index can be defined for the mode where

$$n_{eff} = \frac{\beta}{k_0} \quad (14)$$

and β is the propagation constant, $k_0 n_1 < \beta < k_0 n_2$. If $n_2 \approx n_1$ the effective index is similar to the material index, but the slight non-uniformity in refractive index across the guide is sufficient to maintain total internal reflection whilst making the medium virtually homogeneous as far as polarisation effects are concerned. Hence the wave equation is written in terms of a scalar rather than a vector, i.e.

$$[\nabla_t^2 + k^2 n^2(x,y) - \beta^2]\Psi = 0. \quad (15)$$

The effective index method, like any numerical method, is best appreciated through practise [11]. In fact, there are very few refractive index profiles that submit to exact solution of Maxwell's equations and for circular waveguides with an arbitrary refractive index profile numerical methods are the norm [ref 9, p 336]. Moreover, if the refractive index profile can be approximated to a recognised form such as a gaussian profile then the resulting electric field distribution also has a recognisable form. An example of a gaussian refractive index profile within a VCSEL will be given later.

Numerical methods do not therefore involve any extra complexity over what might be expected for a well defined refractive index profile. Indeed, the weakly guiding approximation that leads to the use of the scalar wave equation is itself a simplification over the vector wave equation, which is needed if a significant component of the electric or magnetic vector lies in the direction of propagation, but with modern computers even this extra complexity can be handled relatively easily. Examples of the successful application numerical methods to the description of waveguiding in VCSELs can be found in [12,13]. A detailed comparison of the various electromagnetic methods available to the VCSEL modeller can be found in [14].

7.2 Practical VCSEL cavity confinement

Having established that the lowest order waveguide mode is linearly polarised and therefore compatible with the selection rules, it now remains to discuss the definition of the cavity and the specifics of the waveguiding structures that ensue. There are currently three popular techniques, as illustrated in figure 7.13: the etched-post configuration in which the dielectric mirror stack is etched down to the spacer layers between the mirrors, implantation of protons to render the GaAs/AlGaAs outside the cavity semi-insulating, and the oxide-confined structure in which a layer of aluminium oxide containing an aperture is fabricated within the cavity to confine the current. Of these, the etched-post is commonly employed in experimental devices but is not used in commercial

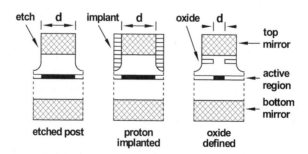

Figure 7.13. Schematic illustration of the common methods used to define the cavity.

fabrication. Oxide confinement is only now, at the time of writing, beginning to be commercialised, but proton implantation is an established technique for rendering conducting GaAs semi-insulating in device fabrication and was the first cavity definition technology to be commercialised in VCSEL production.

Proton isolation

Proton implantation works in two ways. First, protons are highly effective at passivating acceptors. Ionised acceptors are negatively charged so a proton, which is the smallest possible ion and highly mobile through a lattice, can attach itself to the ionised acceptor and render it neutral. It is possible to diffuse protons through a lattice by immersion in a hydrogen plasma but this would not work very well for a VCSEL. The protons would have to diffuse through almost the entire p^+ mirror down to a depth of several microns, and of course they will diffuse laterally at the same time. If the desired cavity width is itself only a few microns across it becomes difficult to define the cavity with sufficient precision. Instead protons are implanted at high energies into the solid to deposit them at the required depth. The incoming ion loses energy through collisions with the ions of the host lattice, which are themselves displaced leaving the material disordered to some extent. Vacancies and dangling bonds are created during the transition through the solid and though many of the vacancies will recombine or form neutral complexes some charged defects will remain which can neutralise dopants of either type, and this is the second mechanism.

Nolte [15] has reviewed proton implantation within the wider context of semi-insulating semiconductors for opto-electronics. Protons implanted at 175 keV produce about 40–50 vacancies per proton, whilst protons implanted at 150 keV remove approximately 3 conduction electrons per proton. Implantation into p-type material requires a higher dose by a factor of about three compared with n-type material. Historically VCSELs were implanted at higher energies than this and figure 7.14 shows the effects of implantation at 300 keV, typical of the energy used in early VCSEL isolation schemes. The upper graphic displays the two-dimensional trajectories of 1000 atoms simulated using the SRIM software package, and the lower graphic shows the distribution of implanted ions expressed as the number of ions per micron. Implantation doses are not normally expressed in this manner but the nomenclature here has been adjusted to reflect the fact that integration under the bottom curve yields the total number of implanted ions. Hence, between 2.4 μm and 2.5 μm there will be approximately 300 implanted ions. These ions will be distributed laterally, perhaps 0.5 μm either side of the centre line. At shallower depths the lateral straggle is even greater and occasionally an isolated proton will be deflected and pursue a path well away from the majority of ions.

Ideally the implantation would leave a well-defined active region but at 300 keV damage was found in the active region caused not only by straggle but

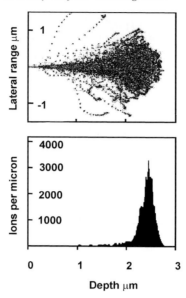

Figure 7.14. SRIM simulation of hydrogen implanted into GaAs at 300 keV.

also by the diffusion of vacancies and other defects. This damage caused a decrease in the luminescence intensity, and although annealing at 450°C for 150 seconds could remove the damage without removing the hydrogen (hydrogen bonded to acceptors will be released at a higher temperature) and therefore restore the luminescence intensity, the lasers were found to suffer from long term stability problems. Consequently lower energies were used for the implantation, which not only maintained the current isolation but also yielded device lifetimes in excess of 10^5 hours. Honeywell was one of the first companies to produce commercially proton implanted devices emitting at 850 nm, and the devices are so stable that Honeywell have stopped publishing reliability data. Failure is defined as a 2 dB drop in output power at 10 mA forward current but Honeywell's proton devices now have lower failure rates than can be reliably measured using the statistical methods normally employed [16].

The random nature of the ion trajectory during implantation precludes definition of small cavities by protonation. Even though lower acceleration voltages may be used leading to implantation depths of less than 2 μm, the protons diffuse beyond the implantation range and also diffuse laterally. In addition the top contacts have to be annealed, which will cause further diffusion. Typically, then, implantation apertures lie in the range 15–20 μm with contact apertures just slightly smaller. Figure 7.15 shows a typical arrangement, with the isolation descending through the mirror into the spacer layer, but stopping short of the active layer. Were the damage to penetrate

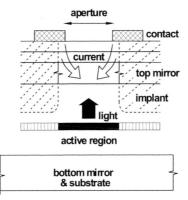

Figure 7.15. A typical implantation isolated VCSEL.

through to the active layer then device performance would be adversely affected, as already described. Even though the current would flow mainly through the undamaged active layer the presence of vacancies within the quantum well adjacent to the active region will lead to recombination and diffusion of carriers away from the gain region.

One of the advantages of the proton isolation scheme is that the devices are planar. There is no need to etch any part of the wafer in order to expose contacts, but again there is no clearly discernible waveguide structure. The lattice disruption due to the proton implants will affect the refractive index slightly, but current conduction in the central region under the aperture will tend to lower the refractive index. There is no waveguiding at all in the bottom mirror so the active region is the only place where the refractive index profile will alter as a result of gain. Although this would hardly seem thick enough to act as a waveguide, it does just this. Bradford et al. [17] have measured the energy distribution within the output beam of 850 nm proton implanted VCSELs with a 17 μm diameter aperture. A scanning near-field optical microscope (SNOM) was used to detect the optical power as a function of the input current. A SNOM is an optical fibre drawn into a tip a few hundred nanometres across and coated with metal to provide an aperture approximately 100 nm across. This aperture is held about 10 nm from the surface and scanned in two dimensions to build up a picture of the optical emission as a function of position. At 6.3 mA the total output power follows a gaussian distribution in two dimensions. Moreover, so does the rate of change of power distribution, which can be measured using a lock-in amplifier in the following way. A Taylor's expansion of the power as a function of the current shows that for a small change in current the differential of the power with respect to current is directly related to the change in current

$$P(I_0 + \Delta I) = P(I_0) + \frac{dP}{dI}(\Delta I) + \frac{1}{2}\frac{d^2P}{dI^2}(\Delta I)^2. \tag{16}$$

Superimposing a small ac voltage onto the dc driving voltage to give a driving current of $I_0 + i\sin(\omega t)$ causes a ripple in the light output. Provided the ripple is small the total light output is given by

$$P(I_0 + i\sin(\omega t)) = P(I_0) + \frac{dP}{dI}[i\sin(\omega t)] \tag{17}$$

so by locking onto the modulated light output the rate of change of output power with current can be monitored. Figure 7.16 illustrates the effect of a small change in current from 6.3 mA to 6.5 mA. At 6.3 mA the output beam has a circular gaussian profile similar to that shown in (a) with the differential power profile also gaussian and centred on the axis of the VCSEL aperture. At 6.5 mA, however, the total power still has a gaussian profile but the differential power shows a bimodal structure with power being shifted to the wings. The power in the centre of the profile is reduced relative to the edges, as indicated by the dashed lines in the contour map, but the current has to be increased to ~ 8 mA before the bimodal structure appears in the total output power.

These structures are definitely waveguide modes and therefore provide clear evidence of waveguiding and the existence of a waveguide structure, despite the lack of any obvious physical mechanism for a refractuive index change throughout the length of the cavity. Bradford *et al.* interpreted their results in terms of a gaussian refractive index profile

$$n^2(r) = n_{inner}^2 \left[1 - \frac{2(n_{inner} - n_{outer})}{n_{inner}}\right]\left[1 - \exp\left(\frac{-r}{\rho}\right)\right] \tag{18}$$

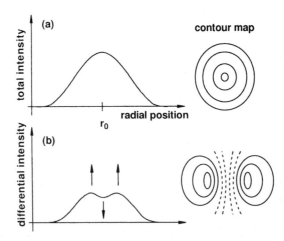

Figure 7.16. Schematic of the total and differential power output of a VCSEL at 6.5 mA with indicative contour maps of the output power. (After Bradford *et al.*)

where ρ is the mode radius, r is the radial position, and n_{inner} and n_{outer} are the refractive indices at the centre of the guide and at the edges, the latter being simply the material refractive index at the edge of the active region. The single mode cut-off point marks the transition from single mode to multi-mode behaviour and is given by the condition

$$2.592 = \frac{2\pi\rho}{\lambda} n_{inner} \left[\frac{2(n_{inner} - n_{outer})}{n_{inner}} \right]^{1/2} \quad (19)$$

with a light intensity profile

$$S(r) = A \exp\left[-1.592 \left(\frac{r}{\rho}\right)^2 \right] \quad (20)$$

where A is a constant. The single mode cut-off is clearly identified as occurring at between 6.3 and 6.5 mA so measurement of the total light intensity gives a mode radius of 4.1 μm, $A = 0.192$, and a refractive index difference between the centre and the outer of 0.0012.

7.3 Oxide confined devices

A small cavity can be defined by converting a layer of AlAs – or at least an AlGaAs layer containing 2% or less of Ga – to high quality aluminium oxide. Heating in steam at about 400°C causes a lateral diffusion of oxidant into the layer so that oxidation occurs from the outside inwards. Stopping the process before complete oxidation leaves a central aperture of unconverted material with high quality alumina outside. The process was discovered by Dallesasse et al. [18] and was found to be highly directional as well as selective on the composition. Fine scale alloys were shown to oxidise slower than large scale alloys such as superlattices, and for relatively thick layers oxidation occurred laterally along the layer rather than perpendicularly. However, for thin layers the rate of oxidation perpendicular to the layer structure becomes similar to the lateral oxidation rate. Typical oxidation rates vary from under a micron per minute at 400°C to ~7 μm/minute at 500°C for AlAs layers ≥ 250 nm thick. The rate does not change much if the films are thicker than this but for thinner films the rate reduces considerably, being about 2 μm/minute for 75 nm thick films at 500°C and about 5 μm/minute for film 150 nm thick at the same temperature [19]. A number of investigations into oxidation kinetics have been conducted but perhaps the most comprehensive summary of experimental findings together with a theoretical treatment is given by Osinski et al. [20].

Oxide-defined devices will out-perform proton implanted devices of similar size not only because of the improved optical confinement but also because of the lack of damage to the active region and contacts. Essentially the advantage of the oxide-defined device over the proton implanted is one of size and therefore threshold current, the latter being well below the milliampere range for a 3μm diameter aperture. Apertures of this size are typical with single mode powers also typically 3–5 mW [21,22]. Smaller apertures than this can of course be made with ease, such is the control over the process, but the waveguide modes become more complicated with a loss of paraxial propagation. The vector wave equation is needed to describe the modes. Moreover the output wavelength blueshifts because as the angle of propagation increases the wavelength of the light in the axial direction of the cavity decreases. Blue shifts of up to 13.4 nm have been reported for photo-pumped oxide confined devices [23]. Calculations using a vector weighted index method as opposed to the scalar effective index method [24] successfully modelled the resonant wavelength of both the fundamental and the first higher order modes (LP_{01} and LP_{11} respectively, where the designation "LP" stands for "linearly polarised"). Blue-shifting of the output wavelength tends to be significant only for apertures below 2.5–3 μm in diameter.

The basic method is illustrated in figure 7.17. The top dielectric mirror has to be etched from the top to expose the high Al-content layer within the cavity. This also exposes the high Al-content layers within the mirror but if the mirror layers have 5% Ga compared with 2% Ga in the aperture layer, the difference in the oxidation rate will allow an aperture to be defined without consuming too much of the mirror layers. If the devices are going to be used in arrays or as part of an integrated circuit the surface must be planarised after etching, but this is a common requirement in semiconductor fabrication and presents no real difficulties [25]. Polyimide is a common planarising material.

Etching the mirrors also changes their properties somewhat because now, instead of a dielectric stack extending out effectively to infinity in the plane, the mirror has a finite dimension. Should the mirror dimension be smaller than the mode field the mirror will diffract the mode. Practical distributed Bragg

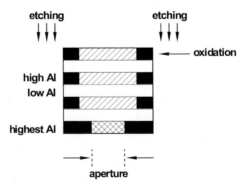

Figure 7.17. The basic mechanism of oxide definition of the cavity. Only a few layers of the mirror are illustrated.

reflectors will diffract in any case, as discussed by Babić et al. [26]. The full properties of real quarter-wavelength semiconductor mirrors can only be calculated by solution of the relevant electromagnetic equations, though some simplifying approximations can be made for quick calculations. The simplification normally revolves around plane waves. In figure 7.6 for example, the reflectivity is calculated assuming a plane wave of infinite extent normally incident on a quarter wavelength stack of infinite extent in the plane. In the oxide-defined VCSEL the mode dimensions themselves can be very small, <3 μm in some cases, and such modes themselves are almost certainly not normally incident. Indeed, the modes can be constructed from the superposition of plane waves at differing angles of incidence so each plane wave component will experience a different reflectivity. Superposing the reflected beams will therefore give a different intensity profile from the incident, which is one way of saying that the mode is scattered (diffracted) from the mirror. Babić et al. treat the diffraction problem more rigorously and show that the diffraction properties depend on the wavelength of the mode, the mode diameter, and the refractive index difference.

Diffraction is a size-related phenomenon and the mode diameter effectively determines the size of the reflecting aperture relative to the wavelength, but there is an additional effect associated with wavelength. In changing from 850 nm to 1.55 μm, say, the wavelength is nearly doubled and all other things being equal the thickness of the layers in the mirror stack will also nearly double, as will the penetration depth into the mirror. In fact, for a given material system in the sub-bandgap normal dispersion regime the refractive index will decrease with increasing wavelength, thereby increasing the thickness of each layer slightly and decreasing the reflection from each interface. The total penetration depth into the mirror will therefore more than double with a doubling of the refractive index, and for small mode diameters and higher order modes with non-normal angles of incidence the scattering out of the mode field will increase. If the material system also changes as a result of changing the wavelength such that the penetration depth into the mirror is increased the scattering losses will increase yet further.

As might be expected from the preceding discussion, the transverse mode behaviour of oxide-defined devices is well explained by waveguiding theory. Huffaker et al. [27] have used 8 μm square devices with high contrast ZnSe/CaF pairs for the dielectric mirrors to demonstrate very clearly four distinct modes of the device; a gaussian profile below threshold, a bi-lobed profile at $1.1 \times I_{th}$ similar to that seen in figure 7.16, a four-lobed output at $2.0 \times I_{th}$, and a complicated eight-lobed structure in the beam at $3.0 \times I_{th}$. The wave-guiding mechanism is quite different from that in proton implanted devices, but again a physical refractive index difference does not extend over the whole of the cavity. Near field scanning optical microscopy on selectively oxidised 10 μm square apertures [28] also shows a complex mode structure at above-threshold currents, and moreover, that each of the modes resonates at a slightly different wavelength. In a four nanometre wavelength range from 846–850 nm some 20 separate modes were found to oscillate, though some of them had a very low intensity. Nonetheless, at least eleven of the modes were strong enough to be considered as competing.

This observation appears to be at odds with the essential idea of the VCSEL as a single mode device but it arises not from the existence of multiple Fabry–Perot modes, but from the fact of waveguiding itself. These are essentially transverse modes of the waveguide which resonate at slightly different frequencies. The fact that the output modes are not true Fabry–Perot modes makes the VCSEL a difficult device to model, though of course the fundamental mode approximates well to Fabry–Perot wavelength because of the near paraxial propagation. For this reason some authors have used the metal contact as a mode filter to suppress the higher order modes [22]. The higher order modes tend to travel in the outer regions of the cavity and are subject to greater diffraction effects, so by aligning a metal aperture with the oxide aperture allows these modes to be blocked. In addition to mode filtering, however, careful control of the cavity temperature is also required for accurate control of the output wavelength. As well as exhibiting multiple output wavelengths, the large aperture devices of Sharma *et al.* [28] also exhibited a red shift in each mode of 0.145 nm/mA ascribed to the effects of temperature in the active region.

The richness of the mode structure might lead one to suppose that the effective guiding in oxide-defined devices is stronger than that in proton-implanted devices. The oxide itself presents a much larger refractive index difference than will occur as a result of gain guiding, but as with the active regions, the positioning within the cavity is crucial if the optical effect of the oxide layer is to be maximised. If the aperture is placed at a node where the optical field is effectively zero the aperture can be expected to have a minimal impact on the optical propagation, though it will of course still limit the current flow. On the other hand, positioning at an anti-node can be expected to maximise the effect. The simplest mechanism is diffraction of a planar wavefront back in to the centre of the cavity (figure 7.18). The low index oxide

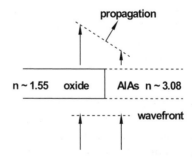

Figure 7.18. Diffraction from an oxide aperture caused by phase retardation in the high index layer.

layer causes an optical phase advance compared with propagation in the high index layer and for an abrupt interface distortion of the wavefront in the vicinity of the interface will act to confine the optical field somewhat. The idea of the oxide aperture as an intracavity lens has been developed by Coldren *et al.* [29] where it was shown that the oxide aperture can be treated either as a uniform waveguide by effectivley distibuting the refractive index change over the whole of the cavity or as a lens which focuses the beam to a point inside the cavity. The uniform waveguide model is perhaps best suited to the abrupt change illustrated in figure 7.17 because the effective focal length is not so clearly defined whereas for a parabolically graded index change that gives rise to a thin spherical lens, the focal length is well defined.

A detailed investigation into the optical effects of the oxide aperture using numerical solutions of the vectorial wave equation [30] compared oxide apertured devices with proton implanted devices. This was a theoretical study in which it is necessary only to restrict the current flow and recognise the existence of gain guiding in order to simulate the device and therefore does not reflect the experimental reality that the statistical nature of the implantation process makes it difficult to define small devices by implantation. The essential properties of the oxide aperture that emerge from this study which make it superior to proton implantation are:

- in proton implanted devices only the gain guiding serves to counteract the effect of beam diffraction within the cavity but within the oxide-defined device the oxide aperture counteracts the effects of diffraction at the mirror;
- therefore the effective mode field diameter is smaller, leading to a better confinement;
- the pumping current is therefore used much more effectively;
- for a given size of device the threshold currents for the oxide defined device is smaller than for the implanted device.

In addition, the importance of the position of the oxide with respect to the standing wave pattern was confirmed with an anti-nodal oxide affecting the optical field much more than a nodal oxide, thereby yielding a smaller threshold current. An interesting aspect of this modelling is an effective plane wave model of the VCSEL that the authors developed. This showed that the reflectivity of the mirrors beneath the oxide is reduced for the fundamental mode, albeit by a very small amount. This has the effect of concentrating the feedback in the central portion of the beam rather than in the tails of the gaussian and effectively confines the mode.

The effect of reducing the aperture below 3 μm has been mentioned in relation to the waveguide modes and the consequent blueshifting, but there are also other effects. It will also be apparent that the diffraction losses from the mirrors will be high and, oddly, an improvement in device performance can be gained by an increase in the number of mirror pairs [31]. However, thermal effects cause the performance to degrade if the number is increased too much, so a much better idea is to reduce the penetration depth into the mirrors. High index-difference DBR's must be used. Deposited materials can be used but the difficulty is to deposit the layers on both sides of the cavity. One very effective way is to grow AlAs/GaAs pairs and oxidise the AlAs, as illustrated in figure 7.17. The large index difference means that a very high reflectivity can be realised with just 3.5 pairs. The requirement on the oxidation rate is now reversed; it is the window layers that must be fully oxidised with the aperture layer only partially oxidised. The thickness dependence of the oxidation rate can be used to advantage here. An aperture layer can be any thickness in principle, but the modelling by Demeulenaere *et al.* showed that 40 nm is the thickness at which the layer begins to be less effective. In the photopumped devices of Shin *et al.* [23] the devices were fabricated by growing a 1λ cavity ($\lambda = 780$ nm) with a four-pair upper mirror and a 5.5-pair lower mirror. This stack was defined by etching 50 μm square mesas which were then oxidised at 450°C in steam. The aperture layer was 30 nm thick whereas the mirror layers were 170 nm thick. The effective penetration depth into the mirrors was 0.06 μm, giving a total cavity length of 0.37 μm. The short penetration depth of the mirrors also means that the stop band is quite broad (figure 7.19) and the requirements for matching the wavelength spectrum of the mirror to the cavity resonance, which is essentially a requirement on the thickness tolerance of each of the layers in the mirror stack, is relaxed.

The photopumped devices mentioned above do not need contacts, of course, but if micro-cavity VCSELs are to prove useful then current has to be supplied. Oxidisation of the top mirror means that the contact has to be made directly onto the cavity itself, which is the essential requirement in the etched-

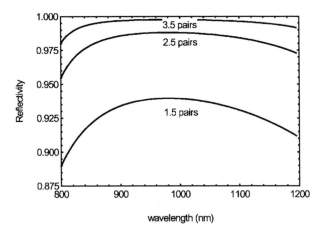

Figure 7.19. The reflectivity of GaAs-oxidised AlAs dielectric mirrors for operation at 980 nm. The greater index difference results in a higher reflectivity for a smaller number of pairs as well as a broad stop band extending over nearly 200 nm.

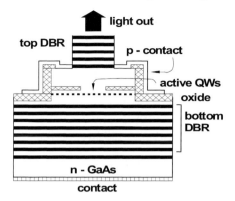

Figure 7.20. Intra-cavity contact in an etched-post structure.

post structure (figure 7.20). The first electrically pumped oxide DBR VCSEL was described in 1995 [32], but the device was not particularly small. Whereas it is possible with photo-pumping to maintain a small cavity length there are inevitably difficulties when it comes to making intra-cavity contacts. Tunnel contacts to a 2λ cavity have been described by Wierer et al. [33] but the fabrication procedure is complicated by the need to align the various contacting layers. Accurate alignment during photolithography becomes an issue when the lateral dimensions of the device approach 1 μm because mis-alignment in such a small device will result in a seriously degraded performance. To this end, a self-aligned device has been described by Choi et al. [34]. Self-alignment is a technique whereby a unified approach to fabrication is adopted and the crucial parts of the device are defined during the fabrication of others. Several photolithography steps are eliminated along with the possibility of mis-alignment, so the technique therefore holds out the promise of high density integration, but in this particular device the mirrors still used multiple pairs of unoxidised AlGaAs/GaAs.

7.4 Long wavelength VCSELs

At the time of writing long wavelength (1.5 μm) VCSELs for telecommunications have not reached the stage of development where commercial production is feasible. The essential difficulty lies with mirrors.

The InGaAsP material system for emission at 1.5 μm is well established and itself presents no problems for the fabrication of active layers. Quantum wells can be grown in InGaAsP lattice matched to InP without difficulty, and in fact for the VCSEL the longer wavelength operation has the potential advantage that the volume of material within the cavity lying closer to the optical anti-node increases so the number of quantum wells that can be incorporated into the active region also increases. Rather than the three wells commonly employed in 850 nm lasers there may be 8 or even 10 wells within a 1.5 μm laser. The longer wavelength operation has the disadvantage that the effective penetration depth into the mirrors is increased. Not only must the thickness of each layer increase to provide $\lambda/4$ phase-matching but the epitaxial material system for 1.5 μm operation has a lower refractive index difference and something like 50 pairs may be needed compared with 30 pairs for emission at 850 nm. The stop band is correspondingly narrower and care must be taken to match the reflectivity maximum to the cavity resonance, and there may be other undesirable consequences such as increased diffraction for higher order modes or smaller apertures.

By far the most serious problem lies with the p-doped mirrors, however. Intervalence band absorption between the light and heavy hole bands and the split-off band becomes significant for this material system [35]. Intervalence band absorption can only occur if there are sufficient hole states, i.e. empty electron states, within the valence band for a transition to occur. The problem doesn't arise in n-type material, or even in low-doped p-type material, because the density of free holes is so low but the p-type mirrors have to be heavily doped in order to increase the conductivity of the contacts. The total mirror thickness will be 11–12 μm compared with 2–3 μm for 850 nm lasers so not only will the resistance be four or five times higher for a similar conductivity but any power dissipated within the mirrors must be conducted away through a greater amount of material. Mirror heating is potentially a difficulty in these lasers but the inter-valence band free-carrier absorption, as the phenomenon is called, adds to the difficulties and, moreover, represents a serious optical loss as well.

AlGaAs/GaAs is an ideal mirror system for this wavelength. The problem still exists that the thickness of each of the layers has to be increased to provide the relevant phase but the refractive index difference means that the number of pairs can be reduced. More importantly, inter-valence band absorption is not a difficulty, and if the mirrors can be oxidised, the penetration depth can be reduced substantially. The difficulty is that AlGaAs is not lattice-matched to this materials system. One way over the difficulty is to grow AlGaAs/GaAs mirrors onto a GaAs substrate and fuse the mirrors onto the top InP layer of the laser structure by heating in vacuum. Typically the wafers are pressed together with a pressure of 15 atmospheres and heated to 630°C in hydrogen for 20 minutes [36]. This results in GaAs/InP junctions that are optically transparent as well as conducting. The GaAs substrate is then etched away. Fusion of the top mirror only was reported in 1996 [37] with continuous operation at 27°C. Double-fused structures in which both mirrors have been made in this way has resulted in 1.5 μm lasers that operate continuously up to 71°C [38]. This temperature is similar to the maximum operating temperatures of commercial 850 nm lasers, but it is by no means certain that wafer fusion will provide a reliable manufacturing technology with a high yield across the wafer.

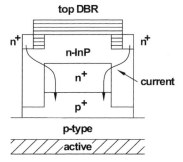

Figure 7.21. The contact structure of the buried tunnel junction (BJT) VCSEL. (After Boehm *et al.*)

An alternative approach is to grow metamorphic, i.e. fully relaxed, AlGaAs/GaAs mirrors onto InP. Although the defect density will be high the optical properties of the mirrors will not be affected, and the conductivity has also been reported as being similar to epitaxial GaAs [39]. On the other hand, epitaxial growth of GaAs on patterned InP without the generation of defects, despite the large lattice mismatch, has been reported and an oxidised-AlAs/GaAs mirror has been made on top of an InGaAsP/InP laser [40]. Selective oxidation of a lattice matched $In_{0.52}Al_{0.48}As$ layer was used to provide a current confining aperture within this structure, thereby showing that similar devices to oxide-defined 850 nm emitters are possible.

Yet another approach is to use a different material system for the wells and mirror. AlGaInAs lattice matched to InP has been shown to be suitable for VCSELs in the range 1.3 μm to 2.0 μm using a buried tunnelling junction [41]. Single mode VCSELs at 1.55 μm show error free transmission at 10 Gbit/s, which shows the potential of this system for communications. The essence of the buried tunnelling junction approach is to fabricate epitaxial *n*-type mirrors

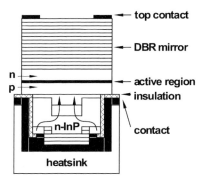

Figure 7.22. The final BJT-VCSEL structure of figure 7.21. (After Boehm *et al.* [41].)

from high index-difference AlInAs-AlGaInAs on an InP substrate on to which the active layers and top contact are fabricated. The device is completed by the deposition of top contact layers and the dielectric mirror (figure 7.21) before being attached to a heat sink (figure 7.22). The substrate is etched off, exposing the underlying DBR, which now becomes the top DBR within the laser. Contacts are attached and light is emitted through what was, within the growth sequence, the bottom set of mirrors. Advantageously, the active layers of the device are mounted very close to the integrated heat sink.

7.5 Visible VCSELs

As with quantum well lasers, visible VCSELs suffer from the difficulty of finding lattice-matched materials that allow both the short wavelength and effective carrier confinement together. AlGaInP alloys lattice-matched to GaAs have direct bandgaps that span the range from deep red to green and several devices emitting single mode powers in the milli-Watt range at wavelengths around 670 nm have been reported (see Chow *et al.* for a review of progress to 1997). The unstrained configuration occurs at an In concentration of 52% irrespective of the Al:Ga ratio (figure 7.23), but at an Al mole fraction of approximately 0.56–0.7 the band changes from direct to indirect as the X band minimum descends below the Γ band minimum. This effectively limits the short wavelength end of the spectrum to 555–570 nm [7] but the practical wavelength limit is 650 nm in this material system because of the need to maintain confinement.

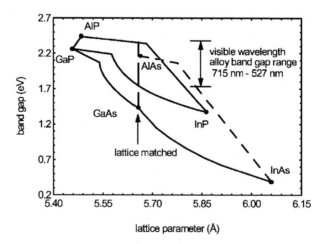

Figure 7.23. The energy gap vs. lattice constant of figure 5.6 expanded to show AlGaInP system.

In addition to the poor confinement demonstrated by this alloy system, two other factors limit the performance of visible VCSELs. First, the high temperature operation is poor. The threshold current can shift markedly with temperature, which is a consequence not only of the low potential barriers confining the carriers, but also a consequence of the cavity resonance detuning [42]. The gain peak shifts to the red with an increase in temperature. So also does the cavity resonance, but this is a consequence of an increase in the refractive index as the temperature increases whereas the gain shift is a consequence of the decreasing band gap. The latter effect is far more pronounced so the cavity resonance detunes itself from the gain maximum. A more stable performance at elevated temperatures can be realised by blue-shifting the gain maximum at room temperature so that the cavity resonance tunes itself in first before detuning if the temperature rise is too high [7]. The room temperature threshold current is increased in this manner, in accordance with figure 7.9, but the change in threshold current with temperature is fairly small. The second effect concerns the mirrors. In a system lattice matched to GaAs the lowest concentration of Al permissible is ~50%, which leads to a relatively small refractive index difference. The number of mirror pairs must increase which leads to higher diffraction, higher resistance, and higher thermal resistance. Oxidised mirrors can be used but then an intra-cavity contact must be made. In general semiconductor DBRs are preferred over insulating dielectric mirrors because an intra-cavity contact is difficult to make in a $1 - \lambda$ cavity and lasers with semiconductor mirrors out-perform lasers with dielectric mirrors.

By way of example, the growth and performance of AlGaInP red VCSELs has been described in detail in a series of papers by Knigge *et al.* [43–45]. The mirrors consisted of 55.5 pairs of *n*-type $Al_{0.5}Ga_{0.5}$ As/AlAs and 35 pairs of *p*-doped $Al_{0.5}Ga_{0.5}$ As/$Al_{0.98}Ga_{0.02}$ As in order to facilitate the formation of an aperture by wet oxidation. The active region consisted of three InGaP quantum wells. Carbon doping was preferred over Zn for the *p*-doping because the series resistance of the mirrors is lower, a fact not fully understood but believed to arise from the increase in doping in the high band gap layers. The *p*-doped mirrors were also graded at the interface. Initially the top of the DBR was coated with a layer of GaAs ~80 nm thick in order to facilitate contacts to the DBR. However, GaAs absorbs at ~870 nm at room temperature so this layer has to be etched off the mirror in the contact aperture, which in itself is difficult to control and exposes the top high-Al layer to the ambient. Operation at an elevated temperature in a humid environment will inevitably lead to some oxidation of the top layer, so a p-$In_{0.48}Ga_{0.52}P$ layer was deposited before the GaAs. This acted as both an etch-stop layer and a passivating layer to protect the top of the DBR. These structures had a threshold current density of ~2.3 kA cm^{-2}, record output powers of 4 mW at 650 nm and 10 mW at 670 nm, and single mode emission up to 65°C for 650 nm emitters and 87°C for 670 nm emitters. Subsequently, the GaAs contact layer was removed and only the etch-stop retained, which resulted in even lower threshold current densities of 1.8 kA cm^{-2}. These devices are still very much under development at the time of writing, and no doubt the structures that will eventually reach the production lines will vary from these devices in some significant respect.

7.6 Summary

The vertical cavity configuration has the dual advantage over edge-emitting devices of both improved quality control during manufacture as well as the possibility of integration with other opto-electronic devices. Devices emitting at 850 nm are the most mature. The materials to emit at 1.3 and 1.5 μm exist and are well characterised because of developments in DH and quantum well lasers, and quantum well systems can be contructed that will emit in the vertical cavity configuration over as wide a range of wavelengths as exists in these other laser structures. The quantum well is preferred over the DH structure because of the enhanced matrix element, but the number of wells is limited by the need to place them at the maximum optical intensity within the cavity.

Key to the performance of these devices is the fabrication of quarter-wavelength pairs for the semiconductor mirrors. Epitaxial growth of approximately 30 pairs is required for mirrors at 850 nm, but for longer wavelength mirrors in other materials more may be needed. The mirrors are therefore several micrometres thick. If these mirrors also form the contacts the composition has to be graded to allow field induced lowering of the barrier, which has the unfortunate effect of lowering the reflectivity slightly. However, this can be offset by the growth of an additional pair or two of layers. The materials and mirror technologies are well established for operation at 850 nm, but for longer wavelength devices the length and conductivity of the mirrors means that free carrier absorption is a problem. Other schemes have been investigated, particularly the use of low index dielectrics to reduce the number of pairs, but these mirrors are then electrically insulating and direct contacts to cavity are required, with the consequence that the surface must be etched away to expose the intra-cavity layers.

Two main types of aperture definition are employed in VCSEL devices; proton implantation and oxide confinement. Proton isolation is the more mature technology but the random nature of the implantation process means that small cavities cannot be defined. Oxide confinement, in which a relatively thick layer of high Al-content AlGaAs is oxidised to provide a current blocking aperture, allows very small cavities to be defined, but the electromagnetic properties of these cavities are hard to model. At cavity diameters below 3 μm blue-shifting of the emission wavelength occurs as the effective wavelength along the cavity is reduced. The vector wave equation needs to be solved for these modes but for larger diameters the ray propagation is paraxial and the fundamental mode approximates well the the Fabry–Perot modes of the cavity.

7.7 References

[1] Iga K 2000 *IEEE J. Sel. Top. in Quant. Electr.* **6** 1201–1215
[2] Towe E, Leheny R F and Yang A 2000 *IEEE J. Sel. Top. in Quant. Electr.* **6** 1458–1463
[3] Koyama F, Morita K and Iga K 1991 *IEEE Journal of Quantum Electronics* **27** 1410–1416
[4] Geels R S, Corzine S W and Coldren L A 1991 *IEEE Journal of Quantum Electronics* **27** 1359–1367
[5] Zhu R, Hargis M C, Woodall J M and Melloch M R 2001 *IEEE Photonics Technology Letters* **13** 103–105
[6] Coldren L A and Corzine S W 1995 *Diode Lasers and Photonic Integrated Circuits* (New York: John Wiley) p 87
[7] Chow W W, Choquette K D, Crawford M H, Lear K L and Hadley G R 1997 *IEEE J. Quant. Electr.* **33** 1810–1824
[8] Babić D I and Corzine S W 1992 *IEEE Journal of Quantum Electronics* **28** 514–524
[9] Snyder A W and Love J D 1983 *Optical Waveguide Theory* (London: Chapman and Hall)
[10] Hadley G R 1995 *Optics Letters* **20** 1483–1485
[11] See for example, Gustavsson J S, Vukušić J A, Bengtsson J and Larsson A 2002 *IEEE J. Quant. Electr.* **38** 203–212
[12] Noble M J, Loehr J P and Lott J A 1998 *IEEE J. Quant. Electr.* **34** 1890–1903
[13] Riyopoulos S A, Dialetis D and Riely B 2001 *IEEE J. Select. Topics in Quant. Electr.* **7** 312–327
[14] Bienstman P, Baets R, Vukušić J, Larsson A, Noble M J, Brunner M, Gulden K, Debernardi P, Fratta L, Bava G P, Wenzel H, Klein B, Conradi O, Pregla R, Riyapoulos S A, Seurin J-F P and Chuang S L 2001 *IEEE J. Quant. Electr.* **37** 1616–1631
[15] Nolte D D 1999 *J. Appl. Phys* **85** 6259–6289
[16] Guenter J K, Tatum J A, Clark A, Penner R S, Johnson R H, Hawthorne R A, Baird J R and Lie Y 2001 *Proceedings of the SPIE* **4826**
[17] Bradford W C, Beach J D, Collins R T, Kisker D W and Galt D 2002 *Appl. Phys. Lett.* **80** 929–931
[18] Dallesasse J M, Holonyak N Jr, Sugg A R, Richard T A and El-zein N 1990 *Appl. Phys. Lett.* **57** 2844–2846
[19] Koley B, Dagenais M, Jin R, Simonis G, Pham J, McLane G, Johnson F and Whaley R Jr 1998 *J. Appl. Phys.* **84** 600–605
[20] Osinski M, Svimonishvili T, Smolyakov G A, Smagley V A, Mackowiak P and Nakwaski W 1999 Simple theory of steam oxidation of AlAs in, *Design, Fabrication and Characterization of Photonic Devices*, Eds. M Osinski, S J Chua and S F Chichibu, *Proceedings of the Society of Photo-optical Instrumentation Engineers (SPIE)* **3896** 534–546

[21] Grabherr M, Jäger R, Michalzik R, Weigl B, Reiner G and Ebeling K J 1997 *IEEE Photon. Tech. Lett.* **9** 1304–1306
[22] Ueki N, Sakamoto A, Nakamura T, Nakayama H, Sakurai J, Otoma H, Miyamoto Y, Yoshikawa M and Fuse M 1999 *IEEE Photon. Tech. Lett.* **11** 1538–1540
[23] Shin J-H, Han I-Y and Lee Y-H 1998 *IEEE Photon. Tech. Lett.* **10** 754–756
[24] Noble M J, Shin J-H, Choquette K D, Loehr J P, Lott J A and Lee Y-H 1998 *IEEE Photon. Tech. Lett.* **10** 475–477
[25] Jung C, King R, Jäger R, Grabherr M, Eberhard F, Michalzik R and Ebeling K J 1999 *J. Opt. A: Pure Appl. Opt* **1** 272–275
[26] Babić D I, Chung Y, Dagli N and Bowers J E 1993 *IEEE J. Quant. Electr.* **29** 1950–1962
[27] Huffaker D L, Deppe D G and Rogers T J 1994 *Appl. Phys. Lett.* **65** 1611–1613
[28] Sharma A, Yarrison-Rice J M, Jackson H E and Choquette K D 2002 *J. Appl. Phys.* **92** 6837–6844
[29] Coldren L A, Thibeault B J, Hegblom E R, Thompson G B and Scott J W 1996 *Appl. Phys. Lett.* **68** 313–315
[30] Demeulenaere B, Bienstman P, Dhoedt B and Baets R G 1999 *IEEE J. Quant. Electr.* **35** 358–367
[31] Jungo M, Monti di Sopra F, Erni D and Baechtold W 2002 *J. Appl. Phys.* **91** 5550–5557
[32] MacDougal M H, Dapkus P D, Pudikov V, Zhao H and Yang G M 1995 *IEEE Photon. Tech. Lett.* **7** 229–231
[33] Wierer J J, Evans P W, Holonyak N Jr and Kellog D A 1998 *Appl. Phys. Lett.* **72** 2742–2745
[34] Choi W-J and Dakus P D 1998 *Appl. Phys. Lett.* **73** 1661–1663
[35] Joindot I and Beylat J L 1993 *Electronics Letters* **29** 604–605
[36] Margalit N M, Piprek J, Zhang S, Babić D I, Streubel K, Mirin R P, Wesselmann J R, Bowers J E and Hu E L 1997 *IEEE J. Selected Topics in Quant. Electr.* **3** 359–365
[37] Ohiso Y, Amano C, Itoh Y, Tateno K, Tadokoro T, Takenouchi H and Kurokawa T 1996 *Electronics Letters* **32** 1483–1484
[38] Black K A, Abraham P, Margalit N M, Hegblom E R, Chiu Y-J, Piprek J, Bowers J E and Hu E L 1998 *Electronics Letters* **34** 1947–1949
[39] Goldstein L, Fortin C, Starck C, Plais A, Jacquet J, Boucart J, Rocher A and Poussou C 1998 *Electronics Letters* **34** 268–270
[40] Gebretsadik H, Bhattacharya P K, Kamath K K, Qasimeh O R, Klotzkin D J, Caneau C and Bhat R 1998 *Electronics Letters* **34** 1316–1318
[41] Boehm G, Ortsiefer M, Shau R, Rosskopf J, Lauer C, Maute M, Köhler F, Mederer F, Meyer R and Amann M-C 2003 *J. Cryst. Growth.* **251** 748–753
[42] Sweeney S J, Knowles G and Sale T E 2001 *Appl. Phys. Lett.* **78** 865–867
[43] Knigge A, Franke R, Knigge S, Sumpf B, Vogel K, Zorn M, Weyers M and Tränkle G 2002 *IEEE Photon. Tech. Lett.* **14** 1385–1387
[44] Knigge A, Zorn M, Sebastian J, Wenzel H, Weyers M and Tränkle G 2003 *IEE-Proc.-Optoelecton.* **150** 110–114
[45] Zorn M, Knigge A, Zeimer U, Klein A, Kissel H, Weyers M and Tränkle G 2003 *J. Cryst. Growth* **248** 186–193
[46] Sfigkakis F, Paddon P, Pacradouni V, Adamcyk M, Nicoll C, Tiedje T and Young J Y 2000 *J. Lightwave Technol.* **18** 199–202

Problems

1. Derive an expression for the mode spacing of a short cavity m_λ wavelengths long, where m_λ is an integer, first neglecting material dispersion and then with material dispersion taken into account. Hence calculate the mode spacing for a one wavelength cavity at 850 nm consisting of $Al_{0.2}Ga_{0.8}As$ where $n = 3.445$.

2. The following table shows the refractive index of AlGaAs at various wavelengths. In order to demonstrate the trade-off between band offset and mirror thickness, calculate the Fresnel coefficients for a mirror stack composed of $GaAs-Al_{0.9}Ga_{0.1}As$ pairs, $GaAs-Al_{0.8}Ga_{0.2}As$ pairs, and $Al_{0.1}Ga_{0.9}As-Al_{0.8}Ga_{0.2}As$ pairs at 850 nm, 1.3 μm, and 1.5 μm. Hence calculate the number of pairs required for a power reflection coefficient of 0.96 and the total mirror thickness.

	$\lambda = 0.85\ \mu m$	$\lambda = 1.3\ \mu m$	$\lambda = 1.5\ \mu m$
$x = 0$	3.596	3.413	3.392
$x = 0.1$	3.518	3.357	3.333
$x = 0.2$	3.445	3.298	3.276
$x = 0.8$	3.088	2.985	2.971
$x = 0.9$	3.035	2.938	2.925
$x = 1.0$	2.982	2.892	2.881

3. Oxidised AlGaAs has a refractive index of 1.61 [46]. Calculate the Fresnel coefficient for the oxide-GaAs interface at the three wavelengths in (2) and hence calculate the number of pairs and total mirror thickness for a power reflectivity of 0.96.

4. Estimate the free carrier absorption coefficient at 1.5 μm in a GaAs-AlGaAs stack doped to 5×10^{17} cm^{-3}.

Chapter 8

Diode laser modelling

Rate equations constitute one of the simplest forms of laser model. They are essentially one-dimensional in so far as the transport of carriers from the contact to the active region is modelled but no account is taken of the complex dynamics of light generation, recombination, and carrier diffusion in the plane of the active region, be it a DH laser or a quantum well, and as chapter 10 will show, for high power broad area lasers these effects are significant. The purpose of rate equation modelling is essentially to predict the temporal and output characteristics of a particular diode laser structure using known material parameters and equations for the rate of change of carrier and photon densities. Rate equation models therefore offer the possibility of being able to understand which material and structure dependent properties most affect the diode laser response and therefore which may be optimised in the design of new lasers. The approach taken in this chapter is to define some general equations applicable to all lasers but especially to the idealised DH laser. These rate equations will then be used, expanding them where necessary, in specific examples of diode laser models. All the models considered have appeared in the literature and therefore represent real cases.

8.1 Rate equations; the idealised DH laser

The rate of change of carrier and photon densities are coupled by necessity, as photons are generated through the loss of carriers by recombination. The carrier density N follows the simple expression

$$\frac{dN}{dt} = R_i - R_{out} \qquad (1)$$

where R_i is the rate of carrier generation and R_{out} is the rate of carrier dissipation. The only generation mechanism of any significance is the constant supply of carriers provided by the current. Therefore

$$R_i = \frac{I_s}{qV} \qquad (2)$$

where V is the volume of the active region under consideration.

Several dissipative mechanisms exist, however. Spontaneous emission, non-radiative recombination, leakage from the active region, and of course stimulated emission. Often the recombination term is expressed by a polynomial

$$R(N) = AN + BN^2 + CN^3 \tag{3}$$

where A, B, and C are respectively the surface and defect, radiative, and non-radiative recombination coefficients. For simplicity, however, these three will be lumped together and characterised by a single time constant τ. Therefore

$$R_{out} = \frac{N}{\tau} + R_{stim}. \tag{4}$$

Hence we have the first of the coupled equations

$$\frac{dN}{dt} = \frac{I_s}{qV} - \frac{N}{\tau} - R_{stim} \tag{5}$$

which is coupled to the photon density via R_{stim}, the rate of stimulated recombination.

The photon *density* generation rate is given by

$$R_{ph} = \frac{V}{V_{ph}} R_{stim} = \Gamma R_{stim} \tag{6}$$

where V_{ph} is the volume occupied by the photons. This is always larger than the active carrier volume in guided structures and this ratio is known as the optical confinement, Γ. Hence

$$\frac{dS}{dt} = \Gamma R_{stim} + \Gamma \beta_{sp} R_{sp} - \frac{S}{\tau_p} \tag{7}$$

which takes into account the spontaneous emission and the loss of photons from the ends of the cavity via the cavity lifetime τ_p. Spontaneous emission, by its nature, is random and does not necessarily contribute to laser output but there is always the possibility that spontaneous emission into the lateral modes of the laser will occur. This is expressed by the spontaneous emission coupling factor β_{sp}. The last term uses the cavity decay time τ_p which essentially describes how fast the photon density in the cavity would decay if left to circulate between the mirrors without any regeneration by stimulated emission.

This is the second of the coupled equations, but before either can be solved we need to express R_{stim} in terms of carrier and photon density. Looking at the gain, therefore, we can express the increase in photon density in the following way

$$S + \Delta S = S e^{g\Delta x} \tag{8}$$

so

$$S \approx S(1 + g\Delta x) \tag{9}$$

for very small Δx (figure 8.1). Therefore

$$\Delta S \approx g\Delta x S \tag{10}$$

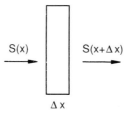

Figure 8.1. Gain in a small volume element.

and

$$\frac{\Delta S}{\Delta t} \approx g \frac{\Delta x}{\Delta t} S \qquad (11)$$

so

$$R_{stim} = \frac{dS}{dt} = g \frac{dx}{dt} S = g v_g S \qquad (12)$$

where v_g is the group velocity.

Hence

$$\frac{dS}{dt} = \Gamma v_g g S + \Gamma \beta_{sp} R_{sp} - \frac{S}{\tau_p}. \qquad (13)$$

8.2 Gain compression

The actual variation of gain with time is complicated by the fact that the carrier density depends in some manner on the photon density, so photon gain is a non-linear function of photon density also.

Imagine a very simple model of a gain medium. A small volume of space (for the sake of argument assume unit cross-sectional area) is fed by a constant current injecting carriers at a sufficient rate to cause gain. In order to get continued gain from this region some form of optical feedback is needed, so let this element be placed in the centre of a cavity (figure 8.2). If the width of the

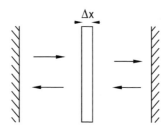

Figure 8.2. A small gain element in an optical cavity.

element is Δx such that $g\Delta x$, where g is the gain, is small, then the photon number on traversing the gain element once is

$$S_2 = S_1 \exp(g\Delta x) \approx S_1(1 + g\Delta x) \qquad (14)$$

i.e. as in equation (10)

$$\Delta S \approx g\Delta x S.$$

Ignore the details of the interaction of the photons with the electrons in the gain medium and make the assumption that the increase in photon number is instantaneous. Then the rate of increase in photon number is then given by the rate at which the photons traverse the element, i.e.

$$R_T = \frac{L_c}{v_g} \Delta S \qquad (15)$$

where L_c is the cavity length and v_g is the group velocity of the photons. Therefore

$$\frac{dS}{dt} = \frac{(1-\alpha)Sg_0\Delta x L_c}{v_g} \qquad (16)$$

where the term $1-\alpha$ represents the fractional loss of photons on reflection from each mirror. For this treatment it is not strictly necessary to add this term since it is only a coefficient, but physically it is necessary to imagine a loss of photons from the ends of the cavity. For simplicity we make this loss symmetrical. Then, by simple rearrangement,

$$\int \frac{1}{S} dS = \int \frac{(1-\alpha)g\Delta x L_c}{v_g} dt. \qquad (17)$$

All the terms on the right hand side with the exception of gain are independent of time. If we assume the gain is constant then we arrive at the conclusion that

$$S \propto \exp(gt) \qquad (18)$$

which is what we would intuitively expect. The photon number increases exponentially in a step-wise manner because of the discrete nature of the gain process in this example (figure 8.3).

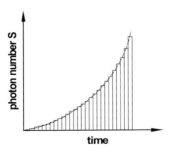

Figure 8.3. Repeated passes through the cavity leads to an exponentially increasing photon number with time.

However, the photons are created at the expense of electrons through stimulated emission, so if all other contributions to the photon density and electron depletion are ignored, and it is assumed that all photons traverse the gain medium, i.e. the optical confinement is unity, then

$$\frac{dN_{stim}}{dt} = -\frac{dS}{dt} \qquad (19)$$

where N_{stim} is the number of electrons lost to stimulated emission. Therefore

$$\frac{dN}{dt} = \frac{I}{q} - \frac{(1-\alpha)Sg\Delta x L_c}{v_g}. \qquad (20)$$

The first term on the right is constant but the second leads to an exponentially increasing photon density under the assumption of constant gain (equation 18). This is unphysical as the photon number would rapidly exceed the electron number (in this instance numbers are preferred to densities since the photon density does not have much meaning in an essentially two dimensional system) and the increase in photons would reach a limiting value where all the electrons were depleted on a single traversal. We would rapidly reach a situation where the increase in photon number would be small compared with the number of incident photons so the gain would be very small, and possibly close to zero. This can be realised by putting the time-averaged rate of change of electron number to zero, i.e.

$$\frac{I}{q} = \frac{(1-\alpha)Sg\Delta x L_c}{v_g} \qquad (21)$$

and under these conditions

$$g \propto \frac{I}{S}. \qquad (22)$$

For a given current applied to the system the gain will decrease with increasing photon density.

It should be emphasised that this is a very simple model. It does not physically represent many laser structures. Possibly only a quantum well VCSEL (chapter 7) will approach this physical construction, but nearly every other design will have an extended gain region which fills almost the entire cavity. Furthermore, the details of the interaction between the photons and the electrons have been entirely omitted. The above equations tell us only that in these idealised circumstances the gain must decrease with increasing photon number and, for very high photon densities, will tend to zero. For a working model of the gain compression, as the phenomenon is called, a quantum mechanical treatment of the electron interaction with the photon field is needed, included within which is a treatment of the intra-band relaxation times as well as the distribution function for the electrons. In the very simplified treatment given above the fact that electrons occupy certain energy levels, some of which might not participate in the stimulated emission, was not considered. According to Agrawal [1] the linear form of the saturated gain, i.e.

$$g_c = g_0(1 - \varepsilon S) \tag{23}$$

where ε is the gain compression factor and S is the photon density, can be derived using the third order perturbation theory within the density-matrix approach, but that clearly is beyond the scope of this book. Therefore I will simply use the results that others have derived. There is in fact no great loss associated with this approach because gain compression is usually introduced phenomenologically into the rate equations in any case. Furthermore, there is not as yet any firm consensus on the mathematical form of the gain compression.

The linear form given above has the advantage that the gain can go to zero at high photon densities, satisfying the intuitive picture that the gain decreases and at least tends towards zero as carrier depletion effects (spectral hole burning) become significant. Unfortunately the gain can also become negative and does if some artificial constraint is not introduced to prevent it. One of the earliest models of gain compression [2] used the expression

$$g_c(S) = \frac{g_0}{(1 + \varepsilon S)} \tag{24}$$

which was derived for a homogeneously broadened two-level system. In such a system, ε has a specific dependence on the lasing material but for the diode laser the parameter was regarded as simply being a convenient variable to describe the gain saturation. As this form is not generally valid for semiconductor lasers Agrawal [1] modelled the gain medium as an ensemble of two-level systems with different transition frequencies distributed in accordance with the joint density of states (see chapter 3) and arrived at the expression

$$g_c(S) = \frac{g_0}{\sqrt{1+S}}. \tag{25}$$

This has the physical advantage of Channin's expression (equation 24) in that the gain does not become negative but at low photon densities approximates by means of the binomial expansion to

$$g_c = g_0\left(1 - \frac{S}{2}\right). \tag{26}$$

All of the above expressions have been used at some stage by different authors in various device models. For this reason it is appropriate to use the symbol g_c to refer to the compressed gain and use the particular form appropriate when describing particular laser models.

8.3 Small signal rate equations

The rate equations as so far given, including the effects of compressed gain, are applicable only to static conditions of operation. Very often the static characteristics are of limited interest compared with the dynamic response of lasers, that is, the manner in which the output power varies with a time-dependent input current. In order to model transient or time dependent effects a modulation of some sort must be superimposed onto the rate equations. This is the so-called small signal analysis. The method outlined here is essentially applicable to most lasers and will be developed to deal with specific lasers in turn, starting first with the idealised DH laser.

Essentially, we want to know how the output power can be expressed in terms of the basic laser parameters. The output power can be expressed as the rate of transmission of photons through the end of the cavity multiplied by the energy per photon, i.e.

$$P_o = R_{ph} h\nu \qquad (27)$$

where

$$R_{ph} = \frac{SV_p \alpha_m}{\tau_r} \qquad (28)$$

and S is the photon density, V_p is the volume occupied by the photons, $\alpha_m = 1 - R$ is the reflection loss of the mirror and τ_r is the round trip time of the cavity. This last can be expressed as

$$\tau_r = \frac{2L_c}{v_g} \qquad (29)$$

where v_g is the group velocity. Therefore

$$P_o = \frac{\alpha_m}{2L_c} S v_g V_p h\nu. \qquad (30)$$

We have made the implicit assumption that there are no transmission losses either at the other end of the cavity or during the round trip. That is, the reflection losses at the output end of the cavity are the sole contributor to the cavity losses. Formulating the rate of photon decay in the absence of stimulated emission in much the same manner as the increase in photon number was formulated to show that the gain must saturate we would find,

$$\frac{v_g \alpha_m}{2L_c} = \frac{1}{\tau_p} \qquad (31)$$

where τ_p is the cavity lifetime. Hence

$$P_o = \frac{SV_p h\nu}{\tau_p}. \qquad (32)$$

Small signal rate equations

For the small signal analysis each term in the rate equation is assumed to have the form

$$A = A_0 + a \cdot e^{j\omega t} \quad (33)$$

where A_0 refers to the static component and $a \cdot e^{j\omega t}$ is a small sinusoidally oscillating component. Thus the photon and carrier densities, the gain, and also the driving current are assumed to have this form. Since the current is the driver this is essentially a problem of forced harmonic oscillation, and we should not be surprised to find that the solution has in fact the classic features of the forced, damped harmonic oscillator.

Consider the electron density equation first. From before

$$\frac{dN}{dt} = \frac{I_s}{qV} - \frac{N}{\tau} - g_c v_g S. \quad (34)$$

Substituting in the small signal forms of N, I, and S,

$$\frac{dN_0}{dt} + j\omega n e^{j\omega t} = \frac{I_{s0}}{qV} + \frac{i_s e^{j\omega t}}{qV} - \frac{N_0}{\tau} - \frac{n e^{j\omega t}}{\tau} - g_c v_g S_0 - g_c v_g s e^{j\omega t}. \quad (35)$$

Under steady state conditions

$$\frac{dN_0}{dt} = 0 \quad (36)$$

and the algebraic sum of the static components on the right must also equal zero. Eliminating the time dependent exponentials also yields

$$j\omega n = \frac{i_s}{qV} - \frac{n}{\tau} - g_c v_g s. \quad (37)$$

Similarly, for the photon density the large signal equation is (equation (13))

$$\frac{dS}{dt} = \Gamma v_g g_c S + \Gamma \beta_{sp} R_{sp} - \frac{S}{\tau_p}. \quad (38)$$

The spontaneous emission terms be can ignored for simplicity so that

$$\frac{dS}{dt} = \Gamma v_g g_c S - \frac{S}{\tau_p} \quad (39)$$

and under steady state

$$\Gamma v_g g_c = \frac{1}{\tau_p}. \quad (40)$$

Making the small signal approximations, in particular the gain

$$g_c = g_{c0} + g e^{j\omega t}. \quad (41)$$

The terms in the time constant τ_p cancel out to arrive at the final expression

Diode laser modelling

$$j\omega s = \Gamma v_g g S_0. \tag{42}$$

The small signal gain can be derived from the first order Taylor expansion of the gain expressed as a function of carrier density.

$$g_c(N_0 + n) = g_c(N_0) + n \frac{\partial g_c}{\partial n} \tag{43}$$

so that

$$g = n \frac{\partial g_c}{\partial n}. \tag{44}$$

Therefore

$$j\omega s = \Gamma v_g \left[\frac{\partial g}{\partial n} \right] n S_0 \tag{45}$$

or

$$j\omega s = \Gamma v_g g_n n S_0 \tag{46}$$

where g_n is the differential gain. Solving for the small signal carrier density,

$$n = \frac{j\omega s}{\Gamma v_g g_n S_0}. \tag{47}$$

Turning to the injection current, and expanding the gain,

$$i = qV \left[j\omega n + \frac{n}{\tau} - \frac{s}{\Gamma \tau_p} - v_g g_n n S_0 \right] \tag{48}$$

and substituting for n

$$i = qV \left[\frac{j^2 \omega^2 s}{\Gamma v_g g_n S_0} + \frac{j\omega s}{\Gamma v_g g_n S_0 \tau} - \frac{s}{\Gamma \tau_p} - \frac{j\omega s}{\Gamma} \right]. \tag{49}$$

Recognising that

$$\Gamma = \frac{V}{V_p} \tag{50}$$

we have, after some further manipulation,

$$i = \frac{qV_p s}{v_g g_n S_0}\left[j\omega v_g g_n S_0 + \frac{j\omega}{\tau} + \frac{v_g g_n S_0}{\tau_p} - \omega^2\right] \quad (51)$$

$$\frac{V_p s}{i} = \frac{v_g g_n S_0}{q\left[j\omega v_g g_n S_0 + \frac{j\omega}{\tau} + \frac{v_g g_n S_0}{\tau_p} - \omega^2\right]} \quad (52)$$

$$\frac{V_p s}{i} = \frac{v_g g_n S_0}{q\left[\frac{v_g g_n S_0}{\tau_p} - \omega^2 + j\omega\left(\frac{1}{\tau} + v_g g_n S_0\right)\right]}. \quad (53)$$

Now, let

$$\omega_r^2 = \frac{v_g g_n S_0}{\tau_p} \quad (54)$$

so that

$$\frac{V_p s}{i} = \frac{\omega_r^2 \tau_p}{q\left[\omega_r^2 - \omega^2 + j\omega\left(\frac{1}{\tau} + \omega_r^2 \tau_p\right)\right]} \quad (55)$$

and

$$\frac{V_p s}{i\tau_p} = \frac{1}{q\left[\left(1 - \frac{\omega^2}{\omega_r^2}\right) + j\omega\left(\frac{1}{\omega_r^2 \tau} + \tau_p\right)\right]}. \quad (56)$$

In order to get the small signal output power we simply take the static output power and substitute the small signal photon density. Therefore

$$\frac{p}{i} = \frac{h\nu}{q} \frac{1}{\left[\left(1 - \frac{\omega^2}{\omega_r^2}\right) + \frac{j\omega}{\omega_r}\left(\frac{1}{\omega_r \tau} + \omega_r \tau_p\right)\right]}. \quad (57)$$

This is the equation derived by Coldren and Corzine [3] though the treatment is different. Following Coldren and Corzine, the solution to these equations for a typically idealised DH laser of cavity dimensions 200 μm $\times$ 5 μm $\times$ 0.25 μm emitting at 1.5 eV and with a cavity decay time of 2 picoseconds is given in figure 8.4, where the labels 1 to 6 represent output powers increasing by a factor of ten each time. Coldren and Corzine used output powers ranging from 10 μW to 1 W.

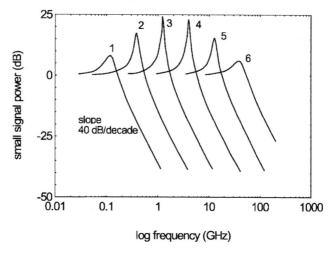

Figure 8.4. Frequency response of an idealised DH laser. (After Coldren and Corzine [3].)

Two things are immediately noticeable. First, the response is, as expected, characteristic of a forced, damped harmonic oscillator with a frequency independent output below the resonant frequency, rapidly falling off to zero above. Second, the resonant frequency and the maximum output power both vary with the driving current, a phenomenon which has had a severe impact on lightwave communications technology.

8.4 Modelling real laser diodes

The modelling presented in the previous section is general and non-specific in so far as it is intended to serve as an introduction to the field of laser modelling. It is nonetheless real for all that but there are a number of idealisations involved which mean that the model as it stands does not apply to a number of structures of interest. It is the purpose of the next section to look at some real experimental laser diode structures and their models.

8.4.1 InGaAsP/InP quantum well lasers

The first structure of interest is the InGaAsP quantum well laser as described by Bernussi *et al.* [4]. InGaAsP/InP lasers exhibit a strong temperature dependence in their performance and Bernussi *et al.* modelled the static characteristics in order to determine which parameters are responsible. According to Bernussi, the introduction of lattice matched and strained quantum wells has reduced the threshold current in lasers based on this material system but the temperature sensitivity remains unaffected. Several factors have been considered as possible reasons for this sensitivity, but the consensus is that the more intrinsic properties of the laser are involved; properties such as gain, differential gain, barrier recombination, and thermionic emission. The temperature sensitivity of the static properties of InGaAsP layers can be described fairly accurately because of the extensive experimental work undertaken by many workers (see Bernussi for references), so a steady state phenomenological model of a compressively strained quaternary multi-quantum well laser was constructed in order to account for the properties of lasers incorporating these layers.

The experimental structures modelled were 1.3 μm strained (0.75%) multi-quantum well buried heterostructures with room temperature threshold currents I_{th} of 8–11 mA and slope efficiencies η_{ext} of 0.17–0.21 mW/mA per facet. In the temperature range 20–100°C the temperature variation of the threshold current is described by a value of the characteristic temperature, T_0, defined in chapter 3, of ~60 K. To recap, the smaller the value of T_0 the more sensitive to the effects of temperature. For example, Coldren and Corzine [3, p 58] describe values of T_0 greater than 120 K near room temperature in GaAs/AlGaAs DH lasers emitting at 850 nm and even higher values (150–180 K) in quantum well lasers based on the same material system. For strained InGaAs/AlGaAs quantum wells, $T_0 \geq 200$ K but in InGaAsP/InP quantum well and DH lasers, the system under consideration here, typical values lie on the range 50–70 K. Such a temperature sensitivity results in poor laser performance at higher temperatures and practically necessitates the use of thermo-electric coolers to ensure stable operation.

The single mode rate equations

$$\frac{dN}{dt} = \frac{\eta_i I}{qV} - v_g a(N - N_0)S - R(N) \tag{58}$$

and

$$\frac{dS}{dt} = v_g[\Gamma a(N - N_0) - \alpha(N)]S + \Gamma \beta BN^2 \tag{59}$$

can be written. Here η_i is the internal efficiency, i.e. the fraction of the current which results in carriers being injected into the active region, a is the differential gain, $R(N)$ is the recombination term from equation (3), $\alpha(N)S$ represents the cavity loss (which can also be written as S/τ_p), Γ is the optical confinement, and β is the spontaneous emission coupling coefficient. These equations are in all respects similar to those already considered. These parameters were measured experimentally on the laser diode structures.

In order to solve these equations to find the threshold current and slope efficiency experimental values of many of the parameters are required. The differential lifetime, τ_d, defined as

$$\tau_d = \frac{\partial N}{\partial R} \tag{60}$$

was measured at 10°C and 50°C by varying the current. Experimentally there was no detectable difference between the two measurements, showing that this parameter is essentially temperature independent. Bernussi *et al.* concluded that of the coefficients describing the recombination rate in equation (3), A and C are temperature independent, C also being negligibly small, and B is inversely proportional to the temperature. Subthreshold gain measurements were performed as a function of photon energy at different temperatures and bias currents to derive the temperature dependence of both dg/dN and N_0, the carrier density at transparency. In the temperature range 20–80°C dg/dN decreased from 1.25×10^{-15} cm^2 to 6.8×10^{-16} cm^2, and N_0 increased from 1.5 to 2.2×10^{18} cm^{-3}. The remaining parameters, the internal losses and the internal efficiency were found to be in the range 16 cm^{-1} (20°C) to 30 cm^{-1} (100°C) for the loss and ~60% for the efficiency.

The solutions evaluated by Bernussi *et al.* are shown in figure 8.5. The theoretical model matches very well the external efficiency but is a little out in the description of the threshold current. Nonetheless it provides a reasonable description of the variation of these properties with temperature. Using this model, Bernussi *et al.* were able to assess systematically which parameters most affected the solution by setting their dependence on temperature to zero in turn. The conclusion is that two parameters are principally responsible; the internal efficiency η_i and the losses α. Absorption losses mainly affect the differential gain, in turn affecting the differential efficiency, and carrier emission from the well mainly affects the threshold current. Carrier emission

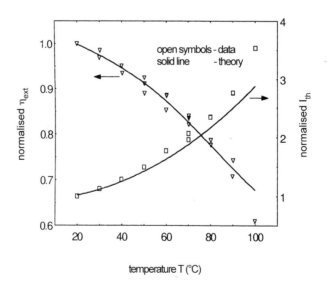

Figure 8.5. The efficiency of quantum well lasers as a function of temperature. (After Bernussi *et al.* [4].)

can be reduced by using larger barriers but there are limitations imposed by the material system so to some extent the temperature sensitivity of the InGaAsP/InP system would appear to be intrinsic to the materials.

8.4.2 Separate confinement heterostructure quantum well laser

Still on the topic of quantum wells, the next model of interest is the separate confinement quantum well laser, as studied by Nagarajan *et al.* [5]. Characteristic responses of two of the In0.2Ga0.8As-GaAs strained quantum well lasers investigated by Nagarajan *et al.* are illustrated in figures 8.6 and 8.7. The first has an optical confinement layer 76 nm in width and the second a confinement layer 300 nm wide.

Two things stand out in these results. First, as expected from the previous models, the lasers exhibit a power dependent resonant frequency but it is very obvious in the case of the larger confinement layer that the resonant frequency is actually sitting on an edge, and unlike the case of the idealised DH laser the response is not flat below resonance. Second, the smaller confinement layer gives rise to a larger bandwidth. This is especially important. It is evident from the previous models, and will be demonstrated here again, that the resonant frequency is directly proportional to the differential gain (equation 54) which is higher in quantum well lasers. In the early 1980's it was not apparent why experimentally quantum well lasers were not exhibiting this increased bandwidth, and experiments such as Nagarajan's and the associated modelling, led to an improved understanding. Given that the shorter device has the higher

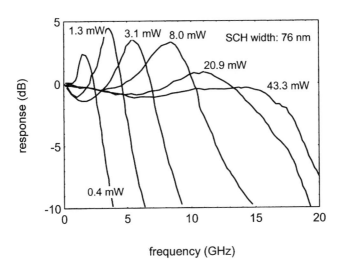

Figure 8.6. The modulation response of narrow SCH quantum well lasers as studied by Nagarajan *et al.* (After Nagarajan *et al.* [5].)

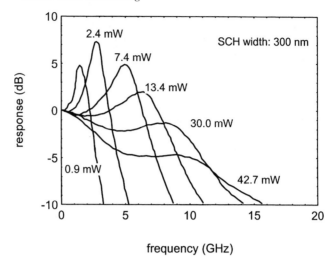

Figure 8.7. Frequency response of a wide SCH laser as studied by Nagarajan *et al.* (After Nagarajan *et al.* [5].)

bandwidth, you might guess that transport across the confinement layer is significant. This is so, and will be demonstrated in Nagarajan's model.

We start, however, with the rate equations of which there are three rather than the two we have so far dealt with. We need to describe the rate of change of carrier density in the confinement layer, the rate of change of carrier density in the well, and also the rate of change of the photon density. Therefore the carrier density in the confinement layer is given by,

$$\frac{dN_{sch}}{dt} = \frac{I_s}{qV_{sch}} - \frac{N_{sch}}{\tau_s} + \frac{1}{\tau_e} \cdot \frac{N_w V_w}{V_{sch}} \tag{61}$$

where τ_s represents the ambipolar transport time including the time constant for carrier capture into the well from the confinement layer and τ_e represents the time constant for thermionic emission out of the well into the confinement layer. The concept of ambipolar conduction has been met in chapter 6 and will be justified further, but for now we simply use the fact that we are not considering the electrons and holes as separate carriers, or even the fact that they are injected from opposite ends of the confinement layer, but treat them as one carrier. Nagarajan *et al.* have argued that the well capture time is small (<1 ps) and can be ignored as a contribution to τ_s, an assumption in keeping with the discussion on carrier capture in the last chapter. The ratio of volumes for the well and the confinement layer allows for the conversion of densities from one region to the other. In the well,

$$\frac{dN_w}{dt} = \frac{1}{\tau_s} \cdot \frac{N_{sch} V_{sch}}{V_w} - \frac{N_w}{\tau_n} - \frac{N_w}{\tau_e} - \frac{v_g G(N) S}{1 + \varepsilon S} \tag{62}$$

where τ_n is the bimolecular recombination lifetime, $G(N)$ is the carrier dependent gain and ε is the phenomenological gain compression factor. The photon density is as before,

$$\frac{dS}{dt} = \frac{\Gamma v_g G(N) S}{1+\varepsilon S} - \frac{S}{\tau_p} \qquad (63)$$

and the usual small signal substitutions can be made.

$$j\omega n_{sch} = \frac{i_s}{qV_{sch}} - \frac{n_{sch}}{\tau_s} + \frac{1}{\tau_e} \cdot \frac{n_w V_w}{V_{sch}} \qquad (64)$$

$$j\omega n_w = \frac{1}{\tau_s} \cdot \frac{n_{sch} V_{sch}}{V_w} - \frac{n_w}{\tau_n} - \frac{n_w}{\tau_e} - \frac{v_g g_n n_w S_0}{1+\varepsilon S_0} \qquad (65)$$

$$j\omega s = \frac{\Gamma v_g g_0 S_0 n_w}{1+\varepsilon S_0} - \frac{s}{\tau_p(1+\varepsilon S_0)} - \frac{s}{\tau_p} \qquad (66)$$

with

$$\frac{\Gamma v_g G_0}{1+\varepsilon S_0} = \frac{1}{\tau_p} \qquad (67)$$

given by the steady state photon density equation. Eliminating n_{sch} and n_w yields

$$M(\omega) = \frac{s(\omega)}{i} = \frac{1}{(1+j\omega\tau_s)} \frac{A}{(\omega_r^2 - \omega^2 + j\omega\gamma)} \qquad (68)$$

where

$$A = \frac{\Gamma(v_g g_n/\chi) S_0}{qV_w(1+\varepsilon S_0)} \qquad (69)$$

$$\chi = 1 + \frac{\tau_s}{\tau_e} \qquad (70)$$

$$\gamma = \frac{(v_g g_n/\chi) S_0}{(1+\varepsilon S_0)} + \frac{\varepsilon S_0/\tau_p}{(1+\varepsilon S_0)} + \frac{1}{\chi\tau_n} \qquad (71)$$

and

$$\omega_r^2 = \frac{(v_g g_n/\chi) S_0}{\tau_p(1+\varepsilon S_0)} \left(1 + \frac{\varepsilon}{v_g g_n \tau_n}\right). \qquad (72)$$

The functional form of the modulation response $M(\omega)$ appears simple enough but when the quantities A, ω_r, γ, and χ are substituted back in then the truth of modelling is revealed; the more accurate the model the more complex it must become. Even this is not the most complex form of the model derived by Nagarajan et al. An alternative expression was derived using a frequency dependent τ_s which contains terms in ω^3 and which cannot be solved analytically. That is to say, the so-called transport factor of equation (70) then becomes

$$\chi = 1 + \frac{\tau_s}{(1+j\omega\tau_s)\tau_e}. \tag{73}$$

The modulation response must be solved for numerically under these conditions. However, there is no particular advantage in going to the extra complexity involved, especially as the physical interpretation of a frequency dependent time constant for transport is not immediately obvious. Figure 8.8 shows a comparison of the two solutions.

Concentrating on $M(\omega)$ in equation (68) it can be seen that the pre-factor $(1+j\omega\tau_s)^{-1}$ is responsible for the low frequency roll-off, as the phenomenon of the decrease in response prior to the resonance is called. Further, it can be seen that the transport factor, which is in effect a measure of the relative time the carriers spend in the well and the confinement layer, effectively reduces the differential gain to g_n/χ and increases the bimolecular recombination time to $\chi\tau_n$. In short, the more time the carriers spend in the confining layer rather than in the well the smaller the recombination rate and the lower the differential gain, which is what you would expect on physical grounds.

The thermionic emission time from a quantum well has been derived by Schneider *et al.* [6] assuming the carriers have bulk-like properties, which is not the case but is justified as a necessary simplification,

$$\tau_e = \left[\frac{2\pi m^* L_w^2}{k_B T}\right]^{1/2} \exp\left(\frac{E_B}{k_B T}\right). \tag{74}$$

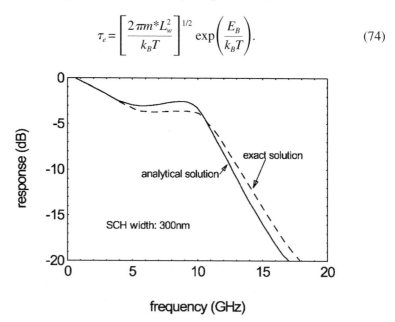

Figure 8.8. Frequency response the SCH quantum well laser. (After Nagarajan *et al.* [5].)

This varies with the quantum well width and barrier height, but in the choice of well width a trade-off with other parameters is necessary. A large well width will increase the thermionic emission time and hence the fraction of time spent in the well but in so doing the current into the well will be increased in order to provide the same carrier density. Alternatively, the transport time across the confinement layer can be decreased. The transport time is given by

$$\tau_s = \frac{L_s^2}{2D_a} \tag{75}$$

where D_a is the ambipolar diffusion coefficient. Hence, a smaller confinement layer width L_s will decrease the fraction of time spent in the confinement layer and will consequently reduce χ, which in turn will lead to an increased resonant frequency, as observed.

The current flowing in the device can be examined in more detail. Following Nagarajan we consider the structure shown in figure 8.9. Electrons are injected from the n^+ cladding from the right and holes from the p^+ cladding from the left. The heterojunctions for the optical confinement are situated at $x = \pm L_s$ and the quantum well is at $x = 0$. The first aim is to show that the current flowing into the well, I_w, has the form

$$I_w = I_s \mathrm{sech}\left(\frac{L_a}{L_s}\right) - qV_w \frac{L_a}{L_s} \tanh\left(\frac{L_a}{L_s}\right) \frac{p_w}{\tau_a} \tag{76}$$

where L_a is the ambipolar diffusion length, V_w is the volume of the quantum well, p_w is the hole concentration in the well, and τ_a is ambipolar carrier lifetime, and I_s is the current entering the SCH from the cladding. All these terms will be defined as we go.

We start with current continuity. The current continuity equation expresses the reality that charge entering a particular volume element of a semiconductor

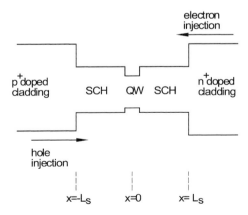

Figure 8.9. Band diagram of SCH laser.

(which in this case will be the quantum well) must either emerge at the other side, remain trapped within the region, or be neutralised by recombination. Therefore,

$$divJ = \frac{d\rho}{dt} - qU(n, p) \qquad (77)$$

where $divJ$ is the divergence in current, ρ is the charge density, and $U(n, p)$ is the net recombination rate. Therefore

$$\frac{\partial n}{\partial t} = \frac{1}{q}\frac{dJ}{dt} - U(n, p). \qquad (78)$$

We need to know the specific form of J. Nagarajan *et al.* have entered into a fairly detailed discussion of the current flow and the possible effects of the structure. A purely diffusive model of carrier transport can be assumed and an ambipolar diffusion coefficient derived, but the lack of other conduction mechanisms, such as ballistic transport, needs to be justified. Ballistic transport is the term given to the passage of energetic carriers through a medium. In bulk material it will occur at high fields such that the carriers have an average energy well away from the band extrema. In quantum systems, or deep submicron structures, ballistic transport can occur simply because the devices are too small for the energy loss mechanisms which thermalise the electrons to apply. In the SCH laser electrons and holes injected into the optically confining layer from the cladding layers are by definition "hot", i.e. energetic, with an excess energy equal to the band offset. Therefore they have a different wave-vector, and hence momentum, from carriers in quasi-thermal equilibrium with the lattice. By definition, the presence of current flow means that the system is not strictly in thermal equilibrium but the Fermi distribution of carriers will be only slightly perturbed and diffusion will take place at the conduction band minimum because in the SCH structure the carrier injection level need not induce transparency in the wide band-gap layer, which by its nature is already transparent to radiation emitted from the quantum well. The quasi-Fermi levels will therefore lie within the band gap. A similar argument applies to drift conduction provided the electric field is not too large. Multi-phonon emission or absorption processes must occur, therefore, in order to change the momentum of the injected carrier.

Nagarajan *et al.* have used the reported intervalley scattering times of ~2 ps in GaAs (see [5] for references) to conclude that hot carriers will have thermalised long before the quantum well is reached in the longest of their confining structures (300 nm). Furthermore, in the high carrier concentrations present in lasers electron-electron scattering cannot be neglected and this occurs on time scales of ~10 fs. In short, Nagarajan concluded that diffusive transport of electrons is justified in this particular laser model. Hole transport might be another matter, however. As will be justified shortly, this is an ambipolar diffusion model and as such averages the diffusion times of the carriers. Nagarajan has argued that since holes diffuse more slowly than electrons in the GaAs-AlGaAs structure these are the limiting carrier, and the ambipolar diffusion can be approximated by hole diffusion with a diffusion coefficient of twice the equilibrium value. Hot electrons injected into the structure may well cause deviations from the strict diffusion model but since the electrons effectively diffuse much faster than the holes their effect will hardly be noticeable. Hot holes, on the other hand, could have a significant effect. The valence band offset in this material is lower than the conduction band offset and so the excess energy is reduced, making thermalisation easier. Nagarajan concluded that ballistic transport of holes at room temperature is not significant, so the current can be taken to be diffusive.

The current consists of a drift component and a diffusion component. Taking the electron current first

$$J_n = \sigma \varepsilon + qD_n \frac{\partial n}{\partial x}. \qquad (79)$$

The conductivity can be written as

$$\sigma = qn\mu \qquad (80)$$

where

$$\mu = \frac{qD_n}{kT} \qquad (81)$$

so that

$$J_n = qD_n \left[\frac{qn\varepsilon}{kT} + \frac{\partial n}{\partial x} \right]. \qquad (82)$$

Hence

$$\frac{\partial J_n}{\partial x} = qD_n \left[\frac{qn}{kT} \frac{\partial \varepsilon}{\partial x} + \frac{\partial^2 n}{\partial x^2} \right] \qquad (83)$$

for electrons and for the holes,

$$\frac{\partial J_p}{\partial x} = qD_p \left[\frac{qp}{kT} \frac{\partial \varepsilon}{\partial x} - \frac{\partial^2 p}{\partial x^2} \right]. \qquad (84)$$

Above threshold, we consider the laser to be operating in the high injection regime, i.e. $n \approx p$, so that the active region is electrically neutral. Any electric field existing within the active region must therefore be constant, i.e.

$$\frac{\partial \varepsilon}{\partial x} = 0. \qquad (85)$$

Substituting this back into (83) we are able to express the time derivatives of the carrier concentrations as

$$\frac{\partial n}{\partial t} = D_n \frac{\partial^2 n}{\partial x^2} - U(n, p) \qquad (86)$$

and

$$\frac{\partial p}{\partial t} = D_p \frac{\partial^2 p}{\partial x^2} - U(n, p) \qquad (87)$$

and under steady state conditions

$$\frac{\partial p}{\partial t} = \frac{\partial n}{\partial t} = 0. \qquad (88)$$

Hence

$$\frac{\partial^2 p}{\partial x^2} + \frac{\partial^2 n}{\partial x^2} = \frac{U(n, p)}{D_p} + \frac{U(n,p)}{D_n} = U(n, p)\left(\frac{D_n + D_p}{D_n D_p}\right). \qquad (89)$$

It is convenient, however, to express the current in terms of one carrier only, rather than both. We can write

$$\frac{d^2 p}{dx^2} = \frac{1}{f}\left(\frac{D_n + D_p}{D_n D_p}\right)U(n, p) = \frac{U(n,p)}{D_a} \qquad (90)$$

where f is some factor that takes into account the electron density. We call D_a the ambipolar diffusion coefficient. It is effectively an average of the two diffusion coefficients and can be determined experimentally.

The recombination rate $U(n, p)$ can be described in terms of the net hole density and a characteristic lifetime, called the ambipolar lifetime. Hence

$$\frac{d^2 p}{dx^2} = \frac{p}{D_a \tau_a} = \frac{p}{L_a^2} \qquad (91)$$

where L_a is the ambipolar diffusion length. This is a second order differential equation with solutions of the form

$$p(x) = A e^{x/L_a} + B e^{-x/L_a} \qquad (92)$$

where A, B are constants to be determined, leading to knowledge of the steady state carrier concentration in the active region. The boundary conditions will determine the specifics of the solution.

Let the concentration of carriers in the well be p_w. The well is set at $x = 0$ so

$$p_w = A + B. \qquad (93)$$

Now let us look at the current injected into the SCH region. There is a constant supply of carriers injected into the SCH region at a rate given by I_s.

$$I_s = -qA_sD_a \frac{dp}{dx}\bigg|_{x=-L_s} \tag{94}$$

where A_s is the cross sectional area at $x=-L_s$.

$$\frac{dp}{dx}\bigg|_{x=-L_s} = \frac{1}{L_a}(Ae^{-L_s/L_a} - Be^{L_s/L_a}). \tag{95}$$

Hence

$$\frac{L_aI_s}{A_sqD_a} = Be^{L_s/L_a} - Ae^{-L_s/L_a}. \tag{96}$$

Using the combination

$$A = p_w - B \tag{97}$$

and

$$B = p_w - A. \tag{98}$$

We find that

$$\frac{L_aI_s}{A_sqD_a} = B[e^{L_s/L_a} + e^{-L_s/L_a}] - p_w\frac{e^{-L_s}}{L_a} \tag{99}$$

from which it follows that

$$B = \frac{p_w\dfrac{e^{-L_s}}{L_a} + \dfrac{L_aI_s}{A_sqD_a}}{e^{L_s/L_a} + e^{-L_s/L_a}} \tag{100}$$

and similarly

$$A = \frac{p_w\dfrac{e^{-L_s}}{L_a} - \dfrac{L_aI_s}{A_sqD_a}}{e^{L_s/L_a} + e^{-L_s/L_a}}. \tag{101}$$

The current into the well is I_w, where

$$I_w = -qA_sD_a\frac{dp}{dx}\bigg|_{x=0} \tag{102}$$

and similarly solving for the differential, i.e.

$$p_w = A + B. \tag{103}$$

Substituting for A and B yields

$$I_w = I_s\operatorname{sech}\left(\frac{L_s}{L_a}\right) - \frac{qD_aA_sp_w}{L_a}\tanh\left(\frac{L_s}{L_a}\right) \tag{104}$$

which can be rearranged to give

$$I_w = I_s \text{sech}\left(\frac{L_s}{L_a}\right) - qV_w\left(\frac{L_a}{L_w}\right)\left(\frac{p_w}{\tau_a}\right)\tanh\left(\frac{L_s}{L_a}\right). \tag{105}$$

Note that

$$\text{sech}(z) = \frac{2}{e^z + e^{-z}} \quad \tanh(z) = \frac{e^z - e^{-z}}{e^z + e^{-z}}. \tag{106}$$

This expression for the current provides a further insight into the importance of transport effects. We can define a steady state transport factor, α_{sch}, different from the transport factor already defined in the solution to the rate equations. In a direct analogy to the bipolar transistor,

$$\alpha_{sch} = \frac{\partial I_w}{\partial I_s} = \text{sech}\left(\frac{L_s}{L_a}\right). \tag{107}$$

For efficient transport from the injection at the SCH barrier to the well $L_s/L_a \ll 1$ (figure 8.10). Nagarajan *et al.* have also derived a small signal transport factor utilising a frequency dependent diffusion length. The derivation, however, is not straightforward, but the result is. Again analogous to the microwave response of the bipolar transistor, [7] where for effective operation charge must be transported across the the base of the transistor in response to the frequency, charge must also be transported across the SCH layer in order to be captured by the quantum well. Therefore

$$\alpha_{sch, \text{small signal}} = \frac{1}{(1 + j\omega\tau_s)} \tag{108}$$

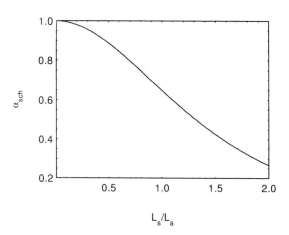

Figure 8.10. The steady state transport factor falls off rapidly for $L_s/L_a > 1$.

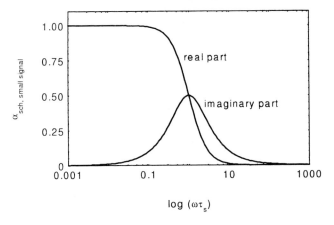

Figure 8.11. The small signal transport factor.

which is nothing other than the pre-factor in $M(\omega)$ in equation (68). This is in essence the same functional form as the dielectric response function. The imaginary part, representing carrier transport out of phase with the oscillating driver, is a peaked function with a maximum value at $\omega\tau_s = 1$ (figure 8.11). The real part, representing lossless (in phase) current falls to a value of $\frac{1}{2}$ at this point and physically represents the fact that carriers cannot be transported across the SCH layer in time with the oscillating signal. Nagarajan reported values of $\tau_s \approx 50$ ps at room temperature for the larger of his devices (SCH layer 300 nm) suggesting a transport cut-off frequency of 2×10^{10} rad s^{-1}.

8.4.3 Three level rate equation models for quantum well SCH lasers

Nagarajan's model suffers from the disadvantage that capture into the quantum well is not treated explicitly but is lumped together into the SCH transport time τ_s. This is justified if the quantum well capture time is small compared with the transport time, which is the argument used by Nagarajan, and which can be justified to some extent by the discussion on carrier capture in chapter 6. At the largest device used by Nagarajan, $\tau_s \approx 50$ ps, with the capture time was quoted as being less than 1 ps, so the approximation is justified. For the smallest devices used by Nagarajan, however, which are approximately a factor of 4 down in SCH width, the transport time will be halved assuming purely diffusive transport. Quantum well capture times may well be important.

McDonald and O'Dowd [8] have compared two- and three-level rate equations which specifically account for carrier capture into the quantum well. The McDonald-O'Dowd (M-O) model is actually an extension of that presented by Nagarajan and provides a direct comparison. The mechanism of carrier capture is considered to be by means of *gateway* states, which are virtual bound states of the quantum well. The existence of such states –

localised in space to the quantum well and lying just below the continuum of levels – was originally proposed by Brum and Bastard [9]. The precise nature of the states is not clear but a very convenient way to envisage them is to think of states that are neither continuum states or bound states. Recalling the physics of the quantum well in chapter 6, the virtual bound states lie close to or just above the barrier and are neither fully bound nor continuum states. They are called gateway states because they help to change the carrier from 3-D to 2-D. The calculations performed by Brum and Bastard of the capture times in GaAs single quantum wells showed that the capture time can be several tens of picoseconds and oscillates with quantum well width as either a new state is bound or a virtual bound state within a longitudinal optical phonon energy of the continuum edge appears. Phonon emission is one of the principal means by which an electron can lose energy so if the state is separated from the continuum by more than a phonon energy the electron cannot easily be captured by the gateway state, but once captured it is then relatively simple to transfer from the gateway state to a lower lying quantum level by a radiative transition.

The scheme envisaged by M-O is illustrated in figure 8.12. Carriers enter the barrier region and lose energy in the gateway states via phonon emission; some carriers are lost to recombination while others are gained through emission from the gateway states. Similar mechanisms apply throughout the process from injection to lasing. Population densities and time constants can be assigned to all the processes, but not all of those listed necessarily apply to single quantum well devices under consideration here, for example inter-well tunnelling. Nagarajan's model contained a description of multiple quantum

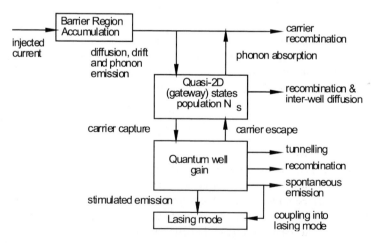

Figure 8.12. Schematic quantum well laser model. (After McDonald and O'Dowd [8].)

well devices and McDonald and O'Dowd's implementation can be extended to other wells. It will not concern us here.

The specific single quantum well laser treated by M-O has a separate confinement structure where the *n*- and *p*- cladding regions are assumed to be highly doped so that the cladding layers are in intimate electrical contact with the confining layers and transport up to this interface can be neglected. The quantum well and confining regions are assumed to be undoped with electrons and holes being injected from either side of the active region. We therefore have four coupled equations to describe carrier densities in the laser structure and the photon density. Within the SCH region, denoted by the subscript *sch*, carriers are gained through injection and by emission from the gateway states and are lost through recombination and after transport across the SCH region. Hence

$$\frac{dN_{sch}}{dt} = \frac{I}{e} - \frac{N_{sch}}{\tau_n} - \frac{N_{sch}}{\tau_d} + \frac{N_g}{\tau_g} \qquad (109)$$

where N_{sch} is the number of carriers in the SCH region, τ_n is the recombination time, τ_d is the diffusion time across the SCH region, N_g is the number of carriers in the gateway states, and τ_g is the carrier lifetime in the gateway states. In Nagarajan's model the diffusion time incorporates capture into the quantum well; the M-O model similarly incorporates capture into the gateway states within the diffusion time across the SCH region. Within the gateway states

$$\frac{dN_g}{dt} = \frac{N_{sch}}{\tau_d} - \frac{N_g}{\tau_{capt}} + \frac{N_w}{\tau_{esc}} - \frac{N_g}{\tau_g} - \frac{N_g}{\tau_n} \qquad (110)$$

where τ_{capt} is the capture time from the gateway states into the quantum well and τ_{esc} is the time for the reverse process of escape. Within the well,

$$\frac{dN_w}{dt} = \frac{N_g}{\tau_{capt}} - \frac{N_w}{\tau_{esc}} - \frac{N_w}{\tau_n} - v_g G \frac{S}{1 + \varepsilon \Gamma_w S}. \qquad (111)$$

The last term is slightly different in form from those previously given, with the optical confinement factor for the quantum well appearing in the denominator. Otherwise the general form of the expression is identical and the symbols have their previous meaning. For the photon number *S*,

$$\frac{dS}{dt} = \Gamma_w v_g G \frac{S}{1 + \varepsilon \Gamma_w S} - \frac{S}{\tau_p} \qquad (112)$$

where τ_p is the photon lifetime.

There are a number of assumptions implicit in the model in respect of the photon number and the capture mechanism. First, only the dominant cavity mode is considered so there is only one photon number rate equation. Second, the effect of spontaneous emission coupled into the laser mode is ignored as the laser is assumed to be biased well above threshold. Under these circumstances the stimulated emission is far more intense than spontaneous emission. Third, the capture process is assumed to be ambipolar and describable by an ambipolar time constant. Such a situation can arise if the holes are the first carrier type to be captured by the quantum well thereby increasing the electron capture rate through Coulomb attraction. At the same time other holes are repelled from the well so the hole capture rate is reduced. According to M-O the ambipolar capture time is approximately twice the hole capture time and the dynamics of the quantum well laser are described by a set of rate equations applicable to a single carrier. Finally, only the lowest lying levels in the quantum well are considered. Capture from, and emission to, higher lying sub-bands can play a part in the capture process, especially for narrow quantum wells, but the effect is ignored here.

The small signal modulation transfer function $s(\omega)/i(\omega)$ can be derived as before, but this will not be done here. As you would expect, the increased complexity arising from having four coupled equations rather than three renders the expression very long and complex, and the essential physics is hard to discern. The reader is referred to the original publication for the details; only the results are given here.

The principal finding of the three level rate equation model is that under the same conditions used by Nagarajan et al. for a lattice matched InGaAs quantum well laser designed to operate at 1.55 μm and consisting of a single well 8 nm thick surrounded by 300 nm thick SCH regions of InGaAsP, the three level model exhibits greater low-frequency roll-off and reduced damping at resonance resulting in a higher resonance peak. The reason for this is that the carrier population in the gateway states can replenish the carrier population in the well at a faster rate than is possible assuming only two levels where it is implicit in the model that capture is accompanied by transport across the SCH region. In the three level model no transport need take place for capture into the well to occur so there is reduced damping. However, the time taken for capture

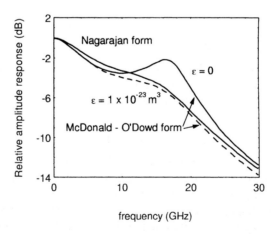

Figure 8.13. Response of the quantum well laser. (After McDonald and O'Dowd [8].)

into the gateway states is an additive factor so there is an increased roll-off. An additional interesting feature of the model is that it is possible to reproduce the frequency response of the two-level model if a large phenomenological gain suppression factor is assumed. In the two-level model transport effects alone will result in a low frequency roll-off and high damping rate while in the three-level model a combination of transport effects plus gain suppression are required.

8.5 Electrical modelling

An alternative approach to the previous models of the modulation transfer function is to model the electrical properties of the diodes lasers. Weisser et al. [10] have developed a simple model of the electrical impedance of quantum well lasers where the emphasis is not on the light output characteristics (L-I) but on the electrical impedance (V-I) characteristics. These models naturally concentrate on the transport properties and the capture time into the quantum well.

The rate equations are somewhat simpler and expressed in slightly different form to reflect the fact that the impedance is the important property. Thus

$$\frac{dS}{dt} = S\left(\Gamma G - \frac{1}{\tau_p}\right) \tag{113}$$

$$\frac{dN_w}{dt} = -N_w\left(\frac{1}{\tau_{eff}} + \frac{1}{\tau_{esc}}\right) + \frac{N_c}{\tau_{capt}} - SV_{qw}G \tag{114}$$

$$\frac{dN_c}{dt} = \frac{I}{q} - \frac{dV_c}{dt}\frac{C_{sc}}{q} - \frac{N_c}{\tau_{capt}} + \frac{N_w}{\tau_{esc}} \tag{115}$$

where N_c and N_w represent the number of unconfined carriers in the core and the number of confined carriers in the well respectively, V_c is the voltage across the core and V_{qw} is the volume of the quantum well. The gain is implicitly taken to be

$$G(N_w, S) = G(N_w)(1 - \varepsilon S). \tag{116}$$

The term in the space charge capacitance of the diode, C_{sc}, represents the fact that under small signal conditions charge will flow into the diode by virtue of its capacitance. The time constants τ_{capt} and τ_{eff} are effective time constants which represent respectively capture into the well, including transport across the core region, and effective lifetime in the well. All other symbols correspond to their previous meanings.

In essence this model represents a return to the simpler ideas of Nagarajan et al. without the complexity of the gateway states. The steady state form of these expressions can be found by putting the time-dependent terms to zero. Sub-threshold, i.e. $S=0$, this yields

$$N_{w0} = \frac{I_0 \tau_{eff}}{q} \tag{117}$$

and

$$N_{c0} = \frac{I_0 \tau_{capt}}{q} \left(1 + \frac{\tau_{eff}}{\tau_{esc}}\right). \tag{118}$$

Above threshold the number of carriers in the well is deemed to be clamped at the threshold value because the material gain is also clamped to its threshold value. Therefore,

$$N_{w0} = N_{w,th} \tag{119}$$

and

$$N_{c0} = \frac{1}{q}\left(I_0 + \frac{\tau_{eff}}{\tau_{esc}} I_{th}\right) \tau_{capt} \tag{120}$$

with

$$I_{th} = \frac{q N_{w,th}}{\tau_{eff}}. \tag{121}$$

Crucially, because the laser is essentially a forward biased diode, the number of carriers can be related to the voltage drop by

$$N_c \propto e^{qV_{c0}/mkT} \tag{122}$$

where k is the Boltzmann constant and m is the ideality factor so that the differential diode resistance

$$R_d = \frac{dV_{c0}}{dI_0} \tag{123}$$

can be evaluated, i.e.

$$R_d(I_0 < I_{th}) = \frac{mkT}{qI_0} \tag{124}$$

and

$$R_d(I_0 > I_{th}) = mkT \cdot \left(I_0 + \frac{\tau_{eff}}{\tau_{esc}} I_{th}\right)^{-1}. \tag{125}$$

At lasing threshold the diode resistance should drop by a factor

$$f_{th} = 1 + \frac{\tau_{eff}}{\tau_{esc}} \qquad (126)$$

from which an estimate of the ratio of time constants can be made. If negligible carrier re-emission from the well occurs this corresponds in the above analysis to $f_{th} \sim 1$.

As yet no estimate of the frequency dependent impedance has been made. This is derived by making the usual small signal assumptions, i.e.

$$S(t) = S_0 + s e^{j\omega t} \qquad (127)$$

and

$$\omega_r = \sqrt{\frac{G_d S_0}{\tau_p}} \qquad (128)$$

for the relaxation frequency. The differential gain G_d is given by

$$G_d = V_{qw} \frac{\partial G}{\partial N_w} \qquad (129)$$

and the damping factor by

$$\gamma = \frac{1}{\tau_{eff}} + \omega_r^2 \left(\tau_p + \frac{\varepsilon}{G_d} \right). \qquad (130)$$

Finally, a bias dependent electrical diode time constant is defined

$$\tau_0(I_0) = \tau_{capt} + \xi R_d(I_0) C_s c(I_o) \qquad (131)$$

where

$$\xi = \frac{\tau_{esc}}{(\tau_{eff} + \tau_{esc})}. \qquad (132)$$

The frequency dependent impedance is then given by

$$Z(\omega) = \frac{v_c(\omega)}{i(\omega)}. \qquad (133)$$

which again is a complicated expression in which it is difficult to discern the physics. Following Weisser *et al.* we will limit the cases considered to weak carrier re-emission from the wells, i.e.

$$\frac{\tau_0}{\tau_{esc}} \ll 1 \qquad (134)$$

and

$$\varepsilon S_0 \ll 1 \tag{135}$$

so that below threshold

$$Z(\omega) = R_d \frac{1}{(1+j\omega\tau_0)} \cdot \frac{(1+j\omega\tau_1)}{(1+j\omega\tau_{eff})} \tag{136}$$

with

$$\frac{1}{\tau_1} = \frac{1}{\tau_{eff}} + \frac{1}{\tau_{esc}} \tag{137}$$

and above threshold

$$Z(\omega) = R_d \frac{1}{(1+j\omega\tau_0)} \cdot \frac{\omega_r^2 - \omega^2 + j\omega\gamma_1}{\omega_r^2 - \omega^2 + j\omega\gamma} \tag{138}$$

where

$$\gamma_1 = \gamma + \frac{1}{\tau_{esc}}. \tag{139}$$

These equations were fitted to measurements on $In_{0.35}Ga_{0.65}As$ quantum lasers consisting of four 5.7 nm wells separated by 20 nm GaAs barriers, with 48 nm GaAs confinement regions and $Al_{0.8}Ga_{0.2}As$ cladding layers [11]. The lasers had Be doped regions 4.5 nm wide placed above each quantum well where the doping varied from highly doped (2×10^{19} cm^{-3}), low-doped (5×10^{18} cm^{-3}), to undoped (i.e. unintentional). The lasers were all cleaved to a cavity length of 200 μm but the width varied from 3 μm to 40 μm. Figure 8.14 shows the data for undoped ($\bigcirc$), low-doped ($\triangledown$), and highly doped ($\square$) samples for a current

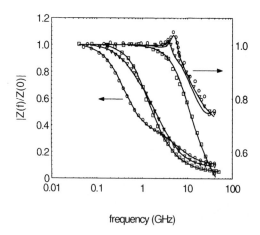

Figure 8.14. Normalised impedance below threshold (left) and above threshold (right). (After Esquivias *et al.* [11].)

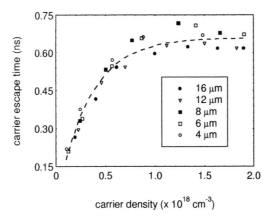

Figure 8.15. The carrier escape time. (After Esquivias *et al.* [11].)

of 1 mA (left) and 4 mA above threshold (right). The fit below 20 Ghz is excellent and allowed extraction of values for τ_{eff} which were found to be in good agreement with those obtained from the frequency dependence of the spontaneous emission.

Also derived were values for the escape time from the quantum well, shown below. The carrier escape time plotted as a function of doping density for devices of widths 4 μm $d \leq 16$ μm shows excellent consistency and lies between 200 ps and 700 ps. The effective carrier escape time is seen to be reduced dramatically with the addition of doping to the active layers, an effect not entirely understood by the authors. However, it is a factor to be considered in the design of high speed *p*-doped lasers. A further feature of this electrical model is the derivation of the diode time constant τ_0 related to the capture time and the RC time constant of the diode capacitance. This time constant was found to decrease with increasing current but under conditions where the capture time is small compared with the space-charge time constant so that

$$\tau_0 \approx \tau_{sc} \tag{140}$$

the time constant should scale with diode width. This is in fact the case. However τ_0 increases with increased doping so while Esquivias *et al.* concluded that high doping may have other benefits the low-frequency roll-off is enhanced, as shown by the fequency response of the impedance above threshold.

8.6 Circuit level modelling

A natural extension of impedance modelling is circuit level modelling. This is the representation of the laser diode as an equivalent circuit of electrical components. An impulse to the circuit simulates the injection current and the voltage across a component represents the light output. The advantage of this type of modelling for devices such as diodes is that stray capacitances and inductances arising from the packaging and mounting can be included easily in the circuit so that a *system* performance is more readily evaluated. Incidentally, there is a neat irony in this type of modelling. It is usually performed on digital computers using highly developed device modelling packages such as p-SPICE, but at one time analogue computers represented one of the most powerful methods of computation. Analogue computation is the process of representing mathematical functions, such as integrators, differentiators, adders, subtractors, etc, by circuit components (analogue of the mathematical function) and physically building a circuit to model the process. The computation occurred in parallel and was therefore exceedingly quick; switch on the current and monitor the output voltage. The principal disadvantage is that the more complex the problem the bigger the computer needed to solve it, and such computing went out of fashion more or less by the late 1970's. With circuit level modelling we now have analogue computation on a digital computer, but without the speed of the original.

There have been many papers published on the subject of electrical modelling and it is still an active topic of research. It is not the intention to review them all or discuss the finer details of circuit models. Rather, the intention is only to illustrate the basic principles of such circuits and the interested reader can pursue the models in the literature. Following Kibar *et al.* [12], we start, as always, with the rate equations expressed in a slightly different form for a single mode laser.

$$\frac{dP(t)}{dt} = GP(t) - \gamma P(t) + R_{sp}. \tag{141}$$

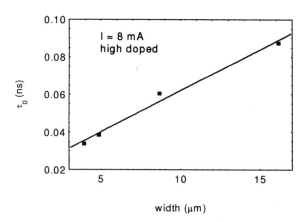

Figure 8.16. Diode time constant against width. (After Esquivias *et al.* [11].)

Here P is the total photon number and the right hand terms of this equation represent the stimulated emission rate (GP), the total photon loss rate (γP), and the spontaneous emission into the lasing mode (R_{sp}). The electrons in the excited state are given by

$$\frac{dN(t)}{dt} = \frac{I_{bias}}{q} - GP(t) - \gamma_e N(t) \tag{142}$$

where the right hand terms represent the electron injection, the loss through stimulated emission, and the total loss rate through spontaneous emission and non-radiative mechanisms. These terms are generalised expressions of phenomena already discussed earlier.

Under steady state conditions

$$\frac{dP(t)}{dt} = 0 \tag{143}$$

and

$$G - \gamma = -\frac{R_{sp}}{P(t)}. \tag{144}$$

Making the appropriate small signal assumptions and eliminating the large signal components leads to

$$\frac{d(\delta P)}{dt} = P\delta G + G\delta P - \gamma \delta P + \delta R_{sp} \tag{145}$$

$$\frac{d(\delta P)}{dt} = (G - \gamma)\delta P + \delta R_{sp} + P\delta G. \tag{146}$$

δR_{sp} depends only on N so we can write

$$\delta R_{sp} = \frac{dR_{sp}}{dN} \delta N \tag{147}$$

and G, which depends on both N and P can be written as

$$\delta G = \frac{\partial G}{\partial N} \delta N + \frac{\partial G}{\partial P} \delta P. \tag{148}$$

Collecting terms in δN and δP together and writing

$$\Gamma_P = \frac{R_{sp}}{P} - \frac{\partial G}{\partial P} P \tag{149}$$

$$\sigma_{N \to P} = \frac{\partial G}{\partial N} P + \frac{\partial R_{sp}}{\partial N}. \tag{150}$$

We have

$$\frac{d(\delta P(t))}{dt} = -\Gamma_P \delta P(t) + \sigma_{N \to P} \delta N(t). \tag{151}$$

Similarly

$$\frac{d(\delta N(t))}{dt} = -\Gamma_N \delta N(t) - \sigma_{P \to N} \delta P(t) + \frac{i_m(t)}{q} \qquad (152)$$

where i_m is the small signal current and

$$\sigma_{P \to N} = \frac{\partial G}{\partial P} P + G \qquad (153)$$

$$\Gamma_N = \gamma_e + \frac{\partial \gamma_e}{\partial N} + \frac{\partial G}{\partial N} P. \qquad (154)$$

These small signal equations form the basis of the equivalent circuit. The modulation current i_m can be represented as a current source within the circuit so clearly the equation in $d(\delta N)/dt$ is a current equation. If the output voltage from the circuit represents the light intensity then we want the equation in $d(\delta P)/dt$ to be an equation in voltage. That is,

$$v_1(t) = \frac{d(\delta P(t))}{dt} = -v_3(t) + v_2(t) \qquad (155)$$

and

$$i_1(t) = q \frac{d(\delta N(t))}{dt} = -i_2(t) - i_3(t) + i_m(t). \qquad (156)$$

Setting the modulation current i_m to zero for convenience we must have an equivalent circuit as shown in figure 8.17. This is the only circuit that will result in the coupled equations in both current and voltage. The algebraic sum of all the currents is zero and the voltages v_1 and v_3 add to v_2. The issue then is to determine the nature of the impedances Z_{1-4}.

Clearly

$$\delta N(t) = \frac{1}{q} \int i \cdot dt = \frac{Q(t)}{q} \qquad (157)$$

i.e. δN simply represents the number of excess electrons. For the photon number δP we have the simultaneous requirement that $i_3 \propto \delta P$ and $v_1 \propto d(\delta P)/dt$, which is satisfied if

$$\delta P = L \cdot i \qquad (158)$$

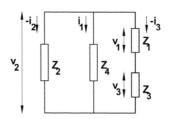

Figure 8.17. The basic equivalent circuit of a diode laser.

Circuit level modelling

which is the flux linkage through an inductor ($v = L di/dt$). Therefore Z_1 is an inductor. Substituting this in the expression for i_3 we have

$$i_3 = q \cdot \sigma_{P \to N} L \cdot i_3 \tag{159}$$

i.e.

$$L = \frac{1}{q\sigma_{P \to N}}. \tag{160}$$

We also have

$$v_3 = \Gamma_p \delta P = \Gamma_p L \cdot i_L \tag{161}$$

which can be satisfied by equating Z_3 to a resistance of value $\Gamma_p L$. Substituting for δN into v_2 and i_2 give Z_4 and Z_2 respectively as a capacitor of value $q/\sigma_{N \to P}$ and a resistor of value $1/(\Gamma_N C)$. Including the modulation current i_m as a current source of infinite impedance we have the circuit shown in figure 8.18, where

$$R_p = \Gamma_p L \tag{162}$$

and

$$R_n = \frac{1}{\Gamma_N C}. \tag{163}$$

This is an exact equivalent circuit for the single mode laser under small signal conditions. The inductor represents the optical element of the laser and the capacitor represents the carriers. It is relatively easy to imagine the frequency response of such a circuit; it is that of a classic damped harmonic oscillator slightly modified by the presence of R_n. It is also easy to imagine the response to a pulsed input, as might occur during amplitude modulation of the laser. Charge will oscillate around the circuit from the inductor to the capacitor and vice versa, the magnitude being reduced on each circuit by the power dissipation in the resistors. Multi-mode lasers can be modelled similarly by adding additional series inductor-resistor combinations in parallel.

While the above approach to modelling has the advantage that the circuit is simple once derived, its derivation from the rate equations is not straightforward. Others have taken a different approach. For example, Lu *et al.*

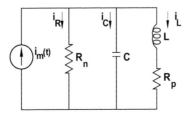

Figure 8.18. The equivalent circuit of figure 8.17 with the components identified.

[13] have modelled quantum well lasers directly from the rate equations proposed by Nagarajan *et al.* [5], making most of the active elements current sources in order to represent the rate of change of charge. Whilst it has the advantage that derivation of the circuit is easier, being linked more closely with the rate equations, it has the disadvantage that the response of the circuit to input signals is less intuitive. Rate equation modelling of this sort seems in some respects more complicated than simply solving the rate equations but it has the advantage that sophisticated software packages exist for modelling complex circuits, so if the rate equations can be expressed in a form representative of circuit elements it is a fairly straightforward matter to solve the equations. Examples can be found in the literature of a wide range of devices, including VCSELs [14].

8.7 Summary

Rate equation modelling lies at the lowest level of laser modelling as the transport of carriers is essentially one-dimensional. Carrier and photon densities are assumed to be uniform under a contact and within this approximation rate equations for the interplay between the two can be established and solved. In particular such models reveal the factors that influence the small signal modulation response, which is of great importance in communications. The more complex the laser, the more complex the rate equations, and within this chapter the evolution from the DH laser, with just two coupled equations, to the separate confinement heterostructure quantum well laser, with four coupled equations to take into account transport to the well and capture into it, has been shown. Gain compression has been treated phenomenologically in common with most of the models in the literature. Finally, circuit level modelling has been introduced as an alternative way of solving these equations, and some examples have been mentioned.

8.8 References

[1] Agrawal G P 1988 *J. Appl. Phys.* **63** 1232–1234
[2] Channin D J 1979 *J. Appl. Phys.* **50** 3858–3860
[3] Coldren L A and Corzine S W 1995 *Diode Lasers and Photonic Integrated Circuits* (New York: Wiley)
[4] Bernussi A A, Pikal J, Temkin H, Coblentz D L and Logan R A 1995 *Appl. Phys. Lett.* **66** 3606–3608
[5] Nagarajan R, Ishikawa M, Fukushima T, Geels R S and Bowers J E 1992 *IEEE J Quant. Elect.* **28** 1990–2008
[6] Schneider H and von Klitzing K 1988 *Phys Rev B* **38** 6160–6165
[7] Sze S M 1981 *Physics of Semiconductor Devices* 2nd edn (New York: Wiley) 159
[8] McDonald D and O'Dowd R F 1995 *IEEE J Quant. Elec.* **31** 1927
[9] Brum J A and Bastard G 1986 *Phys Rev B* **33**(2) 1420

[10] Weisser S, Esquivias I, Tasker P J, Ralston J D, Romero B and Rosenzweig J 1994 *IEEE Photonics Technology Letters* **6** 1421
[11] Esquivias I, Weisser S, Romero B and Ralston J D 1996 *IEEE Photonics Technology Letters* **8** 1294–1296
[12] Kibar O, Van Blerkom D, Fan C, Marchand P J and Esener S C 1998 *Applied Optics* **37** 6136–6139
[13] Lu M F, Deng J S, Juang C, Jou M J and Lee B J 1995 *IEEE J. Quant. Elect.* **31** 1418
[14] Mena P V, Morikuni J J, Kang S M, Harton A V and Wyatt K W 1999 *J. Lightwave Technol.* **17** 2612–2632

Problems

The rate equations described in this chapter generally have to be solved numerically, so for these problems some simplifying assumptions have been made in order to allow you to get a feel for the nature of the equations.

1. Consider an AlGaAs DH laser emitting at 850 nm. The active region has 3% aluminium ($n = 3.571$) and the confinement layers have 30% ($n = 3.378$). The active region is 150 nm thick. Ignoring the spontaneous recombination estimate the current and the current density required to produce 5 mW from a stripe 5 μm and 500 μm long. Assume that only one facet emits light.
2. Estimate the carrier density in the above laser by using the differential gain to express the rate of stimulated recombination in terms of the carrier density.
3. Assuming the carrier density to be pinned at transparency, or very close to it, estimate the life time of the carriers required for the rate of recombination to equal R_{stim}. Hence estimate the maximum density of spontaneous photons in the mode assuming a spontaneous emission factor of 1.2×10^{-5}, which is typical of experimentally measured values. Why is this a maximum? Estimate a realistic spontaneous photon generation rate.
4. You want to modulate a SCH quantum well laser at 2.5 GHz. You need the optically confining layer to be as thick as possible. Estimate the maximum thickness before the transport time across this layer begins to affect the modulation frequency by plotting the small signal transport factor as a function of layer thickness. Assume the ambipolar diffusion coefficient is 2.5 cm^2 s^{-1}.
5. Typical ambipolar lifetimes range from 1 ns to 10 ns, resulting in ambipolar diffusion lengths of about 500 nm to 1.7 μm. Plot equation (chapter 4, 105) assuming a current density of 400 A cm^{-2} entering the optically confining layer of a quantum well laser for both these lifetimes and calculate the thickness of the optically confining layer for 80% of the current to reach the well assuming a hole density of 10^{17} cm^{-3}. Calculate the current for a confining layer thickness of 100 nm.

Chapter 9

Lightwave technology and fibre communications

9.1 An overview of fibre communications and its history

It is difficult to say exactly when the idea of transmitting information down a silica fibre in the form of light first occurred but the first communication-grade fibre was produced by Corning in 1970, just 10 years after the invention of the laser and 8 years after the diode laser was invented. Although double heterostructure lasers had been invented they had not been developed sufficiently for commercial operation at this time. The attenuation at the HeNe wavelength of 633 nm was about 20 dB/km (the dB is defined $10 \cdot \log[P_{in}/P_{out}]$ so the signal decreases in intensity by a factor of ~100 every kilometre). Clearly at this rate of attenuation, signals could not be transmitted very far, but improvements in fibre technology rapidly led to a reduction in the attenuation and the development of CW AlGaAs lasers led to transmission at 850 nm where the attenuation was about 2 dB/km. Subsequently, lasers and fibres have both developed and transmission at both 1.3 μm and 1.55 μm for long haul communications systems rapidly developed.

It is often not immediately clear in modern texts why transmission at 850 nm was important. In fact there were three feasible wavelengths for the transmission of light; 850 nm, 1310 nm, and 1550 nm, as shown in figure 9.1 where the attenuation properties of fibres is depicted schematically, though not to scale. There are clear minima in the attenuation due to impurity absorption peaks super-imposed on to the intrinsic infra-red absorption of the glasses or the Rayleigh scattering processes. The main impurity peak is associated with the OH$^-$ ion derived from water during the manufacturing process. In many modern depictions of fibre absorption the first peak is not illustrated – quite possibly because the fibres are so pure that the absorption in this region is negligible – and the attenuation decreases smoothly down to 1310 nm. Therefore the historical reason for transmission at 850 nm is not always obvious but in the late 1970s, it was abundantly clear that the lowest attenuation achievable within the commercial laser wavelength ranges then available lay at

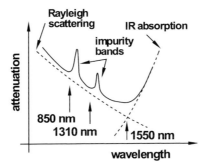

Figure 9.1. Schematic loss in silica based fibres.

850 nm. The invention of the AlGaAs DH laser made transmission at this wavelength a commercial reality, whilst the drive to exploit the lower losses at other minima pushed the development of the ternary InGaAs and quaternary InGaAsP systems.

The UK was a leader in the field at this time. The first non-experimental fibre-optic link was installed in the UK by Dorset police after lightning knocked out their communication system (*for the full history of fibre optics, see* [1,2]). However, research was being conducted around the world. Early in 1976, Masaharu Horiguchi (NTT Ibaraki Lab) and Hiroshi Osanai (Fujikura Cable) made the first fibres with low loss, 0.47 dB/km, at the longer wavelength of 1.2 μm. At the same time the lifetime of lasers was being extended at Bell Labs to about 100,000 hours (10 years) at room temperature, and experiments at the third window at 1.55 μm were being conducted. A characteristic feature of the history of fibre communications is the rapid advance within the laboratories compared with the technological implementation. To some extent this is inevitable; it is one thing to demonstrate the feasibility of a given technology within an experimental environment, it is quite another to show that the technology is reliable enough to be commercially viable. Moreover, considerable commercial inertia exists. As technology develops it is not always economically sensible to dispense with existing systems in favour of the new, and new technology is implemented as opportunities arise.

While research into the next generation of fibre communications was being conducted (as well as fibre development in Japan, InGaAsP lasers emitting continuously at 1.25 μm were also demonstrated in 1976) Bell System started sending live telephone traffic through fibres at 45 Mb/s in Chicago using 850 nm technology. Indeed, in late 1977 AT&T and other telephone companies settled on 850 nanometer gallium arsenide light sources and graded-index fibres as a standard for commercial systems operating at 45 Mb/s. Meanwhile, the low loss at long wavelengths led to renewed research interest in single-mode fibre and in August 1978 NTT transmitted 32 Mb/s through a record 53 kilometers of graded-index fibre at 1.3 μm. A consortium consisting of AT&T,

The Post Office and STL committed themselves to developing a single mode transatlantic fibre cable using the new 1.3 μm window. This was intended to be operational within 10 years (by 1988) and by the end of 1978, Bell Labs abandoned further development of new coaxial cables for submarine systems. This commitment was made public by Bell Labs in 1980 with the first transatlantic fibre-optic cable, TAT-8, designed to operate with single mode fibre at 1.3 μm. Even in 1978, however, the NTT Ibaraki lab made single-mode fibre with a record 0.2 dB/km loss at 1.55 μm (figure 9.2), which is as low as it is likely to get. The theoretical limit imposed by Rayleigh scattering is ~0.16 dB/km.

Rayleigh scattering occurs because the optical materials are not homogeneous. The fibre core is composed of a silica/germania alloy to increase the refractive index, but the germania is not uniformly distributed. Rather, the distribution is inhomogeneous and gives rise to minute variations in refractive index. Light is reflected at each interface and is scattered in all directions because of the irregular shape of the inhomogeneities (figure 9.3). Rayleigh scattering therefore imposes a limitation on the amount of germania that can be incorporated into the core. It is desirable to increase the refractive index difference between the core and the cladding of the fibre, but if the refractive index of the core is increased by adding germania the Rayleigh scattering will increase also. One way to increase the index step size is to depress the index of the cladding layer by the addition of fluorine to the glass.

Commercial second-generation systems emerged in 1981, operating at 1.3 μm. first through graded-index fibres, but very soon afterwards through single mode fibres. By 1985, single-mode fibre had spread across America to

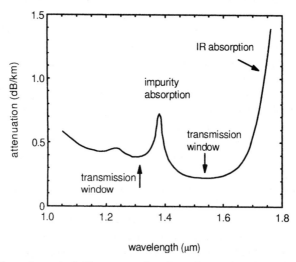

Figure 9.2. Loss in optical fibres as a function of wavelength. (After Miller and Kaminow.)

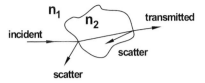

Figure 9.3. Rayleigh scattering at index fluctuations.

carry long-distance telephone signals at 400 Mb/s and above. The following year AT&T sent 1.7 Gb/s through single-mode fibres originally installed to carry 400 Mb/s. TAT-8, the first transatlantic fibre-optic cable, began service in December 1988 using 1.3 μm lasers and single-mode fibre. Long-haul systems such as these obviously could not transmit signals over the entire distance without some form of boosting and repeaters – electronic devices designed to receive signals, reshape the pulses, recover the timing, and re-launch the signals – were incorporated into the systems. It was significant, therefore that in 1987 David Payne at the University of Southampton developed the erbium-doped fibre amplifier (EDFA) operating at 1.55 μm. This device was to prove crucial in the development of wavelength division multiplexing (WDM) systems.

By the early 1990s optical fibre communications had therefore progressed through four generations; 850 nm using multi-mode fibre of core diameter ~ 50 μm, 1310 nm transmission through multi-mode fibre, 1310 nm transmission through single mode fibre of core diameter ~ 10 μm or less, and 1550 nm transmission through single mode fibre. There was much talk of the fifth generation of coherent transmission using optical heterodyning. In essence this is the optical equivalent of phase modulation. Theoretical work had shown that detection of phase modulated signals required about 500 photons whereas detection of amplitude modulated signals required 10 times as many. Coherent transmission therefore held out the promise of greater spacing between repeaters and therefore potentially large savings in system installation. The development of the EDFA changed this. EDFAs are simpler devices in which the only active element is the laser that pumps the fibre (either 1440 nm or 980 nm), whereas the repeater has very complicated receiving, pulse-processing (signal recovery and timing), and transmitting circuitry that must be powered throughout the lifetime of the system. Moreover, EDFAs exhibit gain over a relatively wide bandwidth and can be used to boost signals at several wavelengths simultaneously, and hence are suited to WDM, whereas a repeater is specific to a particular wavelength. WDM systems therefore constitute the fifth generation optical fibre communication system, with the sixth perhaps being based around soliton transmission in the future. Solitons are short single pulses that do not experience dispersion and can therefore be transmitted over phenomenal distances. However, there are no commercial soliton based

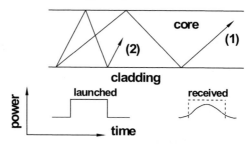

Figure 9.4. Modal dispersion blurs the temporal profile of the pulse as it propagates down the fibre.

systems, and the optical sources required for such systems will not be discussed further.

Dispersion is an important consideration in long haul systems, especially as the transmission rates increase. In amplitude modulated transmission (also known as IM/DD – intensity modulation, direct detection) where the 1's and 0's characteristic of digital signals are represented respectively by high and low intensities of light, dispersion smooths out the intensity modulations (figure 9.4) and the signal can become undetectable in the background long before attenuation will have reduced the total intensity to unmeasurably small levels. There are three principal types of dispersion:

1. *Modal dispersion.* In multi-mode fibres the optical power is partitioned among the different modes that can propagate. Higher order modes are incident on the core-cladding interface at a higher angle of incidence and therefore take a longer path length down the fibre. Clearly in figure 9.4 mode (2) takes a longer time to travel a given length down the fibre than mode (1). The pulse spreads as a result.
2. *Wave-guide dispersion* (figure 9.5). The transition to single-mode fibre does not eliminate dispersion, only the multi-mode component. The launch optical power is never truly monochromatic (this is a relative term) so there will be components at slightly different wavelengths. The longer wavelengths (labelled red) travel slower than the shorter wavelength components (labelled blue) because the angle of incidence changes to satisfy the

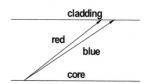

Figure 9.5. Waveguide dispersion.

An overview of fibre communications and its history

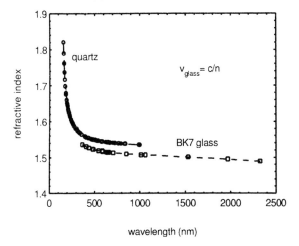

Figure 9.6. The refractive index of two different glasses; quartz and borosilicate (BK7) as a function of wavelength. The dispersion is similar but the absolute value is offset slightly.

eigenvalue equation for the mode (see appendix II). This type of dispersion is called "waveguide" dispersion because it arises from the very fact of waveguiding.
3. *Material dispersion.* The refractive index of a material is a measure of the velocity of the light within the medium; the larger the refractive index the lower the velocity. In glasses the principal electronic resonances (see chapter 4) lie well into the UV so the visible and near infra red wavelengths correspond to normal dispersion and the refractive index idecreases with increasing wavelength (figure 9.7).

The wave-guide dispersion and material dispersion cancel each other out at 1310 nm in the conventional step index fibre. By happy coincidence this is also one of the loss minima, so transmission at this wavelength has the double advantage of being dispersion free and low loss. Although fibres exhibit lower loss at 1550 nm the non-zero dispersion can negate this advantage, and in order to transmit at this wavelength the refractive index profile must be changed. This is relatively easy to do. Optical fibres are produced by depositing doped layers by CVD onto the inside of a fused silica preform, so by adding different dopants to each layer the desired refractive index profile can be generated once the preform is heated and drawn into a fibre. Figure 9.7 shows three common designs, though others are possible.

The simplest profile is the matched cladding step-index; so-called because the refractive index of the first layer is matched to the preform and the refractive index change is abrupt. The refractive index of the cladding can be lowered by the addition of fluorine so that a smaller fibre core can be used. It

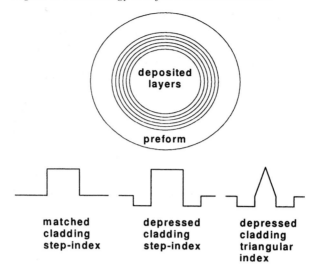

Figure 9.7. The formation of different refractive index profiles by depositing layers on the inside of a quartz preform.

is possible to achieve zero dispersion at 1550 nm using a step index but the core diameter has to be of the order of 4–5 μm, which does not allow for much alignment error when splicing or connecting cables. A triangular core, generated by grading the layers, can shift the zero-dispersion point to longer wavelengths with a core diameter of 6–7 μm and a fractional change in refractive index from the apex of the triangle to the cladding of between 0.8 and 1%. In general, Maxwell's equations have to be solved numerically to determine the propagation properties of a particular refractive index profile. It is apparent that the triangular core forms a self-focusing guide and has a relatively small mode-field diameter. For a detailed discussion on the above, see [3]. The use of a small core diameter (< 10 μm) allows for greater tolerance to bending losses.

Operating at the dispersion minimum therefore allows for the highest bit-rate-distance product. The theoretical transmission distance for single mode fibre with 1mW of launched power at an attenuation of 0.2 dB/km is illustrated in figure 9.8 [4]. The loss-limit is clearly identified, but dispersion imposes its own limits depending on the spectral width of the laser. The case of $\Delta\lambda = 0$ corresponds to the case where the bandwidth of the signal is greater than the spectral width of the source. A spectral width of 0.1 nm corresponds to 12.5 GHz, which is greater than the transmission rate. At 2 Gb/s, for example, loss limited transmission over 200 km is possible, but with a source spectral width of 0.1 nm this is reduced to just over 100 km, and for 0.2 nm width, the maximum distance becomes ~ 60 km.

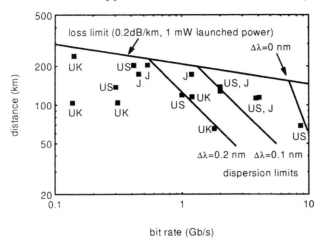

Figure 9.8. Theoretical limits and experimental transmission system performance in the UK, US, and Japan upto 1987. (After Li.)

The current standard for fibre communications is a transmission rate of 10 Gb/s with the promise of 40 Gb/s in the near future. Applications for such technology include long haul communications systems and local area networks (LAN's) but the primary interest in this chapter is the laser technology used in long-haul systems. These are inter-continental and inter-state trunk routes, and in countries like Britain, which are relatively small compared with the USA or even some European countries, the systems that link the major cities. Long haul optical communications systems run under the SONET/SDH (synchronous optical network/synchronous digital hierarchy) framework but local area networks (LANs) run under the ethernet. The SONET standard came about because early systems used proprietry formats. Synchronous means in this context that all timings are taken from a master atomic clock for the whole network. The basic frequency is 51.84 Mb/s and transmission rates are multiples of this frequency. A data rate is referred to as OC-n, where $n = 1, 2, 3, 4$, etc. OC stands for Optical Carrier, and n is the carrier number. Thus OC-1 is at 51.84 Mb/s, OC-3 is at 155 Mb/s (3×51.84) and 10 Gb/s corresponds to OC-192.

The ethernet is the system used to link PC's to the internet and is the standard format for local area networks. Standard Ethernet data rates are 10 Mb/s, 100 Mb/s, and 1000 Mb/s, but ethernet signals are not synchronised. Two or more computers on a line can start to send signals at the same time, so ethernet systems detect collisions between data rates and send instructions to both the sources to wait for a random time. Not all ethernet traffic is carried over optical fibre. Twisted copper pairs are still used, but there is increasing demand for high speed ethernet connections, especially in the so-called "first mile". This is the connection between the subscriber and the networks and can run under Mb/s or even kb/s rates. With the demand for high bandwidth connections to the home for video data and such, the ethernet will itself have to provide broadband communications, and so will the metropolitan area networks (MAN). Consequently much of the laser technology described in this chapter which is currently used in long haul systems will eventually find itself employed in LANs and MANs.

It is clear, therefore, that there are several important requirements of laser sources:

1. *Wavelength*: 850 nm, 1310 nm, or 1550 nm are the standards, but for WDM systems other wavelengths around these will be important. The ITU (Internation Telecommunications Union) has specified six channels for WDM; O-band (1260 nm–1310 nm), E-band (1360 nm–1460 nm), S-band (1460 nm–1530 nm), C-band (1530 nm–1565 nm), L-band (1565 nm–1625 nm), and the U-band (1635 nm–1675 nm).
2. *Spectral width*: laser sources have to be single mode and stable against the effects of high speed modulation. Specifically, chirp is caused by variations in the output wavelength caused by carrier induced changes in the refractive index during a pulse, and relaxation oscillations can lead to multiple modes being transmitted.
3. *Modulation frequency*: OC-192 comprises four channels of a WDM system, each channel of which operates at OC-48 (2.5 GHz). The lasers have to be modulated at this rate in order to transmit the signals [5].
4. *Noise*: laser relative intensity noise (RIN) arises from fluctuations within the laser and is crucial to system performance. Amplitude modulated (AM) systems are more demanding in this respect than digital systems, and for AM OC-48 RIN figures of –150 dB/Hz are necessary [6].

9.2 Materials and laser structures

Historically, transmission at 850 nm has employed AlGaAs DH lasers, and the properties of these devices has been described extensively by Thompson [7]. AlGaAs lasers emitting at 850 nm are still used in some communications applications, particularly where the distances are short, such as in LANs. For long-haul systems, lasers at 1.31 μm and 1.55 μm are used.

The first generation of optical fibre systems used broad area Fabry–Perot laser structures coupled to multi-mode fibres. However, these exhibit complex spatio-temporal dynamic behaviour characterised by filamentation of the near-field output (see chapter 5).The principle disadvantage that this causes for fibre transmission systems, apart from the difficulty in coupling the light to the fibre, is the instability of the brightness-current (L-I) characteristic as optical power is transferred from one lateral mode to another. Although the filamentary behaviour itself may be quite stable at a given current (equally, there are circumstances where it is not) the lasers have to be modulated, which means

Materials and laser structures 269

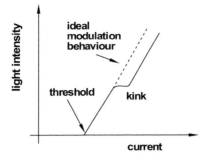

Figure 9.9. Schematic kink instability in the L-I characteristic.

changing the current. As the current is changed and new lateral modes are illuminated there is a transfer of optical and electrical power to the new mode, which results in a so-called "kink" in the L-I curve (figure 9.9). The ideal characteristic is a linear change in intensity with current and the kink causes difficulty in the modulation of the laser.

Stabilisation of the laser output was achieved by means of the stripe geometry. Originally the stripe geometry was used to limit the active area and improve the thermal conduction away from the active region, but it has the advantage that it also results in a stable lateral single mode for stripe widths of the order of 10 μm. Although the use of the stripe gives rise to stable L-I characteristics, modulation of stripe lasers at high frequency is not without problems. Figure 9.10 shows the response of a stripe laser (AlGaAs DH) to a current pulse [7]. The DC current is set 6 mA below threshold so in the two cases here the current is pulsed 2 mA and 10 mA above threshold. Several things stand out:

1. There is a delay of ~1 ns between switch-on and the maximum current. This is due to the RC time-constant of the device, as charge has to be injected into the active area.
2. There is a delay between switch-on and the onset of light emission. The delay decreases with power input. Thompson has shown a series of measurements at current intervals of 2 mA above threshold between the two cases shown here.
3. The period of the oscillation decreases with power input.
4. The maximum light output is between 2 and 3 times steady state.
5. The envelope of the decay is approximately exponential.

Clearly such a laser cannot be modulated at high frequency. For pulses less than 3 ns duration there will be no light emission at all, and for pulses of <10 ns duration the average light output will depend on the pulse duration. If the laser were to be modulated with a sinusoidal signal, the frequency response would be similar to a classic damped harmonic oscillator. The maximum output will occur at the resonant frequency, which is the same as the relaxation oscillation

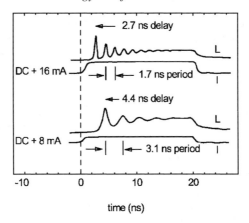

Figure 9.10. Relaxation oscillations in a 20 μm wide stripe laser of cavity length 500 μm. (After Thompson.)

frequency. The output spectrum is also very much broader than in steady state, as described in chapter 3. All lasers tend to exhibit relaxation oscillations – they are a fundamental feature of the operation and arise from the complex interplay between carrier and photon densities at start-up – but not all lasers suffer from the spectral instabilities associated with Fabry–Perot cavities in DH stripe lasers. Changing the optical feedback mechanism can help to overcome these difficulties.

Apart from the increased spectral content due to the relaxation oscillations, the output can also be chirped. The refractive index varies very slightly with the carrier density (chapter 4), but the carrier density does not follow the voltage pulse instantaneously and rises instead according to the RC time-constant. The basic F-P condition of fitting a whole number of half wavelengths into the cavity can be manipulated to show that

$$\Delta\lambda = \lambda \frac{\Delta n}{n} - \frac{\lambda^2}{2.n.L} \qquad (1)$$

where n is the refractive index, λ is the wavelength, and L is the cavity length. As an example, consider a laser with a cavity length of 500 μ emitting at 850 nm and with a refractive index of 3.50. The mode spacing $\Delta\lambda$ is the absence of any change in refractive index, Δn, is about 0.2 nm, rising to 0.44 nm for a change in refractive index of 0.001. More likely, the change will be of the order of 0.01, so the mode spacing will be even larger. This change in mode spacing is occurring over the duration of the change in carrier density, and so is significant for short pulses. In essence the output wavelengths change during the pulse.

For emission at 1.3 μm and 1.5 μm the quaternary InGaAsP has emerged as the material of choice [8]. To recap, the band gap and the lattice constant can be varied independently, so it is possible to obtain a range of compositions $Ga_xIn_{1-x}As_yP_{1-y}$, where $y \approx 0.22x$, all lattice matched to the InP substrate, corresponding wavelengths ranging from 0.92 μm to 1.65 μm, given by the band gap

$$E_g = 1.35 - 0.72y + 0.12y^2. \qquad (2)$$

In InGaAsP emitting at 1.55 μm the split-off valence band and the conduction band are at approximately the same energy from the valence band maximum in the Γ valley, so Auger losses are large.

Stripe geometries have been employed to control the lateral modes, but index-guided structures are preferred. A variety of designs have been put forward for index-guided heterostructure lasers, but all essentially involve interrupting the growth of the laser structure at some point, etching away well defined regions, and then regrowing epitaxial layers onto the etched regions. The regrown layers can either constitute the index guiding layers or the active region of the laser itself. The double channel planar buried heterostructure laser is an example of the first type of structure, the basic sequence of steps being illustrated in figure 9.11. The laser structure, including the active region, is grown first and two channels are etched into the substrate to isolate a small region of active material. Overgrowth of first p-InP and then n-InP into these channels followed again by p-InP provides not only an index step around the active region but also a reverse biased p–n junction which blocks the current. The channelled substrate buried heterostructure is an example of the type of laser where the active region is grown later onto an etched substrate (figure 9.12). This is not always a re-growth technique as p–n junction substrates of the type illustrated can be purchased. A stripe contact helps to define the current, though in the form shown a reverse biased p–n junction still exists outside the channel and limits the current.

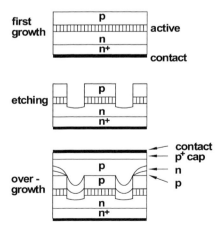

Figure 9.11. Double channel planar buried heterostructure laser.

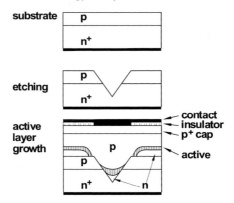

Figure 9.12. V-groove laser, also called channelled substrate buried heterostructure with oxide- or nitride-defined stripe contact.

These are not the only designs possible, however. Ridge waveguides and inverted rib structures have been described by Miller and Kaminow, but the two shown here have the advantages that both the index guiding and the current confinement are strong, so for typical communications lasers the threshold currents are of the order of 15–20 mA with single stable mode output. Dixon and Dutta [9] also describe these two basic structures as being among the favoured sources offered by AT&T in the mid to late 1980s, though the DCPBH laser incorporated a grating for wavelength selective feedback. The active layer cross-section is typically 1.5 μm × 0.2 μm so butt-coupling to single mode fibre is relatively straightforward with an acceptable margin for alignment error.

9.3 Laser performance

9.3.1 Mode selectivity

Although these buried heterostructure devices exhibit stable single modes, the output of the device is still subject to the effects due to the Fabry–Perot cavity, such as multimode emission and chirp. Amann [10] has considered the side-mode suppression ratio (SSR), defined as the the ratio of the number of photons within the principal mode, S_0, to the number of photons in the next strongest mode, S_1. For equal reflectivities of the facets, r_1 and r_2, the SSR is given by

$$SSR = \frac{2P\delta g}{h\nu v_g n_{sp} \alpha_{tot} \alpha_m} \qquad (3)$$

where P is the output power per mirror in the principal mode, δg is the modal gain difference between the two strongest modes, v_g is the group velocity, $h\nu$

is the photon energy, n_{sp} is a spontaneous emission factor ~ 2, α_{tot} is the total loss, including the mirror loss, α_m. The mirror loss is an effective loss expressed in cm^{-1} and is evaluated from the round trip loss according to equation (chapter 3, 43). Consider, for example, a cavity 500 μm long in which the reflectivity at each facet is typically $r = 0.35$. Assuming no loss within the cavity, the reduction in intensity due to the facet reflectivity alone after 1 round trip is $R^2 = 0.123$, giving $\alpha_m = 21$ cm^{-1}.

The modal gain is given by the product of the optical confinement, Γ, and the active region gain, g_a, and the gain difference can be found by expanding the active region gain around the maximum using the Taylor expansion,

$$g_a(\lambda) = g_a(\lambda_0) + \delta\lambda \frac{dg}{d\lambda} + \frac{1}{2}(\delta\lambda)^2 \frac{d^2g}{d\lambda^2}. \qquad (4)$$

Recognising that $dg/d\lambda = 0$ as the gain is a maximum, and putting

$$\frac{1}{2}\frac{d^2g}{d\lambda^2} = -b_2 \qquad (5)$$

then

$$\delta g = \Gamma b_2 \Delta\lambda^2. \qquad (6)$$

Using typical values for 1.55 μm lasers, $b_2 = 0.15$ cm^{-1} nm^{-2}, $\Delta\lambda = 0.6$ nm, $\Gamma = 0.2$, $R_1 = R_2 = 0.35$, the SSR is 17 dB [10], and is commonly no more than 20 dB in F-P lasers. High performance data transmission systems require SSR's of 30 dB or more, so more complex laser structures are required.

9.3.2 Modulation response

The response of a diode laser to a small sinusoidal signal of frequency ω can be determined by small signal analysis (chapter 8), which generally yields an expression of the form

$$\frac{s(\omega)}{i(\omega)} = \frac{s(0)}{i(0)} \frac{\omega_0^2}{\omega_0^2 - \omega^2 + j\omega\omega_d} \qquad (7)$$

where s and I are the small signal photon density and current density respectively, and ω_0 is a resonant frequency given by

$$\omega_0 \approx \sqrt{\frac{v_g g_c S_0}{\tau_p}}. \qquad (8)$$

Here g_c is the small signal (differential) compressed gain,

$$g_c = \frac{g_0}{1 + \varepsilon S_0}. \qquad (9)$$

This behaviour is illustrated in figure 9.13.

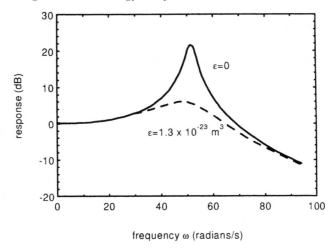

Figure 9.13. Small signal response function for a diode laser. (After Miller and Kaminow.)

The frequency ω_d is given by

$$\omega_d \approx \frac{\varepsilon S_0}{\tau_p}. \tag{10}$$

The response function corresponds to that of a damped, harmonic oscillator (chapter 3) with ε, the phenomenological gain compression factor, acting as the damping term. At resonance $\omega = \omega_0$ and the amplitude of the oscillation reaches a maximum. In an ideal oscillator without damping the amplitude would become infinite, but in reality no oscillator is ideal. There is always some damping present, but even so the amplitude can be so large that in a mechanical system the oscillator can literally shake itself to bits. In a laser the amplitude is equivalent to the photon density, which can only continue to increase if the gain does not saturate. In the phenomenological gain compression terms commonly employed in diode laser modelling, i.e. $g \propto (1 - \varepsilon S)$ or $g \propto (1 + \varepsilon S)^{-1}$, the gain becomes zero at high photon densities, and the stronger the gain compression represented by a higher value of ε, the lower the magnitude of the photon density at which zero gain occurs. This means, of course, that there can be no increase in photon density beyond this value, so the greater the gain compression the lower the magnitude of the photon density at resonance.

The peak frequency occurs at

$$\omega_p^2 = \omega_0^2 - \frac{\omega_d^2}{4} \tag{11}$$

Laser performance 275

so in order to increase the resonant frequency it is necessary to:
- increase the differential gain;
- increase the photon density via optical confinement and a higher output power;
- decrease the photon lifetime by shortening the cavity.

Experimentally, a dependence of the resonant frequency on the square root of the output power has been observed, and this tallies with the relaxation oscillations shown in figure 9.10.

The fundamental laser parameters are but one factor in the frequency response. These set only the fundamental limitations, but other influences may limit the response in practise. Packaging, for example, which is associated with stray capacitances and inductances can contribute to the delay between current and voltage waveforms, and broaden the electrical pulse.

9.3.3 Gain switching

The small signal transfer characteristic of figure 9.13 does not directly represent the performance of a laser in practical communications systems where the laser is modulated instead by a large square pulse. Were the resonant frequency ω_0 to be independent of the current it would be possible to evaluate the response to a step function simply by taking the Fourier transform of the pulse, but the fact that the resonance frequency changes with current means that a different approach is needed. The rate equations can provide an insight into the phenomenon of gain-switching, as the change from a sub-threshold condition to a super-threshold condition is known.

The time interval of interest is the delay between the application of the pulse and the switching on of the laser as the photon density reaches a threshold value and seriously begins to deplete the carrier density. Let this time be known as t_{on}. Neglecting any recombination, whether non-radiative or leading to spontaneous emission, the rate of change of the carrier density in the cavity is given by the current, i.e.

$$\frac{dN}{dt} = \frac{I(t)}{qV} \tag{12}$$

where V is the volume. Straight forward integration yields the solution

$$N(t) = N_0 + \frac{1}{qV} \int_0^t I(t) \, dt \tag{13}$$

i.e.

$$N(t) = N_0 + \frac{Q(t)}{qV} \tag{14}$$

as the integral of the current is just the total charge. Strictly, this only applies for $0 < t < t_{on}$, since for $t > t_{on}$ the photon density affects the carrier density.

Under these conditions, then, the rate equation for the photon density can be rearranged to give

$$\frac{dS}{S} = \left[a\Gamma(N(t) - N_{th}) - \frac{1}{\tau_p} \right] . dt \qquad (15)$$

where a is the differential gain and N_{th} is the threshold carrier density. Recognising the integral of $1/S$ as $\ln(S)$ gives the solution

$$S = S_0 e^{t/\tau_r} \qquad (16)$$

for $0 < t < t_{on}$, where S_0 is the photon density at $t \leq 0$, and

$$\frac{1}{\tau_r} = a\Gamma \left(\frac{Q_t}{qV} + N_0 - N_{th} \right) - \frac{1}{\tau_p}. \qquad (17)$$

Equation (15) has been used to relate the charge to the carrier density.

For a laser biased at threshold the steady state carrier density at $t = 0$, N_0, is found by setting $dS/dt = 0$ so that

$$N_0 = N_t + \frac{1}{a\Gamma\tau_p} \qquad (18)$$

and

$$\tau_r = \frac{qV}{a\Gamma Q_t}. \qquad (19)$$

In other words, equations (16) and (19) show that in response to a square current pulse the photon density rises exponentially with a characteristic time delay τ_r dependent inversely on the amount of charge injected. For high injection levels, then, the time delay decreases, as expected. If the pulse is short so that a definite, but small amount of charge is injected into the laser then after the electrical pulse has ended the photon density will also decrease exponentially, as

$$\frac{dS}{dt} = -\frac{S}{\tau_p}. \qquad (20)$$

Strictly, the lifetime in the equation above is not the lifetime characteristic of steady state behaviour, which is determined by the optical losses in the cavity, but has to be modified by the fact that the carrier density is also driven below threshold at the same time, so increasing the losses through absorption. Single optical pulses of half-width a few pico-seconds, and total duration ~60 pico-seconds are possible [8, p 547]. If a communications system were to be operated in this mode this would set an upper frequency limit of ~15–20 GHz.

9.3.4 Linewidth

As well as the multimodal nature of the output during the relaxation oscillations, the carrier density is fluctuating and from equation (1) this can be expected to affect the output wavelength. Mathematically, chirp can be treated by expanding the gain as a function of both carrier density and photon density. The first step is to express the Fabry–Perot condition in terms of frequency ν.

$$\nu \bar{n} L_c = \frac{mc}{2} \quad (21)$$

where $\bar{n}$ is the average refractive index in the cavity. The average is needed because some of the light propagates within the cladding. Small changes in $\bar{n}$ bring about small changes in ν, but the F-P condition must still be met

$$[\nu\bar{n} + \Delta(\nu\bar{n})]L_c = \frac{mc}{2}. \quad (22)$$

Rearranging, and using equation (21)

$$\Delta(\nu\bar{n})L_c = \frac{mc}{2} - \nu\bar{n}L_c = 0. \quad (23)$$

Expanding the change gives

$$\Delta(\nu\bar{n}) = \nu\Delta\bar{n} + \bar{n}\Delta\nu \quad (24)$$

where there are, additionally, two terms to consider in the refractive index

$$\Delta\bar{n} = \frac{d\bar{n}}{d\nu}\Delta\nu + \frac{d\bar{n}}{dN}\Delta N. \quad (25)$$

The first term is just the rate of change of refractive index with frequency and is therefore the normal material dispersion. The second term is the change in real refractive index caused by a small change in carrier density, which is the effect of interest here. Hence,

$$\Delta(\nu\bar{n}) = \Delta\nu\left[\bar{n} + \nu\frac{d\bar{n}}{d\nu}\right] + \nu\frac{d\bar{n}}{dN}\Delta N \quad (26)$$

where the term in brackets is just the group refractive index n_g. Setting the left hand side to zero (equation 23) and solving for ν gives,

$$\Delta\nu = -\frac{\nu}{n_g}\frac{d\bar{n}}{dN}\Delta N = -\frac{v_g}{\lambda}\frac{d\bar{n}}{dN}\Delta N \quad (27)$$

where v_g is the group velocity, c/n_g. At this point it is necessary to clarify the nature of the refractive index. It is not simply the material refractive index, but the effective index we want in order to account for that part of the optical field that propagates in the cladding layers. The effective index is an approximation arrived at by matching the optical fields at the waveguide boundaries. It can be shown [11] that the *change* in the effective propagation index can be related to the change in the material index by weighting according to the optical confinement,

$$\Delta \bar{n} = \Gamma \Delta n \tag{28}$$

and therefore

$$\Delta \nu = -\Gamma \frac{v_g}{\lambda} \frac{dn}{dN} \Delta N. \tag{29}$$

In order to proceed further with this it is necessary to recognise that a change in the real part of the refractive index due to a change in carrier density is accompanied by a corresponding change in the imaginary part, and this is expressed by the phase-amplitude coupling coefficient, described in more detail in chapter 10 but here simply defined as

$$\alpha = -\frac{4\pi}{a\lambda} \frac{dn}{dN} \tag{30}$$

where a is the differential gain. Hence,

$$\Delta \nu = \frac{\alpha}{4\pi} \Gamma a v_g \Delta N. \tag{31}$$

Part of this expression is recognisable from the photon rate equation, which can be written as

$$\frac{dS}{dt} = \Gamma v_g S g + \Gamma \beta_{sp} R_{sp} - \frac{S}{\tau_p} \tag{32}$$

i.e.

$$\frac{1}{S} \left[\frac{dS}{dt} - \Gamma \beta_{sp} R_{sp} \right] = \Gamma v_g g - \frac{1}{\tau_p}. \tag{33}$$

Expanding the gain about the threshold value, i.e.

$$g = g_{th} + a\Delta N \tag{34}$$

and making use of the fact that

$$\frac{1}{\tau_p} = \Gamma v_g g_{th} \tag{35}$$

leads to the simple result that

$$\Delta \nu = \frac{\alpha}{4\pi} \frac{1}{S} \left[\frac{dS}{dt} + \Gamma \beta \frac{N}{\tau_{sp}} \right]. \tag{36}$$

Clearly the output power P is a linear function of the photon density. Making this assumption and substituting for the photon density in (36), and neglecting the spontaneous recombination term, gives

$$\Delta \nu = \frac{\alpha}{4\pi} \frac{dP/dt}{P}. \qquad (37)$$

There is nothing in equation (37) that depends on the design of laser. Rather, it depends on the material property α and the voltage waveform via dP/dt. In some texts $\Delta \nu$ is written as negative and in others it is positive, as above. It all depends on whether α, the phase-amplitude coupling coefficient, is defined as positive or negative, and this seems to vary from text to text. The linewidth thus defined is the *transient* linewidth. At a steady output power $dP/dt = 0$, so this mechanism is important at the beginning and end of a pulse, and especially so during the relaxation oscillations. For short pulses, of course, the transient linewidth is an important factor throughout the duration of the pulse.

Equation (37) suggests that the frequency change during a long pulse in which steady state is achieved should be zero, but in fact the expression is not complete. Chirp can be understood as arising from the change in refractive index as a function of the carrier density which, from figure 9.10, is oscillatory. However, the oscillatory nature of both the carrier and photon densities is affected by the gain compression factor, which, it has been argued, acts as the damping term. Although it is not obvious from anything that has so far been developed either in the last chapter or here, we can expect, by analogy with critically damped and overdamped systems, that a large gain compression factor will damp out the oscillations in photon density and the carrier density will rise smoothly to the steady state limit. The gain compression factor is structure dependent. It arises essentially because the current supplies carriers at a definite rate, and ultimately the rate of increase in the photon density cannot exceed this. At a large enough photon density the rate of increase is therefore fixed and the gain, ie. the multiplication rate, decreases. The rate at which the gain decreases must depend on the structure through the current, and so equation (37), which contains no structure dependent terms, must be incomplete. The gain has been expanded in the carrier density only (equation 34), but a term must also be included to account for the dependence of the gain on the photon density. An additional term therefore appears in the linewidth that depends on the output power [12].

Expanding the gain as follows,

$$g = g_{th} + \frac{dg}{dN} \Delta N - \frac{dg}{dS} \Delta S \qquad (38)$$

where the negative sign in the third term expresses the physical reality that the gain decreases with increasing photon density, and substituting S for ΔS in the case of a gain switched laser since the initial photon density is so low as to be negligible, leads to

$$\Delta v = \frac{\alpha}{4\pi}\left[\frac{dP/dt}{P} + \gamma S\right] = \frac{\alpha}{4\pi}\left[\frac{d[\ln(P)]}{dt} + \gamma S\right] \quad (39)$$

where γ is a constant. This additional term is decribed by Koch and Linke as an adiabatic term [13] because it doesn't involve variations in the power.

At first sight equation (39) appears to show two additive contributions to the linewidth, but we would expect from the previous discussion that the two terms on the right hand side act in opposition as a large oscillatory term in dP/dt must be accompanied by a small term in γ via the gain compression factor ε. Of the various forms of gain compression in the literature, Coldren and Corzine [11, p 196] choose a form that gives

$$-\frac{\partial g}{\partial S} = \frac{\varepsilon g}{(1 + \varepsilon S)} \quad (40)$$

which makes it clear that the gain compression is responsible for the adiabatic frequency change. A laser that has a high gain compression will exhibit high damping and hence a small contribution from dP/dt, but the adiabatic frequency shift may be large. A laser exhibiting low damping will experience high amplitude relaxation oscillations but low adiabatic frequency shift, as shown in figure 9.17. The frequency shift due to the relaxation oscillations is both positive and negative according to dP/dt, and settles at a value close to

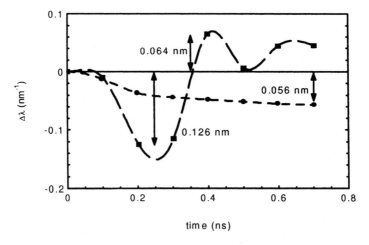

Figure 9.14. Change in wavelength for ridge waveguide (■) and DCPBH lasers (●), both with wavelength selective distributed feedback. The dashed lines are splines fitted to the data. (After Koch and Linke [13].)

zero. The adiabatic shift is very much constant with time. Lasers with high damping, and consequently low but essentially constant frequency shift, are best suited to high transmission rates over long distances, as oscillatory frequency shifts result in a number of different wavelengths being launched into the fibre.

9.4 Single wavelength sources

As the above shows, single wavelength sources stable against both chirp and the effects of relaxation oscillations are very important for long-haul systems. A number of laser structures have been investigated for single wavelength sources, including external cavity devices, coupled cavity devices, and long cavity devices. External cavity devices do not use the facets of the crystal but some other external mechanism of feedback. Coupled cavity devices require the creation of two cavities of slightly different optical length in tandem so that the mode of the laser that propagates must be a common mode of both cavities. This seriously limits the available wavelengths, but both this type of laser and those using an external cavity are difficult to realise commercially because of the potential lack of mechanical stability. Long cavity lasers tend, by nature, to have poor modulation characteristics compared with short cavity lasers, though the linewidth can be very narrow. Such lasers would make good sources for externally modulated signals but at present amplitude modulation (IM/DD) coupled with WDM is providing all the bandwidth needed. By the late 1980s and early 1990s the industry had settled on grating feedback as a means of providing single frequency sources [9]. Indeed, the lasers illustrated in figure 9.14 employed grating feedback.

There are two basic designs of grating feedback laser; the distributed feedback laser (DFB) and the distributed Bragg reflector (DBR). The essential difference between the two is that the DFB laser is pumped over the grating region whilst the DBR laser is pumped over a limited volume with the gratings lying outside.

Historically the DFB laser was developed for reliable single mode operation first, and even today is preferred over the DBR laser for this purpose. The DBR laser can be treated as a Fabry–Perot cavity with a wavelength selective feedback. There are therefore two independent mechanisms at work to select the wavelength; the normal F-P modes, and the Bragg wavelength. These two are not guranteed to coincide, which can degrade the single mode performance of the laser. For this reason the DFB laser is preferred but the DBR structure forms the basis of tunable sources for WDM applications, and is therefore very important. Normally lasers of both types are formed by holographic or e-beam definition of the grating structure in the substrate followed by overgrowth of a GaInAsP waveguide layer and then the GaInAsP active layer of a different composition.

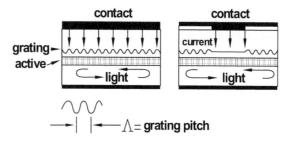

Figure 9.15. Basic DFB (left) and DBR (right) laser structures.

The mathematical treatment of these structures usually involves coupled mode theory, as developed by Kogelnik and Shank [14]. This describes the counter propagating electromagnetic waves within the structure in terms of the refractive index variations, and allows for gain in the grating section. The allowed wavelengths, the modal thresholds, and the side-mode suppression ratios can all be defined using couple mode theory, but the treatment will not be presented here. As with many mathematical treatments, it is easy to lose sight of the important physics of the problem, and though the mathematical treatment is necessary to calculate the precise properties of given structures it is not necessary to develop the theory in order to understand the essentials. Coupled-mode theory is well described by Amann [10] and by Coldren and Corzine [11]. The essential results only will be presented here.

Brillouin [15] has considered the propagation of waves in a continuum with a periodic structure using a perturbation approach. The medium of the DBR is just such a continuum; it is an otherwise uniform structure in which a small periodic fluctuation is introduced. Brillouin has shown that whether a wave propagates or not within such a structure is governed by a condition identical to Bragg reflection, i.e.

$$m\lambda = 2nd \sin \theta \tag{41}$$

where m is an integer, n is the refractive index, and d is the spacing between diffracting planes. For an angle of incidence of 90°, as occurs for the grating lying in the plane of propagation, reflection occurs at a wavelength

$$\lambda = 2n\Lambda \tag{42}$$

where Λ is now the grating pitch rather than the plane spacing.

The DBR laser is physically different from the DFB laser. Within the DBR structure light is generated in a separate gain region and is incident on the DBR, which acts as a reflector with wavelength selective feedback into the active region. This feedback is a maximum at the Bragg wavelength. Within the DFB structure the grating is continuous so the light is generated, propagated, and reflected entirely within a periodic structure, and hence the theory of Brillouin, which deals specifically with such concepts, applies. The

Single wavelength sources

essential idea is that physically a wave at the Bragg wavelength cannot ordinarily propagate in a grating structure because the scattering (diffraction) is symmetrical in both directions and the wave is consequently stationary. It is difficult to quantify these ideas using the concept of diffraction, but the first order perturbation of the wave equation leads directly to the solution

$$\omega^2 = \omega_0^2 \pm 2\omega_0 \Delta\omega \tag{43}$$

with ω_0 equal to the Bragg frequency. The range of frequencies $2\Delta\omega$ is known as the stop band because frequencies within $\Delta\omega$ of ω_0 cannot propagate (figure 9.16).

This result applies strictly to non-lossy propagation – propagation where there is neither loss nor gain – which is not always the case in a laser. Kogelnik [16] has developed the essential equations on which much of the analysis of grating feedback lasers is based. An electromagnetic wave propagating in a medium of relative permittivity ε_r (not to be confused with the gain compression factor) and permeability μ satisfies the scalar wave equation

$$\nabla^2 \xi + k^2 \xi = 0 \tag{44}$$

where ξ is the electric field and

$$k^2 = \frac{\omega^2}{c^2}\varepsilon_r - j\omega\mu\sigma \tag{45}$$

is the propagation constant. The appearance of the conductivity σ in this equation accounts for the loss by absorption, either through the free carriers or through the action of polarised lattice charges. In a semiconductor at photon energies close to the band-edge the lattice charges play no part and only the carriers are important. If these two are modulated such that

$$\varepsilon_r = \varepsilon_{r0} + \varepsilon_{r1} \cos [K \cdot x]$$
$$\sigma = \sigma_0 + \sigma_1 \cos [K \cdot x] \tag{46}$$

where the grating vector is

$$K = \frac{2\pi}{\Lambda} \tag{47}$$

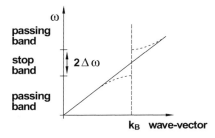

Figure 9.16. The stopping band for Bragg reflection in a periodic structure.

and Λ is the grating period. ε_{r0} is the average dielectric constant and likewise σ_0 is the average conductivity. It is straightforward to show that

$$k^2 = \beta^2 - 2j\alpha\beta + 2\kappa\beta \cos[K \cdot x] \tag{48}$$

with

$$\beta = \frac{2\pi\sqrt{\varepsilon_{r0}}}{\lambda}$$

$$\alpha = \frac{\mu c \sigma_0}{2\sqrt{\varepsilon_{r0}}} \tag{49}$$

and

$$\kappa = \frac{\pi n_1}{\lambda} - j\frac{\alpha_1}{2} \tag{50}$$

where n_1 and α_1 are respectively the refractive index and loss modulation. Although equations (45) and (46) are expressed in terms of the dielectric constant and the conductivity it is normal in semiconductors to talk about the refractive index and the loss, so κ, the coupling constant, is written in these terms. It is a complex number, so modulation of either the refractive index or the loss is sufficient to couple the forward and backward propagating waves to each other so that energy is transferred from one wave to the other. β is what we would normally understand to be the propagation constant in a uniform medium of refractive index $n_0 = \sqrt{\varepsilon_{r0}}$. It is then possible to derive two coupled equations for the wave amplitudes propagating to the left and right

$$-\frac{dA_R}{dz} + (\alpha - j\delta)A_R = j\kappa A_L$$

$$\frac{dA_L}{dz} + (\alpha - j\delta)A_L = j\kappa A_R \tag{51}$$

with

$$\delta = \frac{\beta^2 - \beta_0^2}{2\beta} \approx \frac{n(\omega - \omega_0)}{c} \tag{52}$$

where ω_0 is the Bragg frequency. Equation (52) brings out more clearly the nature of the right and left propagating electromagnetic waves (denoted by the subscripts R and L) coupled via the coupling constant κ.

Kogelnik considered two special cases; index coupling where κ is real, and gain coupling where κ is imaginary. For index coupling stop bands exist as shown in figure 9.16, with the width of the band given by 2κ. For gain coupling, there is no stop band in frequency, but instead forbidden values of the propagation constant exist (figure 9.17), the width of the band being α_1. The notion of gain coupling is the essential difference between Kogelnik and Brillouin. These two extremes of pure index or pure gain coupling lead to an

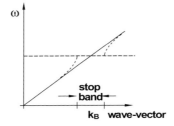

Figure 9.17. Stop bands for gain coupling.

important difference; in index coupled structures there is no mode propagating at the Bragg frequency whereas in gain coupled structures there exists a mode exactly at the Bragg frequency. Potentially this is an advantage. In index coupled devices the two modes either side of the stop band are equally likely to propagate, so the devices are not strictly single mode. In gain coupled devices only the mode at the Bragg frequency propagates and potentially the devices offer better wavelength performance. In reality, real laser structures tend to be index coupled. Within the DFB the grating lies within the gain region whilst in the DBR the grating lies outside the gain region, but in both cases the grating is constructed in a layer adjacent to the active layer and in principle there should be neither gain or loss but pure index modulation within the grating with a coupling coefficient

$$\kappa = \frac{k_0 \Delta n_1}{4} \qquad (53)$$

where Δn_1 is the total refractive index modulation and is therefore equal to $2n_1$. Kapon has described in detail progress in DFB and DBR lasers, and refers to structures that employ pure gain coupling, but in practise index coupled DFB lasers offer a satisfactory single mode performance and there is no need for more complicated structures. The following descriptions of the two types assumes index coupling only.

9.4.1 DBR lasers

The essential difference between the DBR and the DFB has already been described. The DBR acts as a reflector with a maximum reflectivity at the

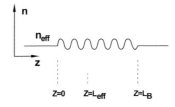

Figure 9.18. Effective mirror model of the DBR.

Bragg wavelength. Therefore the feedback into the active region is wavelength selective. The DBR also increases the effective length of the cavity, so the phase of the reflectivity can be approximated by

$$r = r_0 \exp(-2jk_0 n_{eff} L_{eff}) \quad (54)$$

where R_0 is a wavelength independent reflector [10]. This defines the effective penetration depth of the field into the grating

$$L_{eff} = \frac{1}{2\kappa} \tanh(\kappa L_B). \quad (55)$$

After some manipulation the reflectivity reduces to

$$r \approx -\tanh(\kappa L_B) \exp(-2j\Delta\beta L_{eff}) \quad (56)$$

where L_B is the length of the grating and $\Delta\beta$ is the deviation of the wavevector $k_0 \cdot n_{eff}$ from the Bragg wavevector k_B. Figure 9.19 shows the amplitude reflectivity of the grating at the Bragg wavelength. The maximum value of the tanh(x) function is unity so for long gratings such that $\kappa L_B > 3$

$$L_{eff} = \frac{1}{2\kappa}. \quad (57)$$

Typically κ lies in the range 20–50 cm^{-1}. For a product $\kappa L_B = 2$ the power reflectivity at the Bragg wavelength is 0.93, but this reduces rapidly with reducing κ. Using the identity

$$n_{eff} k_0 = \frac{2\pi}{\lambda} \quad (58)$$

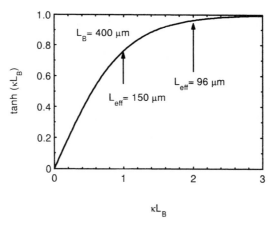

Figure 9.19. The amplitude reflectivity for a fixed $L_B = 400$ μm with different values of the coupling coefficient κ. Note that for lower values of κ the effective mirror length increases.

together with

$$\Delta\beta = \frac{d\beta}{d\lambda}\Delta\lambda \tag{59}$$

it is a straight forward matter to show for $\lambda = 1.55$ μm, $L_B = 400$ μm, and $n_{\mathit{eff}} = 4$, the reflectivity drops to zero for $\Delta\lambda = \lambda - \lambda_B \approx 0.8$ nm.

The DBR laser has excellent wavelength selectivity and can be used to determine the absolute wavelength of the laser. However, great precision in the fabrication of the devices is needed. The effective index of the Bragg grating depends crucially on the details of the guide structure. The effective index is an approximation based on the numerical solution of the wave equation at the boundaries of the guide, and so variations in the guide dimensions will affect this approximation. Hence not only the thickness of the layer but also the width of the stripe is crucial to the operation of this device. Moreover, the frequency comb determined by the F-P condition (remember that the grating acts as a mirror placed L_{eff} away from the cavity) is not necessarily coincident with the Bragg wavelength, which is the condition for optimum performance. If complete de-tuning of the two occurs it is possible that two wavelengths will propagate. At the very least, the side-mode suppression ratio is reduced in normal operation. Finally, the active region and the Bragg grating form two different waveguides coupled together, which is an additional complication. Parasitic losses can result. The upshot of this is that the DBR laser has not been adopted as a commercial technology for single mode sources because of the lack of reproducibility compared with the distributed feedback laser.

9.4.2 DFB lasers

Solution of the coupled mode equations shows that at wavelengths far from the Bragg condition the mode separation is $c/(2L)$, which is identical to the mode spacing in the Fabry–Perot structure. Close to the Bragg condition, however, the modes are pushed out slightly from these values. Moreover, the threshold gain for the modes closest to the Bragg condition is lowest, so these are the modes of the laser that propagate. Effectively the reflectivity of the grating is infinite because the presence of gain ensures that the power returned is greater than the power incident. It is possible to show that the threshold gain and wavelength are given by [10]

$$g_{th} = \frac{2 \cdot Im[\Delta\beta L]}{L} \tag{60}$$

and

$$\delta\lambda = -\frac{\lambda_B}{k_0(\lambda_B)n_{\mathit{eff}}L} Re[\Delta\beta L] \tag{61}$$

where $\Delta\beta$ is complex because of the presence of gain. Figure 9.20 shows the threshold gain-length product as calculated by Amman [10], where

$$\Delta\lambda_m = \frac{\lambda_B^2}{2n_{eff}L}. \tag{62}$$

It is clear from the above that the modes adjacent to the stop band are degenerate and equally likely to propagate. Practically, however, the grating is rarely so long that there is not some reflection from the end facet, and the facet is never in phase register with the grating except by happy accident. This difference is enough to break the degeneracy and the DFB laser operates sucessfully as a single mode laser, with a SSR > 30 dB.

9.5 High bandwidth sources

The limitations on the modulation rates of individual lasers means that bandwidth has been increased by wavelength division multiplexing. WDM can either be dense (DWDM), where the channel spacing is currently 200GHz (1.6 nm) or 100 GHz (0.8 nm), but with the possibility of smaller channel spacings, or coarse, where the channel spacing is 20 nm. DWDM requirements are currently met by DFB laser sources dedicated to a particular wavelength, and a number of sources may be integrated into a single module to increase the bandwidth. CWDM requirements will most likely be met by VCSELs, as the main application is likely to be in the ethernet, the "first mile" connecting the PC to the trunk routes where DWDM will take over.

This is a rapidly developing field. Optical fibres are well developed to deliver a number of wavelengths, and the main technological challenge now is to provide those wavelengths at high speed. A number of commercial laboratories are working actively to develop products that will deliver ever higher bandwidth at lower costs, and what appears in the technical and scientific literature is not necessarily representative of what is on offer. As has been mentioned in several places in this book, a laboratory demonstration is one thing but a commercial device is quite another. The latter is driven by the twin requirements of high yield and low cost manufacture, and the proprietary nature of much of the technology means that it is difficult to get an accurate picture not only of the detail of the products that are already on the market, but also of those that are in the pre-production development stage.

Reports of integrated arrays go back at least as far as 1987 [17] when five wavelengths around 1.3 μm were obtained using DFB lasers with a side-mode suppression ratio of more than 30 dB. A 20-wavelength DFB array was reported in 1992 [18] using tensile strained $In_{0.4}Ga_{0.6}As$ quantum wells designed to emit at 1.5 μm. These took advantage of the shift from the heavy hole to the light hole at the top of the valence band and were therefore expected to demonstrate lower threshold currents. The different wavelengths (3 nm

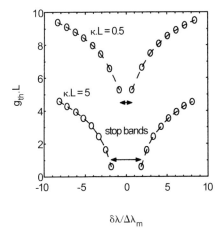

Figure 9.20. Threshold gain-length product and mode spacing for DFB lasers. Each symbol represents a mode. (After Kapon.)

spacing) were achieved by writing gratings shifted in period by 0.5 nm on each of the lasers by electron beam lithography. The same authors reported a compressively strained 18-wavelength array using $In_{0.7}Ga_{0.3}As$ wells [19]. A 10-wavelength 200 GHz channel spacing array was reported in 1999 based on tunable lasers. [20]. Each laser had a 900 μm long active section and a 500 μm long passive Bragg section (figure 9.21). The passive Bragg section can be pumped separately to change the refractive index and provide about 8 nm of tuning. This is not enough to span the wavelength range required for the 10-laser array so each passive Bragg reflector had a different pitch.

These two-section lasers have a limited tuning range of only a few nanometres, and are therefore of limited appeal. Broadly tunable souces are another matter [21–25]. Coldren [26] has reviewed monolithic tunable lasers and reports that after 20 years of being no more than a curiosity such lasers have now attracted serious commercial interest. These devices will increasingly find application in DWDM systems not only as replacements for failed lasers but also as primary sources in re-configurable systems. More than this, however, such lasers may well prove to be the backbone of future all-optical networks. However, Coldren adds a caveat. Broadly tunable lasers do not simply have an additional Bragg reflector with its own current supply but four sections in total. Usually there are two sampled gratings (gratings that have periodic sections removed to modify the reflection spectrum) each with its own current supply and a phase region adjacent to the active region. There are therefore four current connections, and this may well prove unattractive to systems engineers who naturally want ease of control over the output characteristics. That said, a number of telecommunications companies now offer widely tunable DFB laser arrays.

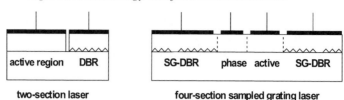

Figure 9.21. Schematic of a two-section and four-section tunable laser. The four-section laser has two sampled grating DBRs as well as a phase region.

The VCSEL is another device that offers potentially very high bandwidth. The VCSEL has a number of advantages over the DH and quantum well lasers when it comes to high speed modulation. The device has an intrinsically smaller area leading to a smaller capacitance and the barrier layers of the quantum well structure can be much thinner. In the edge-emitting devices the barrier layers are also part of the optical confinement structure and must be thick enough to contain the optical field, and as shown in chapter 8 transport across these layers can limit the modulation speed. In the VCSEL a $1-\lambda$ cavity at 850 nm in a material of refractive index ~ 3 is only 280 nm across and each barrier will be less than half of this when the quantum wells are taken into acount. On the other hand the mirrors add to the total resistance of the device and in proton-implanted devices the area of the aperture, while small, is approximately 25 times larger than in oxide confined defines. Current VCSEL devices offer modulation rates at around 2–4 GB/s but reports have started to appear in the literature of 850 nm devices with modulation rates of 10 GB/s [27–32]. Some of these devices use intra-cavity contacts, some use novel confinement techniques not based on either oxides or proton implantation, and others are novel structures suited to planarisation. Intra-cavity contacts essentially present the device manufacturer with two problem. First, the contacts must be accurately aligned, and second, the devices are no longer planar. These are not insurmountable problems but it remains to be seen what device structure the industry settles on for high speed modulation.

9.6 Summary

Lasers used in optical communications systems at 1.3 and 1.5 μm need to meet a number of stringent criteria in respect of the emitted wavelength and the suppression of the side modes. Fabry–Perot lasers based on cleaved cavities have generally proven themselves unsuitable for high bit rate transmission, as not only do relaxation oscillations broaden the output at short pulse lengths, but carrier induced refractive index changes cause the output wavelength to chirp. Moreover, the feedback in such lasers is not wavelength selective and the side-mode suppression at ordinary emitted powers is too small for high performance systems. Wavelength selectivity can be improved by incorporating grating structures into the cavity, which still acts like a Fabry–Perot cavity, but with the additional effect of a modulated reflectivity. Grating based lasers have proven themselves as stable sources for high bit rate transmission.

9.7 References

[1] Hecht J 1999 *City of Light: The Story of Fiber Optics* (New York: Oxford University Press)
[2] Hayes J *Fiber Optics Technician's Handbook* (Albany, New York: Delmar Publishers)
[3] Kalish D and Cohen L G 1987 *AT&T Technical Journal* **66** 19–32
[4] Li T 1987 *AT&T Technical Journal* **66** 5–18
[5] DeDuck P F and Johnson S R 1992 *AT&T Technical Journal* **71** 14–22
[6] McGrath C J 1992 *AT&T Technical Journal* **71** 22–30
[7] Thompson G H B 1980 *Physics of Semiconductor Laser Devices* (John Wiley & Sons)
[8] Bowers J E and Pollack M A 1980 in *Optical Fiber Telecommunications II* (Eds S E Miller and I P Kaminow) (San Diego: Academic Press)
[9] Dixon R W and Dutta N K 1987 *AT&T Technical Journal* **66** 73–83
[10] Amann M-K 1998 in *Semiconductor Lasers II* (Ed. Eli Kapon) (Academic Press)
[11] Coldren L A and Corzine S W 1995 *Diode Lasers and Photonic Integrated Circuits, Wiley Series in Microwave and Optical Engineering* (John Wiley & Sons)
[12] Koch T L and Bowers J E 1984 *Electronics Letters* **20** 1038–1041
[13] Koch T L and Linke R A 1986 *Appl. Phys. Lett.* **48** 614–616
[14] Kogelnik H and Shank C V 1972 *J. Appl. Phys.* **43** 2327–2335
[15] Brillouin L 1953 *Wave Propagation In Periodic Structures* (New York: Dover Publications Inc.)
[16] Kogelnik H 1969 *The Bell System Technical Journal* **48** 2909–2947
[17] Okuda H, Hirayama Y, Furuyama H, Kinoshita J-I and Nakamura M 1987 *IEEE J. Quant. Electr.* **QE-23** 843–848
[18] Zah C E, Pathak B, Favire F, Bhat R, Caneau C, Lin P S D, Gozdz A S, Andreakis N C, Koza M A and Lee T P 1992 *Electronics Letters* **28** 1585–1587
[19] Zah C E, Favire F, Pathak B, Bhat R, Caneau C, Lin P S D, Gozdz A S, Andreakis N C, Koza M A and Lee T P 1992 *Electronics Letters* **28** 2361–2362
[20] Ménézo S, Talneau A, Delorne F, Grosmaire S, Gaborit F and Klempes S 1999 *IEEE Photon. Tech. Lett.* **11** 785–787
[21] Jayaraman V, Cohen D A and Coldren L A 1992 *Appl. Phys. Lett.* **60** 2321–2323
[22] Ishii H, Kano F, Tohmori Y, Kondo Y, Tamamura T and Yoshikuno Y 1994 *Electronics Letters* **30** 1134–1135
[23] Ishii H, Tanobe T, Kano F, Tohmori Y, Kondo Y and Yoshikuno Y 1996 *Electronics Letters* **32** 454–455; Cartledge J C 1996 *J. Lightwave Technol.* **14** 480–485
[24] Rigole P-J, Nilsson S, Bäckbom L, Stålnacke B, Berglind E, Weber J-P and Stoltz B 1996 *Electronics Letters* **32** 2353–2354

[25] Delorme F, Grosmaire S, Gloukhian A and Ougazzaden A 1997 *Electronics Letters* **33** 210–211
[26] Coldren L A 2000 *IEEE J. Select. Topics. in Quant. Electr.* **6** 988–999
[27] Kuo H C, Chang Y S, Lai F Y, Hseuh T H, Chu L T, Laih L H and Wang S C 2004 *Solid-State Electron.* **48** 483–485
[28] Yu H C, Chang S J, Su Y K, Sung C P, Lin Y W, Yang H P, Huang C Y and Wang J M 2004 *Mater. Sci. Eng. B – Solid State Mater. Adv. Technol.* **106** 101–104
[29] Dang G T, Mehandru R, Luo B, Ren F, Hobson W S, Lopata J, Tayahi M, Chu S N G, Pearton S J, Chang W and Shen H 2003 *J. Lightwave Technol.* **21** 1020–1031
[30] Gustavsson J S, Haglund A, Bengtsson J and Larsson A 2002 *IEEE J. Quantum Electron.* **38** 1089–1096
[31] Dang G, Hobson W S, Chirovsky L M F, Lopata J, Tayahi M, Chu S N G, Ren F and Pearton S J 2001 *IEEE Photonics Technol. Lett.* **13** 924–926
[32] Zhou Y X, Cheng J and Allerman A A 2000 *IEEE Photonics Technol. Lett.* **12** 122–124

Problems

1. Satisfy yourself that the side mode suppression ration is independent of cavity length, and estimate the output power equired for a typical Fabry–Perot laser emitting at 1.5 μm to achieve a SSR of 30 dB.
2. Taking the nominal differential gain to be $\approx 4 \times 10^{-20}$ m^2, calculate the resonant frequency in a Fabry–Perot DH buried heterojunction laser with active region dimensions of 1.5 μm $\times$ 0.15 μm $\times$ 300 μm emitting 3 mW at 1500 nm for zero gain compression and a gain compression of 10^{-22} m^{-3}. Assume a refractive index of 3.5.
3. Assuming an effective refractive index of 3.5 calculate the pitch for a Bragg grating at 1.5 μm.
4. Assuming an index modulation of 0.002 calculate the coupling coefficient for a Bragg grating at 1.5 μm. Taking the length of the grating to be 400 μm calculate the effective length of the grating and the power reflection coefficient, and show that the reflection coefficient at the adjacent mode is 0.05 if the active region is 300 μm long.

Chapter 10

High power diode lasers

To a great extent the definition of what constitutes a high power laser is arbitrary. Diode laser technology has been driven in the main by the needs of the telecommunications industry, and before the development of fibre amplifier technology the launch power of the laser was kept below 3 mW to avoid non-linear effects. Compared with this, even 10 mW is a high power, but for many applications 10 mW would be considered low. At power levels exceeding a few tens of mW the lasers are usefully classed as high power, and this will be taken as a working definition for the purposes of this chapter.

Historically the main application for high power diode lasers was in pumping sources for solid state lasers, but high power diode lasers can now be found in diverse applications that require wavelengths ranging from 650 nm to 5 μm. Red lasers emitting at ~650 nm have now become the standard for compact disc (CD) and digital video disc (DVD) writers. Typically outputs of 30–50 mW are required. In the mid infra-red range from 3–5 μm, environmental sensing is one of the most important applications. Many gases and organic compounds have characteristic absorption bands due to molecular vibration within this wavelength range and lasers that can be used in specific detection systems are very important. In the near infra-red, 780 nm emitters have been replaced by red lasers for CD and DVD writing, but these lasers have now become very important sources for photo-dynamic therapy (PDT) and laser surgery, whilst 808 nm emitters are the most common source of optical radiation for diode pumping of solid state lasers. Optical pumping is also important at 980 nm, where the Er ion has a characteristic absorption, and a considerable effort has been expended in recent years in developing long lifetime lasers for pumping erbium doped fibre amplifiers (EDFA) for long haul, high bit rate telecommunication systems. This wavelength has all but replaced 1480 nm for this particular application, but again alternative uses exist for this technology, particularly welding of plastics. Lasers emitting at around 2 μm are also favoured for this type of application.

Diode laser arrays are not themselves ideally suited to materials processing applications, however, principally because of the output beam characteristics. The uses tend to be limited to certain niche applications, such as welding – plastics have already been mentioned in this regard – and laser soldering. The latter requires the application of a short thermal pulse in order to melt the solder down to a well defined depth, and with diode lasers the depth of melting can be varied according to both the power applied and the duration of the pulse. The constraints on spatial resolution and spot size may not be so demanding so the limitations of beam quality that can rule out the diode laser from consideration for many machining applications is not important. The use of a focused beam can allow soldered joints of a millimetre or so in diameter. In fact, diodes are much better suited to this task than solid state lasers, principally because in a solid state laser the pulse duration can be many orders of magnitude smaller, resulting in a much higher pulse power. Instead of melting the solder, such a laser would probably remove it entirely from the surface.

10.1 Geometry of high power diode lasers

There is no such thing as a standard high power laser. Structures vary according to the power required and the application; single, narrow stripes emitting several hundred milli-Watts, broad area lasers emitting a few Watts, and diode laser bars consisting of several broad area emitters, normally on a chip 1 cm long, emitting up to 200 W of optical power (figure 10.1). The cavities may be relatively short with high gain or long with low gain, and of double heterostructure or quantum well construction. Quantum wells may be strained, strain compensated, or lattice-matched. The diversity of device types is extensive, in reflection of the range of applications at various wavelengths and powers.

The simplest way to increase the emitted power is to increase the size of the emitting aperture, but the beam quality is adversely affected. Instead of uniform illumination across the whole facet, the output of a broad area device is usually localised to particular regions which are not in phase with each other.

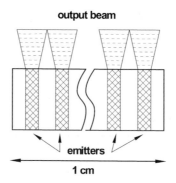

Figure 10.1. Schematic of a diode laser bar.

The laser output is effectively a sum of a series of individual beams, so there may be low spatial and temporal coherence. Nonetheless, apertures around 100 μm in extent are commonly found. High powers can also be achieved from narrow apertures, but the design has to be optimised. Other changes that can be made include reducing the gain and increasing the length of the cavity, and cavity lengths up to a few millimetres can be found. For very high power applications (several tens of W) the output from several emitters integrated onto a single bar has to be combined but the number and spacing of the emitters varies from design to design. Diode laser bars themselves may be stacked to provide an intense high power optical source for pumping solid state lasers, for example.

The application of standard diode bars to materials processing has been described by Treusch et al. [1]. The most common wavelengths are in the near infra-red, from ~ 800 nm out to ~ 1 μm, but in fact any wavelength that is absorbed strongly can be used. A typical product will be a bar 1 cm in length emitting at 808 nm with power outputs from 10 W up to something in excess of 200 W, achieved by adjusting the number and spacing between the emitters. Single emitter devices at ~ 900 nm, usually based on phosphides, are common, as are single emitters in the visible and mid-IR. As an example of a very high power device, figure 10.2 shows the power output and efficiency of a 70 emitter bar. The maximum power efficiency is in excess of 50%, whilst the power is virtually linear with current. At the very highest power levels the rate of increase of power begins to decline, an effect known as thermal roll-over. Thermal roll-over limits the power output in some lasers but in others a phenomenon known as catastrophic optical damage (COD) occurs before the onset of roll-over. COD is a single event that destroys the facet and causes the laser to cease operating, so setting a practical limit on the operating power.

Diode arrays, as the bars are commonly called, may be compact optical sources, but of course they are extended in one dimension. As with low power

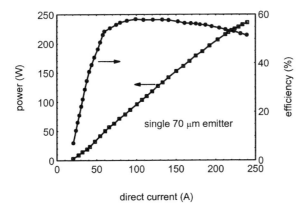

Figure 10.2. Output power and efficiency of a 70 emitter bar. (After Treusch et al.)

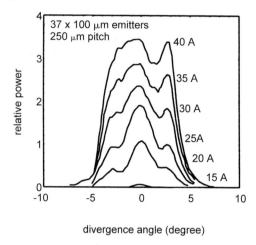

Figure 10.3. Far-field pattern from 37 emitters on a bar. (After Treusch *et al.*)

lasers, divergence in the plane perpendicular to the junction occurs, but the beam in the plane of the bar is by no means well behaved. Figures 10.3 and 10.4 show the divergence in the plane of the junction from bars containing 37 and 70 emitters respectively, where the aperture of the emitters is 100 μm. The peaks are not merely due to the superposition of the various individual beams but the result of non-linearities in the semiconductor material that affect the output. Their existence poses the problem of how to shape the beam and to

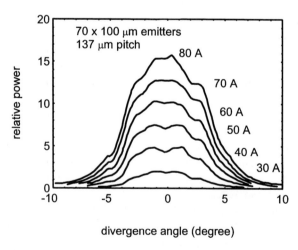

Figure 10.4. Far-field pattern from 70 emitters on a bar. (After Treusch *et al.*)

reduce the spot size and combinations of cylindrical and spherical lenses often need to be employed. Thermal dissipation also needs to be considered, and considerable effort has been expended on the design of micro-channel coolers. These must be stackable so that high power bars can be combined to give multi kilowatt outputs for diode pumping of large solid state lasers, and a summary of the technology can be found in Treusch's paper. For other applications the beam quality has to be improved. How this can be achieved requires a knowledge of the origins of the non-linearities responsible for the characteristic broad area behaviour. This chapter starts with the single emitter as a precursor to arrays.

10.2 Single emitter broad area diode lasers

The single emitter may be used alone as a high power source or as one element in an array. The emitter aperture is usually defined using a stripe contact of some description. As mentioned in chapter 5, the invention of the stripe geometry influenced developments in double heterostructure laser technology, and it is now standard for a stripe contact of some sort to be made in experimental devices of all powers. For low power devices in particular, a stripe geometry allows a low operating current and stable output characteristics. An obvious method of increasing the power output is to widen the stripe so that the emitting aperture is larger, but it is also possible to use narrow apertures and still achieve high powers if the design of the laser is optimised. In order to put these developments in context it is necessary first to be clear about what is meant by the term "broad area laser".

Broad area behaviour is characterised by the appearance of many lateral modes, often called filaments. Seen close up, the light is not emitted uniformly across the facet but from localised regions distributed in an apparently random manner across the stripe. Sometimes these filaments are stable, sometimes not. They can appear to migrate across the output aperture of the laser in a chaotic manner, but the timescales are of the order of nanoseconds. The amplitude of the output can be modulated at frequencies ranging from a few MHz to ~ 1 GHz by a process of "self-pulsation". The complexity of the filamentation process means that it is not easy to summarise it by a few simple rules, or even by a detailed description of the physics. It is possible to predict in general terms the appearance of filamentation using analytical models, but only very complicated numerical modelling will reveal the full details. The only certainty is that filaments appear, and the uses to which the laser may be put are affected. If a high power diode laser were to be required for free space communication the self-pulsation will very seriously interfere with the modulation of the beam. On the other hand, if the laser were to be used as a heat source for welding or soldering, the only important properties are spot size, average power, duration of the pulse, and the pulse power. Of these, only the spot size is affected by the filaments, which are essentially decoupled from each other in phase. The divergence of the beam increases markedly, as illustrated in figures 10.3 and 10.4, but the instabilities have no impact on the other properties, as the duration of the pulse will be many orders of magnitude longer than the characteristic lifetime of the instabilities. Thermal diffusion effects in the processed material will therefore wash out any non-uniformities that might occur on very short time-scales, or over short lateral dimensions.

10.3 Lateral modes in broad area lasers

The appearance of lateral modes can only be understood with reference to the spatio-temporal dynamics of semiconductor lasers as described by Hess and Kuhn [2]. Rate equations, the basis of the modelling presented in chapter 8, are not detailed enough to predict such phenomena. They describe the temporal variations in carrier density as they are transported perpendicular to the junction and density of both photons and carriers in the plane of the junction are considered uniform. Although examples exist where terms describing the diffusion of carriers out of the stripe have been included, these are global terms that take into account a loss of carriers only. The rate equations are therefore essentially one-dimensional, and in the hierarchy of laser modelling they constitute the lowest level. What is needed to model the spatial variations in the output beam in the plane of the junction are models which explicitly include the lateral spatial dependence of both the carrier and photon densities, and analytical models do not exist. The equations must be solved numerically so the insight lies not so much in the description of the model but in the outcome. Hess and Kuhn concluded that stripes wider than ~15 μm typically results in broad area behaviour rather than stripe behaviour, in agreement with the experiments of Thompson [3], which illustrate very clearly the change from narrow stripe to broad area behaviour in AlGaAs DH lasers (figure 10.5). Stripe widths of 10, 15, 20, and 30 μm were investigated and the two narrowest stripes both exhibit a single lateral mode, though the light intensity is distributed asymmetrically about the peak. For broader stripes the appearance of other lateral modes is clear from the intensity distribution along the aperture.

The key to this behaviour lies in the non-linear interaction between carriers and photons. It was shown in chapter 4 that the refractive index of a semiconductor is not a fixed quantity but depends intimately on the carrier density and the gain inside the medium, which in turn depends on the photon density. Regions of high carrier density have a lower refractive index than regions of low carrier density, but regions of high gain have an increased refractive index. Thus the stripe laser operates by gain guiding; the formation of a natural waveguide in the plane of the junction through an increase in the refractive index due to a high gain. If the stripe is narrow enough this

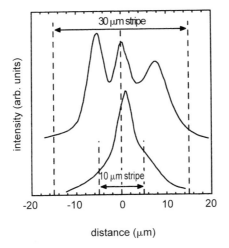

Figure 10.5. Near field radiation pattern for two stripe widths corresponding to "narrow" and "broad area" behaviour. (After Thompson *et al.*)

waveguide causes a single lateral mode centred more or less on the stripe, but for wider stripes higher order waveguide modes propagate. Even in the single mode behaviour demonstrated in figure 10.5 there is evidence of a non-uniform intensity distribution, and considerable emission from outside the immediate confines of the stripe. At a slightly lower current Thompson's data indicates that the mode is symmetric, and it is a feature of stripe lasers that the light intensity can shift as the power increases. Such instabilities often manifest themselves as a characteristic kink in the intensity-current curve (see figure 10.6).

The stripe is intended to confine the current to a localised region of the junction. Proton implantation to form semi-insulating regions, the use of reversed bias *p–n* junctions fabricated by deep diffusion, or the use of a contact window defined by means of a patterned dielectric layer, (silicon nitride or

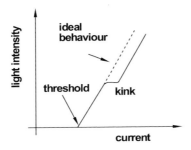

Figure 10.6. Non-ideal (kink) behaviour in the L-I characteristic caused by filamentation.

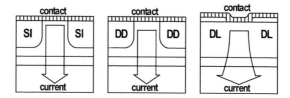

Figure 10.7. Common Stripe formats for diode lasers can use semi-insulating (SI) regions, deep diffusion (DD), or a dielectric layer (DL) beneath the contact. Current spreading is a problem with the DL.

dioxide) directly beneath the contact (figure 10.7) are the three most common approaches. The dielectric layer (DL) is the least efficient structure for confining the contact, as the relatively large separation between the contact and the active junction allows for current spreading. Proton implantation to produce semi-insulating (SI) material and deep diffusion (DD) both give rise to insulating regions closer to the active region and therefore the current flow is better defined. Within the active layer but outside the region of high current flow the carrier density will be relatively low so the effect on the refractive index will be small. Immediately adjacent to the active region the refractive index will be lower than its quiescent level because diffusion will increase the carrier density slightly. Within the stripe the high carrier density will tend to lower the refractive index even further, but at the onset of gain the refractive index will begin to increase. As with the *p–n* homojunction laser, these effects are quite small, and the wave-guiding is weak. Hence a guide several micrometres wide will support only the fundamental mode, as illustrated in figure 10.8.

Hess and Kuhn modelled AlGaAs/GaAs double heterostructure devices emitting at 815 nm, but the physics applies just as well to other materials and structures. The refractive index difference between the core and cladding was 0.25 and the critical thickness of the active layer for single transverse mode behaviour was 0.52 μm. With a modelled thickness of 0.15 μm the structure

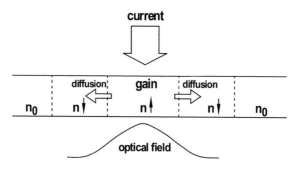

Figure 10.8. Schematic of the electronic and optical properties of the narrow stripe.

Lateral modes in broad area lasers 301

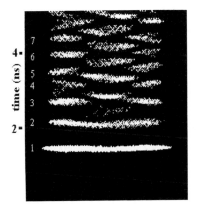

Figure 10.9. Spatio-temporal variations in light intensity in a broad area laser. The time after the initial electrical pulse is shown vertically and the horizontal represents the 50 μm wide stripe with a border of 10 μm either side. Reprinted from Hass O and Kuhn T 1996 Spatio-temporal dynamics of semiconductor lasers: theory, modelling and analysis *Prog. Quant. Electronics* **20** 85–197, © 1996 with permission from Elsevier.

therefore exhibited a single transverse mode. The initial response of a single broad stripe 50 μm wide to a current pulse is dominated by the relaxation oscillations (figure 10.9). The first seven illumination events are shown and numbered in white, the light regions corresponding to high optical field intensities but without the use of grey scale adopted by Hess and Kuhn to represent differences in intensity.

Initially the whole width of the stripe is illuminated because the photon density lags behind the carrier density. Before laser output can occur not only does the carrier density have to reach threshold but the photon density also has to rise from zero. There is a time delay between the two which allows the carrier density to reach a level well above threshold, where it is normally pinned during steady state operation. Consequently the photon density also rises well above the steady state level and depletes the carrier density to a level below threshold, hence the extinction of the emission. The carrier density has to rise again before any further light emission can occur, but it doesn't rise to the same level. Moreover, it is clear from figure 10.9 that the centre of the stripe in event 2 is illuminated slightly before the edges, though essentially the whole stripe is illuminated except for a small strip at both edges. This is probably a consequence of the existence of lateral waveguide modes. The fundamental mode is centred on the stripe and in the first order mode the light intensity distribution is bi-modal. It will only take a few round trips of the cavity to begin to establish these modes, hence the light emission begins in the centre but moves outward as time progresses. Even on the initial illumination event the light output at the edges lasts a little longer than the light output at the centre, which allows the carrier density at the centre to recover quicker than it does at the edges.

Crucially, the centre of the stripe is illuminated on its own by the time event (3) has occurred, and even though the emission is weak, the carrier density in the centre is depleted. Subsequent emission (event 4) is concentrated solely at the edges and thereafter an alternation between emission from the centre and the sides is inevitable. The process thus described is clearly similar to the process behind the relaxation oscillations familiar from narrow stripe lasers, but there are crucial differences. First, variations in carrier density laterally across the stripe are not washed out by diffusion as they are in narrow lasers. The carriers will diffuse at most a few microns during the time interval between successive light emissions so significant lateral variations can be sustained by the pattern of emission and extinction. Indeed, Hess and Kuhn mapped out the carrier densities in a similar manner to the light intensity and it is apparent that periods of high light intensity are followed by periods of low carrier density. Second, the lateral variations in carrier density in the broad area device serve to reinforce the oscillations. Not only is there a tendency for light to be distributed across the stripe according to the waveguide modes, so that the carrier density rises where the stripe is dark, but the increase in carrier density will cause the refractive index to decrease and focus the light into the illuminated regions. The filament thus forms its own waveguide, causing the intensity of light to increase in the filament more rapidly than would normally be expected on the basis of the rate equations. This is called spatial hole-burning. Figure 10.10 illustrates the refractive index changes occurring for both the first mode (a) and the fundamental (b) of the lateral waveguide.

One of the major differences between broad area and narrow stripe lasers, therefore, is the apparent lack of a damping mechanism to iron out the spatial fluctuations. In a narrow stripe in which a stable single mode can propagate the whole width of the stripe is illuminated and there are no regions where the carrier density is allowed to rise to very high levels through the lack of stimulated emission. Filaments simply do not form, but in the broad area device a new filament can form in the adjacent regions of super-threshold carrier

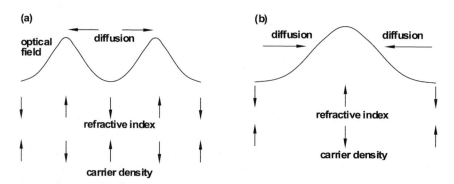

Figure 10.10. Interaction of the optical field and carrier density in the broad stripe.

density and high gain. Moreover, this filament interacts non-linearly with the existing filament and energy is transferred from one to the other. Sometimes this transfer can appear stable but at other times it can seem chaotic.

This non-linear interaction between the carrier density and the optical field is governed by the phase-amplitude coupling coefficient, so called because the phase of a travelling optical wave is governed by the real part of the refractive index and the amplitude is governed by the imaginary part. As has been discussed, the two are intimately linked via the carrier density, so the phase and amplitude are said to be coupled. Hence, a change in the real part of the refractive index alters the propagation time of the photons within the cavity and alters the output wavelength via the Fabry-Perot condition. The spectral linewidth of the laser is enhanced, which is sometimes known as spectral broadening. For this reason, the phase-amplitude coupling coefficient, defined as,

$$\alpha = \frac{dn_r/dN}{dn_i/dN} \qquad (1)$$

where $n = n_r + jn_i$ is the refractive index, and N is the carrier density, is also called the linewidth enhancement factor. It can be shown (Coldren and Corzine, p 209) that this can be transformed to give,

$$\alpha = -\frac{4\pi}{\lambda a}\frac{dn_r}{dN} \qquad (2)$$

where λ is the wavelength of operation and

$$a = \frac{dg}{dN} \qquad (3)$$

is the differential gain. For a fuller discussion on the linewidth enhancement factor, see Yariv [4] where it is shown that the linewidth of a diode laser is enhanced over the theoretical Shawlow-Townes linewidth by a factor $(1 + \alpha^2)$. This arises because fluctuations in the carrier density caused by spontaneous emission events manifest themselves through α as variations in the phase of the optical field and hence as variations in the output wavelength. Indeed, one of the ways of measuring α in diode lasers is to deliberately modulate the carrier density at RF frequencies and measure the resulting phase modulation as well as the ampltiude modulation of the output optical field. Typically $2 < \alpha < 6$.

As might be expected, both the level of pumping and the stripe width are important in filamentation. The filaments illustrated in figure 10.9 occur as a result of pumping at about twice the threshold current. At three times threshold filamentation occurs even within the relaxation oscillations, and the average time interval between each filament at any given point along the width of the stripe is shorter. Increasing the width of the stripe appears to have a similar effect as increasing the pumping power. An example of the experimental observation of the effect of pump power on filamentation is given in

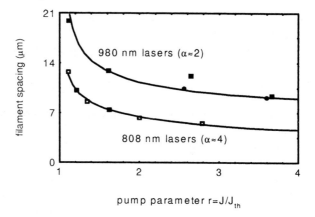

Figure 10.11. Filament spacing as a function of pump parameter. (After Marciante and Agrawal.)

figures 10.11 and 10.12 [5]. Marciante and Agrawal developed a theory that explicitly linked the properties of the filaments to α [6] and found experimentally for 50 μm wide stripes in both AlGaAs lasers (808 nm) and InGaAs lasers (980 nm) that higher values of α lead to smaller values of filament spacing for a given pumping condition relative to the threshold current. The agreement between the model and experiment is shown in figure 10.11 in which the solid lines represent the theoretical model and the symbols represent the experimental data. The chaotic nature of filamentation is reflected

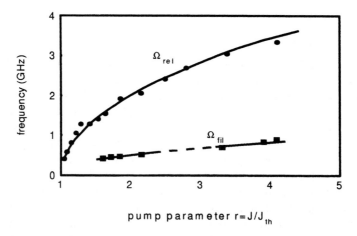

Figure 10.12. Frequency of the output fluctuations as a function of pump parameter. Relaxation oscillations and self-sustained oscillations are clearly distinguishable. (After Marciante and Agrawal.)

in the fact that filaments do not always appear regularly spaced, and hence there is no data for some values of the pump current. Those that are shown therefore correspond to the conditions under which regularly spaced and well behaved filaments were observed.

This is particularly evident in figure 10.12. The output of the laser was analysed with a microwave spectrum analyser so that periodic variations in intensity could be detected. Not surprisingly the relaxation oscillations were readily detected, as shown by Ω_{rel}, but a second component was also detected, labelled Ω_{fil}. Marciante and Agrawal called this "self pulsation". It is a characteristic of stable filaments and can be seen in the steady alternation of the light output from the sides to the centre in figure 10.9 that occurs before the onset of chaotic behaviour. Where stable filaments were not detected the data has been represented as a dotted line but clearly all the data lie on a single straight line describing Ω_{fil}. Experimentally, the observation of filamentation and self-pulsation can mislead as to the true nature of the phenomenon. The filaments appear to migrate laterally across the cavity, some of which can be seen in figure 10.9 just as the filamentation becomes chaotic. This is called "mode hopping", as it appears that the field is hopping from one lateral mode to another. In fact no such thing is occurring. The model of Hess and Kuhn shows that within the round trip time of the cavity one filament can die away as another grows, but the round-trip time is only a few picoseconds in duration. Individual events such as the birth and death of a filament cannot be distinguished. Rather, an average effect is observed, and sometimes that average effect has stable characteristics and at other times under other conditions the effect is chaotic and the filaments appear to migrate.

10.4 Controlling filamentation

10.4.1 Mode filtering

A free running conventional double heterostructure broad area laser will exhibit filamentation at pumping levels just over threshold. The existence of non-uniform optical fields corresponding to multiple waveguide modes coupled with carrier diffusion and phase-amplitude coupling guarantees it. There have been numerous articles published in the open literature describing many different ways in which filamentation can be controlled, but not all these devices will find their way on to the commercial market. Many factors are involved in determining whether a particular design becomes a commercial reality. It is not only necessary to consider whether the device works, but also whether it can be manufactured reliably, whether the cost of changing the process is justified by the return, etc. These are business decisions, but informed by technological considerations. There also has to be a clear reason for adopting a particular solution, and it may well be that expensive, but very effective, solutions are used for specific high technology, but low volume,

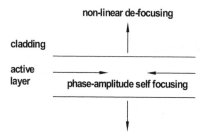

Figure 10.13. Self-defocusing by a layer with a non-linear refractive index to reduce the optical intensity in the active layer.

applications. With this in mind it is not possible to examine all the solutions in the open literature. Those that are given in the following pages are interesting not only because they demonstrate an imaginative approach to the problem but because they also provide further insight into the nature of the phenomenon.

Marciante and Agrawal [7] proposed, in a purely theoretical exercise, inserting an extra pair of epitaxial layers into the structure. The purpose of these layers is to act as self-defocusing layers. Having a slightly larger band-gap than the active layers by about 29 meV, these layers form part of the wave guide and a large fraction of the optical field – somewhere between 80% and 90% – resides within them. The band-gap difference is not so large as to affect current diffusion, however. The layers exhibit a non-linear refractive index which causes a defocusing effect to counter-act the focusing effect of the phase-amplitude coupling coefficient. In effect, as the intensity of the filament increases the refractive index of the extra layers decreases and the transverse confinement is reduced slightly. The intensity of the field within the waveguide is thereby reduced and operation of a 100 μm wide stripe at 810 nm without filamentation was predicted.

Spatial filtering has also been investigated by a number of research groups. Szymanski *et al.* [8] have analysed in detail the theory of a broad area laser with a modal reflector. A modal reflector is a patterned reflector of the resonator which can be used to provide selective feedback for particular modes of the waveguide, and hence suppress the others. Wolff *et al.* [9,10] have investigated the use of a 4f external cavity, so called because the total length of the cavity is 4 times the focal length of the lens employed within the cavity. The basic premise of the technique is that a Fourier transform of an image is produced within the focal plane of a lens. If another lens were to be placed with its focal plane coincident with that of the first the image will be reconstructed from the Fourier transform. A filter, i.e. a physical blockage, placed within the Fourier plane can be used to eliminate part of the fourier transform, so the reconstructed image is altered. In the case of a diode laser the image of the output facet is Fourier transformed and reflected at the focal plane so the same lens is used to form an image back at the output facet. This image then constitutes the

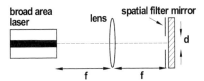

Figure 10.14. A 4-f folded cavity with spatial filter.

optical feedback into the laser. Features corresponding to the higher order modes of the waveguide usually lie further out from the centre of the Fourier transform than those corresponding to the fundamental mode. A single slit of adjustable width is adequate as a filter to eliminate them from the feedback so that only the fundamental mode is excited. It is not necessary to identify these features explicitly, only to adjust the slit width until stable operation is achieved. Furthermore, deliberate excitation of higher order modes is possible by translating the filter in the Fourier plane.

Lang et al. [11] proposed the use of grating confined broad area lasers. This is also a method of spatial filtering. The wave-guide confinement is achieved by the use of an angled distributed feedback (DFB) grating, which can be configured to have a very narrow acceptance angle. As discussed elsewhere, DFB gratings work by reflecting a small amount of radiation from alternate layers $\lambda/4$ wavelengths thick. Each reflection is in phase with each other and if the sequence of layers is thick enough near total reflection ($>99\%$) can be achieved. The acceptance angle is defined as the full-width-half-maximum of the reflectivity as a function of the incident angle. Lang has shown that for a grating configured for normal incidence operation the acceptance angle is 2.5°, i.e. reflection will occur anywhere between an angle of incidence of 88.75° and 91.25°, but this drops to 0.16° for an angle of incidence of 80° and 0.029° for an angle of incidence of 15°. In short, if such gratings are used to define the boundary of a wave-guide rather than the usual refractive index changes, only rays incident at the specified angle will be reflected. All others, including all higher order modes of the waveguide, will be transmitted and only the fundamental mode will propagate.

A schematic of a four-grating laser is shown in figure 10.15, where the grating and cavity are angled from the facet to produce a perpendicular output beam. One of the advantages of such a structure is the self-selection of a stable mode of operation. If the laser mode is incident on the side grating at an angle θ the angle of incidence for the end grating is $90 - \theta$ (figure 10.16). The operating wavelength of the laser is found by solving the simultaneous equations for the allowed wavelengths, i.e.

$$\lambda = 2n\Lambda_1 \sin\left(\frac{\theta}{m_1}\right) = 2n\Lambda_2 \sin\left(\frac{90-\theta}{m_2}\right) \tag{4}$$

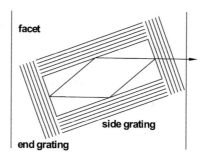

Figure 10.15. Grating confined broad area laser. (After Lang *et al.*)

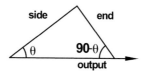

Figure 10.16. Trigonometry of the beam in the grating confined structure.

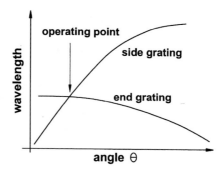

Figure 10.17. Allowed wavelengths as a function of angle.

where Λ_1 and Λ_2 are the grating pitches and m_1 and m_2 are the respective grating orders. The values used by Lang were $\Lambda_1 = 567$ nm, $\Lambda_2 = 304$ nm, $m_1 = 1$ and $m_2 = 2$ giving $\lambda = 980$ nm. This is illustrated schematically in figure 10.17, where the allowed wavelengths are shown as solid lines. The wavelength reflected from the side grating increases with θ whereas the wavelength reflected from the end grating decreases. There is therefore only one solution, which can be varied by changing either of the grating spacings. Provided the operating point lies within the gain spectrum the laser will operate. This type of laser is capable of emitting over 1 W of CW power in a single lobed diffraction limited beam.

10.4.2 Materials engineering

The preceding approaches all involve changing the laser design. Marciante and Agrawal's approach illustrates the importance of the optical intensity in the formation of the filament and their aim is to reduce the intensity by anti-guiding non-linear optical layers. The spatial filtering approach illustrates the importance of higher order modes as a mechanism for initiating the filaments through spatial variations in the optical intensity. In particular the approach of Lang *et al.* allows for very well defined optical propagation at a specific wavelength and a specific angle within the guide. The alternative to changing the laser design is to change the laser material, and this has been done by Kano *et al.* [12], who used strained multiple quantum wells of InGaAsP/InP. Although not specifically concerned with high power laser diodes, Kano *et al.* nonetheless demonstrated significant reduction in the phase-amplitude coupling coefficient α through the use of strain and modulation doping.

The valence band structure is altered in a quantum well. The degeneracy of the heavy hole and light hole bands is lifted and strain can further affect these bands. Kano *et al.* mapped out the valence bands as a function of wave vector in both the [100] and [110] crystallographic planes, both of which lie in the plane of the quantum well, for a 50 Angstrom well sandwiched in barrier layers corresponding to a band gap of 1.1 μm, as shown in figure 5 of appendix IV. In the lattice matched quantum well the heavy hole forms the band gap, being of a higher energy, and the first excited state of the heavy hole is similar to the energy of the light hole ground state at $k = 0$, but in the strained quantum well the heavy hole bands are shifted upward so that both the effective mass and the density of states are reduced. These two are not the same thing. True, the density of states within a particular energy level depends on the effective mass, but the total density of states is affected by the proximity of the levels and both are changed by strain. The conduction band is much simpler in nature than the valence band and the changes to the valence band are more significant.

Strain affects the differential gain, for this appears in the denominator of α. The maximum in gain appears at photon energies corresponding to the band edge. In a lattice matched quantum well where the density of valence band states is high, the requirement for population inversion, that the quasi-Fermi level separation be greater than the band gap leads to a high quasi-Fermi level in the conduction band and a relatively low quasi-Fermi level in the valence band. The hole quasi-Fermi level does not even need to enter the valence band, and in some cases may not do so. For the strained quantum well where the density of valence band states is considerably reduced the distribution of the quasi-Fermi levels is more even (figure 10.18). This means that the gain is more

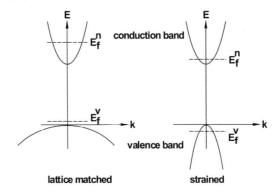

Figure 10.18. A reduced density of states in the valence band leads to a greater penetration of the Fermi-level into the valence band.

sensitive to changes in carrier density, because the differential gain is related to the degree to which the electron and hole densities vary with Fermi level. If the Fermi level is well inside the conduction band the fractional occupancy of states at the conduction band edge is high and changes but little as the carrier density increases. If the Fermi level lies close to the band edge then changes in the Fermi level position can bring about large changes in carrier density and hence large changes in gain. This is illustrated in figure 10.19, where the arrows mark the band edges in relation to a common Fermi level. An increase in the carrier density shifts the Fermi occupation function higher in energy (dotted line) so the change in fractional occupancy for a conduction band in a strained well lying close to the Fermi energy is larger than it is in a conduction band for a lattice matched well lying further below the Fermi level, as indicated by Δf_{lm} and Δf_s. Strain compensation, wherein the wells and barrier are alternately

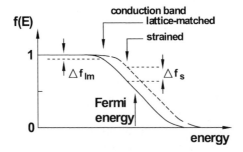

Figure 10.19. The conduction band in a strained quantum well lies closer to the Fermi level than in the lattice-matched case, leading to a larger change in the occupancy function with a slight change in Fermi level.

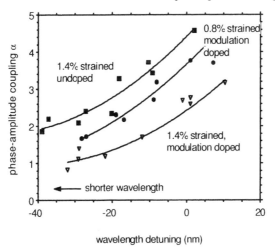

Figure 10.20. Phase-amplitude coupling as a function of wavelength detuning. (After Kano *et al.*)

under tensile and compressive strain similarly affects the valence bands and can result in a much reduced phase-amplitude coupling [13].

This is the principle reason why strained quantum well lasers can exhibit low values of the phase-amplitude coupling coefficient α. Moreover, it is possible to take further advantage of this property of the strained quantum well. The operating wavelength of a traditional Fabry–Perot laser is determined by the point at which maximum gain occurs, which lies close to the band edges, but the gain extends over higher energies. It extends as far as the Fermi levels extend into the bands, as this is the region of transparency. If a wavelength selective method of feedback is employed to shift the operating wavelength to higher energy and hence shorter wavelength, the energy states involved in the transition will lie closer to the Fermi energy in the conduction band. The closer they lie, the greater the differential gain. Kano *et al.* investigated just such a mechanism and recorded significant variations in α, with values as low as 1 in a strained, modulation doped laser detuned from the peak gain by -30 nm (figure 10.20).

10.5 Catastrophic optical damage

Whether a laser is broad area or narrow stripe, the obvious question is, "how much power can the laser emit"? There are two principal limitations; damage and roll-over. Roll-over is reversible but catastrophic optical damage (COD) is a sudden event that destroys the output of the laser. It is recognisable by a sudden deterioration in light-current (L-I) characteristic, and afterwards by inspection of laser facet, which shows unmistakeable signs of damage. Not surprisingly, there is a great deal of work in the open literature on COD as various authors have attempted to identify the exact mechanisms and protect lasers against it. Eliseev [14] has provided a comprehensive review of the subject, including a general discussion on the susceptibility of semiconductor materials to damage from externally generated laser radiation. External damage is in contrast to COD, but it provides a means of rapid heating and an insight into damage mechanisms. Although an enormous amount of data exists on COD, there is in fact as yet no comprehensive picture of the mechanisms. In the following pages the current understanding is discussed in some detail, not only because it provides an insight into the thinking behind laser designs for high power operation, but also because it is an interesting subject in itself.

The main features of COD can be summarised as follows. The external intensity of the laser is used as a measure of the catastrophic damage, and is called the COD intensity. Typical values lie in the range 10–100 MW cm^{-2}. In BAL an average power for COD is given because of beam non-uniformities such as filamentation. In reality it is the internal intensity that is important, and this can exceed the output intensity several-fold. Photons within the laser cavity are strongly absorbed both in unpumped regions, and locally, where absorption is increased by various mechanisms such as a local temperature rise or the existence of absorbing defects. Extended defects such as dislocations are important. In particular absorption occurs at the facet. Partly this is due to surface recombination which, apart from generating heat due to the current flow itself as well as the energy released by the recombination process, causes a decrease in the carrier density at the facets. The laser under population inversion is normally transparent to the emitted radiation, but if the temperature rise is high enough the band gap shrinks and band-gap absorption can occur at the facet. The band gap of a semiconductor decreases with temperature according to

$$E_g(T) = E_g(0) - \frac{\alpha T^2}{T + \beta} \qquad (5)$$

where $E_g(0)$ is the band gap at absolute zero and α and β are constants (in GaAs $E_g(0) = 1.519$, $\alpha = 5.405 \times 10^{-4}$ and $\beta = 204$). Coating the output facet with anti-reflection (AR) layers can give an increase in the measured COD power by up to a factor of three, partly due to the improved coupling of the radiation to the outside and in part due to the protection that the coating gives to the facet.

Once COD has occurred the radiation intensity within the cavity drops within pico-seconds, which corresponds to just a few round trips of the cavity. Whilst the bulk of the laser itself may not be damaged operation will cease due to the lack of optical feedback from the damaged facet. Although the final optical damage event is catastrophic a prior gradual degradation of laser performance occurs over time. Sustained operation at 50% of the COD level degrades the lifetime of laser and the threshold for the catastrophic event can also be reduced by aging. The degradation rate is proportional to the square root of the aging time, which suggests diffusion of some sort as a mechanism. In experiments on aging one of the characteristic features of the degradation is the production of so-called dark line defects (DLD). These are lines of darkness in an otherwise illuminated facet, hence the name, and are strongly indicative of non-radiative recombination. As discussed in chapter 2, non-radiative recombination lifetimes are very short and non-radiative processes tend to dominate over the radiative process. Hence the action of the laser can be destroyed where such recombination centres exist.

The dark line defects are found to contain four characteristic defects; small dislocation loops of diameter ~ 20 nm can be produced which lie along a dark line; larger dislocation loops of 200–500 nm in diameter, elongated along the dark line; compositional segregation characteristic of melting and recrystallisation (that is, the local concentration of Al vastly exceeds the average composition, and adjacent to it is a region containing a far higher concentration of Ga); and defects not immediately visible by electron microscopy but which clearly contain a lot of non-radiative recombination centres. These may be vacancies or interstitials. Environmental oxygen diffusing into the facet has been identified as a factor and indeed initial operation of the laser (called burn-in) in an inert atmosphere improves the lifetime. However, the mechanism is not understood. Thermal decomposition, i.e. As desorption accompanied by group III rich surface regions has also been observed, as has the propagation of dark line defects into the laser from the facet at 200–400 cm s^{-1}. These are all diffusion phenomena of one sort or another.

The observation of dislocations, sometimes in the form of loops and sometimes in tangles, is quite common in lasers subject to COD. Eliseev speculated that a supercritical pulse heats the surface and increases non-radiative recombination, which further heats the surface. A small region near the surface melts and on recrystallisation dislocation loops are formed to compensate the change in volume. In COD in pulsed lasers the optical intensity is proportional to the square root of inverse pulse width, which strongly suggests thermal diffusion as a mechanism. Heat generated in any localised region within a short time will diffuse a characteristic distance determined by the square root of the pulse duration, so the effective volume heated during the pulse varies in like manner. Taking the mass of a small volume of length L and unit area in cross section to be

$$m = \rho L = \rho \sqrt{D\tau} \tag{6}$$

where is the density, D the thermal diffusion coefficient, and τ is the pulse duration, the temperature rise ΔT is related to the amount of heat Q deposited into the volume τ in time by

$$Q = P \cdot \tau = m \cdot C \cdot \Delta T \tag{7}$$

where P is the power and C is the specific heat capacity. Hence

$$P = \rho \cdot C \cdot \Delta T \frac{\sqrt{D\tau}}{\tau}. \tag{8}$$

For a fixed temperature rise ΔT_{cr}, say the critical temperature rise for the onset of COD, then

$$P \propto \frac{1}{\sqrt{\tau}}. \tag{9}$$

This is a simplification as the thermal properties ρ, C, and D can all vary with temperature. Nonetheless, in experiments on laser heating of semiconductors by strongly absorbed radiation such as excimer laser radiation in the UV, simple calculations of surface temperature rises based on very similar ideas agree remarkably well with empirical observation. It is possible to conclude therefore that the source of heat giving rise to the COD event is localised to a volume much smaller than the thermal diffusion length, otherwise the thermal diffusion would not be significant. If the source is optical absorption there must be a strong non-linearity, such as a temperature dependent absorption, that causes the local absorption to increase dramatically within a small volume so that the heat source is localised.

Most investigations into COD have taken place in GaAs based materials. Even where InGaAs and InGaAsP have been examined it has been in material deposited onto GaAs substrates. Other materials have been investigated rather haphazardly, so no systematic data concerning damage exists. In addition, other mechanisms, such as thermal roll-over, may limit the output power so COD is not always observed. In GaAs based devices there is clear evidence of melting in some cases, but not all. The signs of melting are unmistakable, such as the segregation of constituent elements, and the appearance of material around the damage site that seems to have been ejected and re-deposited, recrystallising in the process. Localised heating to temperatures around 1400°C must have occurred, therefore, but it is not always clear whether melting has occurred or not when the damage appears to be simply an array of pits and holes in the surface. COD has not been reported in InGaAsP/InP, and the material system appears to be stable even against degradation. Short DLD's have been produced in 1.3 μm lasers under prolonged high injection at elevated temperatures of 250°C, but it doesn't always occur. No DLD's have been found at all in 1.5 μm lasers but this does not mean they do not occur because mid-IR materials are not inspected for COD as other mechanisms often limit the operational power [14]. There are no DLD's in InGaAs lasers for 980 nm operation and InGaAlAs lasers for 810 nm operation where the In content exceeded 5%. Indium is believed to pin dislocations and to prevent them diffusing into the interior of the laser structure. Aluminium, on the other hand, is deleterious. The threshold for the production of gradual degradation DLD's in InGaAsP/InP is higher by up to a factor of 8 than in AlGaAs, and experiments on identically strained quantum wells of InGaAs with cladding layers of either $Ga_{0.45}Al_{0.55}As$ or $In_{0.49}Ga_{0.51}P$ showed that Al-containing lasers degrade more than Al-free lasers. Dislocations and diffusion are therefore clearly implicated in COD.

Aging in QW lasers decreases the critical COD power, but in general QW lasers give an improvement in COD intensities, which can be up to ~ 80 MW cm^{-2} in pulsed operation. By way of example Eliseev refers to a high power laser consisting of 7 nm single quantum well in GaAs with graded index separate confinement heterostructure layers of AlGaAs. For a stripe 3.1 μm wide single mode powers of 180 mW were measured with a maximal power of 425 mW. This is an enormous power for such a small aperture, but contrary to the normal practise of increasing the aperture to produce broad area lasers, it seems that narrow stripe width lasers may be better suited to high power operation. Lee [15] modelled carrier and thermal diffusion in DH lasers, but as the structures are very similar to SCH QW lasers with a wide optically confining layer the theory should apply equally well. Indeed, the use of narrow stripes in the form of a ridge waveguide has become a recognised technique for achieving high power output in stable single modes, as will be described later. For heat conduction the important factor is the aspect ratio, γ, of the cavity length to stripe width, and for $\gamma < 70$ the output power decreases as the stripe width is increased, but $\gamma > 70$ there is predicted to be dramatic rise in the maximum output power as the stripe width is decreased. Stripe width notwithstanding, single QW lasers in GaAs can exhibit facet overheating by up to 120°C and in multi-quantum well lasers the facet heating increases with the number of wells. Although optical absorption contributes to the facet heating, surface recombination of the injected current is the principle contributor especially at high currents. The facet temperature saturates somewhat at high output power but once the critical facet temperature is reached any increase in power appears to cause COD. The critical facet temperature in GaAs/AlGaAs QW lasers appears to be ~ 150°C.

There are important differences between QW and DH lasers in GaAs/AlGaAs, not all of which are understood, but which nonetheless point to different mechanisms. In double heterostructure lasers the following are generally found:

- stoichiometric oxides grow during operation if the facet is uncoated;
- oxidation occurs even in inert ambients because there is always some residual oxygen;
- oxygen diffuses deep into the laser structure;
- oxidation is correlated with the presence of Al;
- oxidation leads to segregation and the growth of a defective region;
- both laser absorption and electric current at the facet are important;
- thermal runaway of the facet temperature has been proposed but not observed.

In quantum wells, however:

- oxide growth is not well-defined;
- facet oxidation is neither uniform nor extensive;
- facet temperature rise is correlated with current and not the optical power, hence surface recombination is taken to be more important as a mechanism for generating heat at the facet;
- compositional changes at the facet occur within the first few minutes of operation;
- starting facets are variable in composition and oxidation across the facet;
- the near facet region becomes depleted of As and rich in group III elements as the laser is operated;
- there is no clearly identifiable damage state – melting can occur but does not always;
- thermal runaway of the facet temperature has been observed;
- there is critical temperature rise of ~ 120–140°C after which thermal runaway occurs.

Facets are crucial to the process of COD but as yet there is no clear idea as to exactly what is happening. Some of the mechanisms proposed include:

- direct optical absorption both at surfaces and in unpumped regions, especially where heating has reduced the band gap;
- non-linear absorption mechanisms, such as two-photon absorption;
- second harmonic generation within the cavity and consequent absorption of this super-band gap radiation;
- energy dissipation by non-radiative recombination; electron energy is transferred to phonon modes and hence to the lattice;
- formation of shock waves by rapid thermal expansion and the ensuing energy transfer via phonons to the lattice.

In respect of the latter it is interesting to note that the optical guide also acts as an acoustic guide because of the difference in sound velocity between the two materials and acoustic energy is focused within the central region. The direct oxidation of Al in the cladding layer has been mentioned as a factor, but the thermal resistance of the cladding layer also changes with composition. A low thermal resistance helps to stabilise the facet temperature.

The agreement between theory and experimental observation of COD phenomena is patchy. It is clear that a full understanding has not yet been reached, and quite possibly there is no single mechanism. It is often the case that apparently confusing and contradictory data relate to different physical circumstances, in this case different laser structures built in different materials. The two outstanding questions in the models discussed by Eliseev are, first, what is the mechanism for the generation of defects during degradation, and second, what is the precise mechanism of the thermal runaway? The degradation defects are generated during operation under high driving conditions and affect the COD intensity. They must play some part in COD but their origin and precise role remains somewhat of a mystery. The COD models concentrate on four things in particular; the carrier distribution, the current distribution, the temperature distribution, and the absorption properties of the facet. These are not always treated self-consistently. For example, Henry *et al.* [16] postulated an initial absorption coefficient of 140 cm^{-1} at the surface which increases exponentially to about 10^4 cm^{-1} over the first 100°C temperature rise and then increases but slowly to just over 2×10^4 cm^{-1} at about a temperature rise of 600°C. Such a large starting absorption coefficient is hard to justify; it is much lower than the band edge absorption but higher than would ordinarily be expected at sub-band gap wavelengths. There is certainly no experimental data from the lasers to justify the claim, though in low doped GaAs the absorption coefficient at the band edge is close to 10^4 cm^{-1} hence at elevated temperatures the decrease in band gap brings the band edge absorption into coincidence with the photon energy.

The temperature rise at the facet is postulated to arise initially from surface recombination, but again the evidence to justify this claim is not there. As figure 10.21 shows, surface recombination causes a current flow characterised by the surface recombination velocity (SRV), and the associated joule heating. Untreated semiconductor surfaces are well known as centres for non-radiative recombination so there will be an enhanced current flow towards the surface as carriers recombine. The carrier density at the facet will also be depressed, but whether band edge absorption takes place depends not only on whether the carrier density drops below threshold, but also on whether the temperature induced reduction in the band-gap brings the band edges within the range of the photon energy. In a detailed discussion of various thermal balance calculations, Eliseev suggests that SRV alone is insufficient to produce the observed temperature rise in quantum well lasers in GaAs/AlGaAs and InGaAsP/GaAs, so the mechanism for thermal runaway is not clear. Numerical models of thermal balance require only a heat source located at the facet to allow the temperature rise to be calculated, but the physical origin of this heat source is not important. Thus, a temperature rise of 100°C at the surface of an

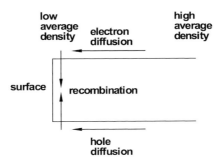

Figure 10.21. Illustration of the currents flowing as a result of surface recombination.

AlGaAs/GaAs DH laser requires a dissipation of 370 kW cm^{-2} [14, p 57] and 5.6 mW cm^{-2} for InGaAs/GaAs single quantum well structures. The differences arise from the thermal properties of the adjacent materials. This explains to a great extent why InGaAsP/InP DH lasers are stable against COD; the heatsinking properties of the layers adjacent to the active region are much better and the average temperature is lower.

The problem identified by Eliseev is to find physical sources to generate such temperature rises, but once the facet temperature has reached a critical value thermal runaway is easy to produce computationally. At elevated temperatures even the cladding layers in quantum well structures can become absorbing because of the reduction in the band gap. For quantum well structures in AlGaAs barriers containing 22% Al a temperature rise of 400°C is all that is needed to induce absorption. Elevated temperatures at the facet therefore strongly increase the optical aborption and optical power dissipation is easy to predict. Below these critical temperatures, however, the optical absorption seems to be unable to generate the power dissipation which will increase the temperature further and increase the absorption, hence the scepticism over Henry's model. The role of the electrical current is therefore considered crucial to the runaway mechanism, though the precise role is again not clear. Electrically induced thermal runaway is not expected to be localised but to occur over the whole of the junction region. This is observed in QW lasers, where the facet damage can often be widespread, but not in DH lasers, leading to the conclusion that in DH lasers the COD is optically induced whilst in QW lasers it is considered to be a result of both optical and electrical effects.

The current distribution within the lasers has been considered in a number of papers reviewed by Eliseev, but the general conclusion seems to be that there is a shortage of mechanisms supplying current to the facet. Photon transport is one mechanism considered. This is the absorption of an internally generated photon at the facet creating electron hole pairs, effectively transporting carriers from the interior of the laser to the facet. SRV is also considered, but in some cases the SRV has to become unfeasibly large (SRV in GaAs is limited to about 2×10^7 cm s^{-1} so anything higher than this becomes unphysical) to generate sufficient heat at the surface. The effect of the shrinkage in band-gap as a function of temperature on the current injection has also been considered, but even this is insufficient to generate the excess current required to produce thermal runaway. For a temperature rise of 100°C above room temperature the band gap shrinks by 50 meV. The current injected into the active layer at the point of the temperature rise increases due to the reduction in barrier. However, it is not a large increase (less than an order of magnitude in the current *density*) given that the band gap of GaAs is ~ 1.4 eV at room temperature, and hence the initial barrier height in the *p–n* junction is of this order.

The effect of temperature on the current flow across the junction is not limited to band-gap shrinkage, however. The electron distribution within the conduction band has an exponential tail at high energies. The total electron distribution is given by the integral over the conduction band of the probability of occupancy multiplied by the density of states., i.e.

$$n(E) = \int_{E_c} D(E) \cdot f(E) \cdot dE. \qquad (10)$$

As the current flow across a *p–n* junction is diffusive, only those electrons with energy exceeding the barrier are available for conduction and for a large barrier this is normally a small fraction of the whole number. The Boltzmann approximation to the electron density is usually valid, and hence the forward bias current density contains a term in $\exp(-V/kT)$. If the barrier is lowered by an amount ΔV for whatever reason, either by applying a voltage or by reducing the band gap, the number of electrons available for conduction across the barrier increases (figure 10.22). However, if the temperature increases from T_1 to T_2 the electron distribution is extended in energy and the number of electrons available for conduction also increases. For a temperature change from 300K to 400 K the increase in current *density* will be several orders of magnitude and certainly much higher than that due to the change in barrier height. Curiously, the contribution to the injected current arising from the temperature alone is not mentioned in the review by Eliseev, and only the effect of the reduction in the barrier is described. Quite possibly, this contribution has been overlooked. If the temperature rise is distributed over approximately 0.1–0.5 μm, as indicated by Eliseev [14, p 57] then the increase in actual current into the surface as a fraction of the whole is very small for barrier lowering. For a cavity 500 μm long, for example, a factor of 7 or 8 increase in current density at the facet results in less than 1% of the current being diverted to the facet, but for an increased injected current density of three or four orders of magnitude it is possible to envisage circumstances in which the whole of the current is diverted to the facet. The precise details of the current redistribution will depend not

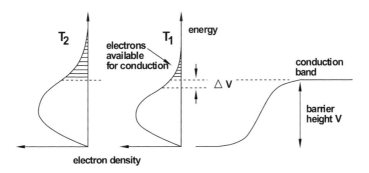

Figure 10.22. Schematic of electron transport across the junction.

only on the laser construction but also on external factors such as the current supply.

There is good reason therefore for supposing that the current contributes to COD, and whilst there is evidence for this in QW lasers it is by no means clear what is happening in DH lasers. As yet there is no microscopic theory of COD and little experimental evidence to support one. The temperature rises in the literature, both measured and predicted, are averages and the degradation mechanisms are not clear. However, cleaved facets are not perfect. They contain several defects, even before laser operation. Vacancies and other defects such as pits have been observed on cleaved surfaces [17] and these will probably act as nucleation centres for degradation. In contrast, where surface have been passivated with sulphur prior to operation the lifetime is significantly enhanced [14, p 69]. One of the problems is that defects can be induced in laser materials and structures at high temperatures, but such high temperatures do not appear to be reached during normal operation. For example, temperatures of 600–800°C are required for rapid thermal oxidation of GaAs [18], but where facets are oxidised uniformly such temperatures have not been observed, and if such temperatures are reached locally the oxidation ought not be uniform. Hence the oxidation of the facets together with the diffusion of oxygen would appear to be a low temperature phenomenon, perhaps an electro-chemical or photo-chemical effect [19]. Theoretical investigations into the diffusion and stability of oxygen in GaAs and AlAs has concluded that the charge state of the oxygen impurity plays an important role. In n-type GaAs the charge state is -2 and the activation energy for diffusion is 1.7 eV, but in p-type GaAs the charge state is $+1$ and the activation energy is 0.77 eV [20]. Diffusion in p-type material under the action of energy supplied by recombination is therefore a distinct possibility. Likewise, Henry *et al.* reproduced DLD's in lasers by melting localised regions with an external laser but the evidence for melting is ambiguous in COD, never mind during degradation.

Some experimental evidence for the atomistic processes at the surface has been gathered in a study of InGaAs/AlGaAs quantum well lasers operated within a scanning electron microscope so the formation of damage could be observed [21]. The temperature of the facet was measured from the peak wavelength shift of the cathodoluminescence (figure 10.23) during the operation and significantly the initial facet temperature rise is lower than after stress testing. This is very clear evidence for the formation of defects which contribute to the surface absorption. After further degradation the quantum well at the facet appears to have diffused and disappeared. From estimates of the diffusion coefficient it is estimated that the temperature must be below 650°C during this phase of the degradation, but this is an upper limit and no estimate of the actual temperature was given. Dissolution of gallium into the anti-reflecting (and also protective) oxide coatings on the facet was identified as the principal cause. Vacancies left behind enhance the interdiffusion of Al and In, and indeed this mechanism is exploited as a means of impurity free vacancy

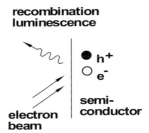

Figure 10.23. Cathodoluminescence in a semiconductor.

disordering of quantum wells in other contexts. This stage of degradation is also identified with the formation of dislocations, so these are clearly identified as arising in the solid state and not as a result of melting. Kamajima and Yonezu [22] also identified the formation of dislocation "knots" in DLD's after pulsed operation, though they ascribed the formation of the knots to the effects of cooling after melting. Each knot is a tangle of dislocations and the number of knots matches the number of pulses. The knots are themselves joined by other dislocations, and the totality forms the DLD.

Again, there is no direct evidence for melting in the above. Rather it seems that the authors have based their conclusions on the supposition that melting is a pre-requisite for the generation of dislocations, but the situation is more complex than this. Hull [23] describes the conditions for homogeneous nucleation of dislocations in solids. In essence a strain of approximately G/30 is required, where G is the shear modulus of the material. The lattice expansion of GaAs at the melting point is not sufficient to produce a strain of this magnitude (it is an order of magnitude too low) so the homogeneous nucleation of dislocations is unlikely. Much more likely is the propagation of dislocations from a pre-existing defect aided by the rise in temperature caused by the local heating.

A fully self-consistent model of COD is probably not possible given the multiplicity of atomistic processes that can occur. Quite clearly facet heating by surface recombination is not a fixed quantity, nor is the photon absorption. Both depend intimately on the nature and density of defects and these change over time. Although band-edge absorption occurs at high surface temperatures, localised absorption at the defects will almost certainly occur, and may be the ultimate cause of COD in DH lasers. Not only are vacancies and dislocations produced during the operation of the laser, but also free As is liberated on the surface. The electronic effects of As on the surface of GaAs are well known in the electronics community concerned with metal-insulator-semiconductor devices. In contrast to silicon, which has a stable oxide, oxidised GaAs is characterised by the existence of a very large density of interface states lying close to mid-gap which pin the Fermi level and prevent the onset of inversion at the surface. GaAs transistor technology has therefore concentrated on

MESFETs and JFETs and heterojunction bipolar devices. The mid-gap states are associated with free As which is liberated during the oxidation, as the oxygen is transferred from As to Ga. The formation of oxides on facets in GaAs based devices therefore has potentially devastating consequences for the electronic properties of the surfaces, and the existence of a large density of states close to mid-gap will provide a ready source of non-radiative recombination centres. The density of these may also be expected to increase as the laser is operated, but whether a limiting density is achieved is an unanswered question. Sulphur passivation creates stable sulphides of Ga, As and Al and prevents the formation of an oxide and improves the lifetime as well as increasing the COD intensity.

Finally on the subject of COD, Eliseev has described at some length the use of various windows over the facets to prevent catastrophic damage. For example, ZnSe deposited as a window has been described [14, p 70], and also non-absorbing mirrors (NAM). These can be constructed in quantum well devices by intermixing the quantum well in the region of the facet. The composition of the well is therefore changed as barrier material is intermixed and the band-gap increases, reducing the absorption at the surface. Output facets in VCSELs are protected by the window-like nature of the mirrors as the Bragg reflectors are buried below the surface.

10.6 Very high power operation

Increasing the total emitting area is an effective way of increasing the total power output, but as figures 10.3 and 10.4 show, the beam quality can be poor and coupling the output into a fibre or achieving a small spot size for materials processing becomes difficult. High power output from single emitters with good beam characteristic is therefore important, and this means optimising the performance of the laser. Assuming that COD can be avoided, thermal roll-over will limit the power output. Temperature rises within the active region cause increased injection over the current confining barriers thereby reducing the carrier concentration in the active region. In the case of lasers using wavelength selective feedback, such as dielectric mirrors or distributed Bragg reflectors, the reduction in band gap with temperature can also shift the gain maximum outside the feedback range, thereby causing the laser to cease to operate. The effects of thermal roll-over can be serious and the phenomenon cannot easily be designed out of the laser beyond the obvious step of optimising the stripe length/width ratio. As discussed, the rate at which heat can be conducted out of the active region depends on the thermal resistance of the materials, but the choice of materials is determined by the wavelength requirements rather than their thermal properties. Other ways have to found of maximising the heatflow out of the active region of the laser. One such way is to mount epitaxial structures epi-side down on the cooling mount so that the active region is as close to the cooler as possible, but this is a purely mechanical step and says nothing about the laser itself.

For a long stripe, the gain has to be correspondingly low in order to keep the output intensity below the COD level. Low gain can be achieved by low confinement, and such devices allow stable high power operation for a number of reasons, not least because the effects of phase-amplitude coupling are minimised. Most important for high power operation, however, a low optical confinement means that a great deal of the optical power is contained in the cladding layers and hence the spot size is increased so that a given output intensity on the facet requires a larger total power output [24]. The key to such a design is to ensure first that the optical confinement is low within the quantum well and second, that the absorption coefficient in the substrate is low (less than ~ 0.03 cm^{-1}). The optically confining layer is therefore relatively thick, unlike a low power device. Normally the active layer thickness is optimised for a minimum equivalent transverse spot size defined as d/Γ, where d is the quantum well width and Γ is the optical confinement (see figure 10.24), but this maximises the intensity at the output facet and leads to a lower total output power. The operating point of a high power laser is chosen at a higher confinement layer thickness such that the spot size is relatively large but the waveguide still operates in the fundamental mode. In fact, the first order mode may well not lase as the minimum in the optical field is located at the quantum well (see figure 10.25). The overlap between the optical field and the heavily doped, and usually absorbing, substrate layers also has to be low, and in symmetric structures [24] the cladding layers can be over 1μm thick to keep the optical field away from the substrate. Modulation speed is compromised, of course, but this is not a problem if the intended application requires CW. If pulsed output is important the response of the laser can demonstrate a delay of the order of microsecond between the electrical injection and the optical output [25,26]. Similar ideas were described in relation to broad area Al-free InGaAsP/InGaP laser diodes emitting at 980 nm [27]. Using lasers with cavity lengths of 1 mm resulted in output powers of 1.5 W/facet (3 W in total) in uncoated devices, and 6 W with the facets coated to 3% and 95% reflectivity.

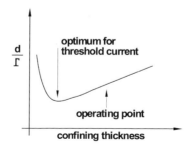

Figure 10.24. Operating active thickness for high power diodes lasers.

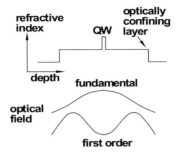

Figure 10.25. Transverse optical modes in high power LOC structures.

The term "aluminium free" has become synonymous with InGaAs sources emitting at 980 nm, although the technology extends to other wavelengths such as 940 nm [28] and 808 nm. However, 980 nm sources are particularly important for pumping erbium ions in fibre amplifiers, and as the output has to be coupled into a fibre high power stable single mode operation from narrow emitters need to be achieved. Although the Al-free technology described above is capable of delivering 1.5 W/facet, this is from a broad area device and therefore is not suitable for coupling into a fibre. Attaining coupled powers in the region of 70–100 mW requires correspondingly high outputs from the lasers, and ideally powers in excess of 200 mW should be achievable as coupling the ouput of the laser into the fibre may be no more than 50% efficient. Erbium has two absorption bands centred on 1480 nm and 980 nm but emits at 1.5 μm, so a length of fibre coiled into a loop (figure 10.26) and inserted into an optical communication network can amplify signal transmitted down the network. The choice between pumping wavelengths has gradually shifted from 1480 nm to 980 nm for several reasons: 980 nm pumping results in a lower noise figure from the amplifier; the drive current requirements for the amplifier are lower; and there is a larger wavelength difference between the pump and the signal. Against this the optical conversion from the pump to emitted light at 1.5 μm is more efficient with 1480 nm pumps. The reliability of 980 nm sources has lagged behind that of 1480 nm sources but increasingly

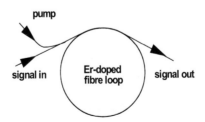

Figure 10.26. Schematic of an Er-doped fibre amplifier.

the former are seen as a viable alternative. It is possible to achieve in excess 250 mW of power in a stable single mode from a ridge-waveguide structure 4–5 μm wide.

As described above, one of the chief difficulties with 980 nm sources has been their reliability, especially against degradation and COD. The basic structure of all 980 nm emitters consists of compressively strained InGaAs quantum wells with a lattice mis-match of ~1.4% on GaInP layers lattice matched to GaAs, but a number of different designs exist. The differences lie mainly in the optically confining layer, which may be uniform GaInAsP, graded GaInAsP, or a superlattice of GaAs/GaInP. Comparing different devices, Pessa et al. [29] describe degradation rates of 0.2 mW COD power per 1000 hours of operation in sources emitting at 50 mW, but this increases dramatically to 10 mW per 1000 hours of operation at 70 mW output. The principle degradation mechanism appears to be the growth of defects at the interface between the laser materials and the facet coating [30], but even so the mean time to failure is estimated to be 18 years at 100 mW operation and 12 years at 150 mW operation [29] based on steady degradation rather than catastrophic single event failure. The importance of the facet has been demonstrated in different designs of laser where the facets have been protected by different methods. NAM's have been fabricated by implantation induced disorder of the quantum wells at the facet [32]. Horie et al. [32] used a combination of very low energy ion bombardment of the facet to remove gallium oxide at the surface, a thin coating of amorphous silicon to act as a diffusion layer, and a coating of aluminium oxide to act as an anti-reflection coating. Oxygen is thereby prevented from reaching the surface and the reliability of the lasers is improved.

The lateral confinement of the optical field in a high power device is usually achieved using a ridge waveguide (figure 10.27). This has become a common design for high power single mode devices since the early 1990's when Jaeckel et al. [33] described high power single quantum well lasers in AlGaAs delivering 425 mW from a 3 μm wide ridge. Analytical models of mode propagation in ridge wave guide lasers do not exist even as yet, and numerical solution of the wave equation is the only method by which an

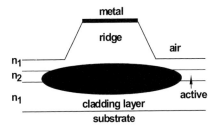

Figure 10.27. Ridge waveguide structure with the optical field shown in black.

accurate description of the mode profile can be obtained. However, the important factors can be identified as the thickness of the low-index layer above the active layer (or optically confining layer in the case of a quantum well laser) and the width of the ridge. The latter is usually a few μm and usually topped by the metal electrode, so the ridge also acts as a stripe contact. Optically it behaves very differently from a stripe, however, where the optical confinement is provided by gain induced refractive index changes. The ridge waveguide works because outside the ridge the optical field can penetrate into the ambient (labelled as "air" for convenience in figure 10.27), forcing the optical power to shift towards the substrate. This not only reduces the gain but it also lowers the effective index of the guide [34], leading to lateral confinement, as the waveguide is symmetrical beneath the ridge and the optical field distribution is different.

This is more complicated than the gain guiding that occurs in the stripe laser but has the advantage that if used in conjunction with low gain devices such as quantum well lasers the optical wave front is kept planar and the effects of carrier induced refractive index changes are minimized. Quantum well ridge wave-guide lasers are often limited in their output power by thermal roll-over. Mawst et al. [27] gave the intensity for COD in Al-free lasers as ~ 15 MW cm^{-2}, and for an effective beam diameter of 5 μ defined by a ridge waveguide this corresponds to an output power of nearly 4 W. This is nowhere near the output powers reported by Pessa et al. [29] and explains why these devices are limited by thermal roll-over rather than COD.

A modification to the design of the ridge finding increasing application is the use of the tapered active region [35]. The taper typically is only a few degrees so over the length of the cavity the width will change from ~ 3 μm at the high reflectivity facet to ~ 30 μm. The advantages of this type of geometry are several fold and include; first, the narrow end serves as a mode filter, so that higher order waveguide modes propagating in the flare are suppressed in the single mode region; second, the broad end allows the beam to spread and the intensity at the facet to be reduced, reducing the probability of COD; and third, the processing is no more complicated than the standard ridge processing and does not involve the use of expensive grating technology. Diffraction limited operation is possible with beam divergences as low 3°–6° [35]. However, these lasers are multi-pass devices and as such are sensitive to variations in the carrier density. A significant depletion in carrier density, otherwise known as spatial hole burning, can occur within the stripe and the problem is exacerbated at high pumping currents. Nonetheless, high performance laser designs can be fabricated using this technology [36], including the integration of a laser with a tapered amplifier system in the so-called "master oscillator power amplifier" (MOPA) configuration [37]. Figure 10.28 shows two configurations of the amplifier arrangement as discussed by Walpole. One of the disadvantages of the MOPA as shown above is the possibility of double cavity effects, the gratings on the end of the single mode active region form one cavity whilst the tapered

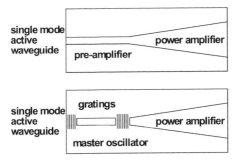

Figure 10.28. Tapered ridge laser structures.

region forms another. Longitudinal modes propagating in the device must be common modes of the two cavities, i.e.

$$\lambda = \frac{n_1 L_1}{m_1} = \frac{n_2 L_2}{m_2} \qquad (11)$$

where L are the lengths of the cavities, m the mode orders, and n, the refractive indices, may be different if the carrier densities within the cavity are different. This effect can be overcome by moving the grating to the other end of the flare to form a single cavity, in which case the active single mode waveguide acts as a mode filter.

10.7 Visible lasers

High Power visible lasers emitting in the red at 650 nm have become the standard for writing optical data storage media. As discussed in chapter 6, AlGaInP lasers are not without their difficulties. Growth of the material by MBE on GaAs substrates has been been considered difficult, but techniques for optimised growth have emerged [38–40]. The principal difficulty is caused by the tendency of GaInP to "order" during growth [41]. Ternary III–V alloys of the type $A_x B_{1-x} C$, where A and B are group III elements and C is a group V element, have been predicted to be thermodynamically unstable [42] and if the surface mobility of the ad-atoms is high enough during growth they allow segregates out into ordered structures of different, but particular, compositions. Facets and terraces can appear on the surface during growth. Ordered in this sense is specific to the phenomenon, and disordered does not mean a lack of crystal structure, merely that the crystal structure and composition are uniform and the surface is smooth. The conditions for disordered growth normally require the GaAs substrates to be cut off-axis, a process which leaves steps on the surface on the atomic scale. Growth on to stepped surfaces helps to ensure disorder.

Eliminating oxygen from the system is also important. Al is highly reactive and the presence of oxygen, either during the growth or afterwards is detrimental. Apart from the role the impurity plays in COD and facet oxidation, oxygen impurities also result in the presence of non-radiative recombination centres. The source of the phosphorus is crucial to the purity of the films during growth, and afterwards capping the films with GaAs or AlGaAs reduces the native oxide at the surface. Nadja et al. [43] have examined a number of sources for MBE and shown how oxygen may be incorporated into the films during growth. Using both gas source MBE with phosphine as the source and solid source MBE with GaP as the source from which P_2 is sublimed, a laser structure consisting of $(Al_{0.7}Ga_{0.3})_{0.52}In_{0.48}P$ cladding layers with a GaInP quantum well was grown, as well as simple ternary or quaternary layers onto a GaAs buffer layer grown onto the substrate. Oxygen was shown to be present in low concentrations in both AlGaInP and GaInP, but in the layers grown using the GaP cell the oxygen concentration was higher. Within the active region of the laser, however, a large concentration of oxygen was recorded. Gas source MBE is therefore better suited to the growth of laser structures despite the toxicity of the phosphine source, but the MBE of these compounds has lagged behind MOVPE.

The principal issue in the lasers is the design of the optical cavity to ensure efficient operation. The band offsets are relatively small for visible lasers, and temperature sensitivity is to be expected. These are limitations associated with the materials, however, and little can be done to overcome them beyond moving to a different material system. For the optical cavity, however, the thickness can be altered to provide optimum confinement while limiting the electrical and thermal resistance of the cladding layers. For GaInP quantum well lasers using AlGaInP cladding grown directly onto GaAs, optical absorption within the GaAs is an important source of loss. The optical intensity can be reduced by increasing the thickness of the cladding layer to ensure a high degree of confinement, and a thickness of ~ 0.5 μm is sufficient [44,45]. Large optical cavities (LOC) have also been used to improve the COD power in 670 nm sources [46].

A variety of designs exist for single emitter structures. Hiroyama et al. [47] have used strain compensated multi-quantum well devices with a buried ridge for emission at 660 nm (figure 10.29). P-type GaInP forms the contact and the current is limited by the existence of a reverse biased heterojunction outside the ridge. Under pulsed conditions (100 ns, 5 MHz repetition rate) this structure was capable of delivering over 100 mW, even at 80°C. Replacing the p-GaAs cap with transparent AlGaAs reduces the operating current [48]. Single emitter sources such as this are finding increasing application in optical storage media. The transition from 780 nm to 650 nm allows for a greater storage density, but the source has to be capable of being focused to a very small spot size in order to take advantage of the potential for increased resolution. Extended sources are therefore of no use, but single mode sources from small

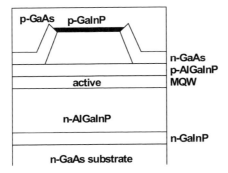

Figure 10.29. A single emitter visible laser of the type constructed by Hiroyama [47].

apertures are ideal. The minimum power required to record optical data is about 30 mW, which is much higher than the power required for reading, typically about 5 mW. In read-write systems, though, using separate lasers for each task simply adds to the complexity of the storage device, and combined use of a high power laser for reading and writing is desirable. This brings with it a particular problem of high speed modulation. Optical read circuits are usually modulated at high frequency (typically 10's of MHz) in order to reduce the noise caused by back reflections into the laser source. If the same laser is required to write this must also be modulated and the circuitry becomes complicated and expensive. One solution is self-pulsation.

The method chosen to achieve self-pulsation is a form of passive Q-switching. The Q (or quality factor) of an oscillator is a measure of the lifetime, and in the context of a laser a low Q means a low photon lifetime within the cavity. Either a high intrinsic loss or a low feedback reflectivity will cause a low Q. In such a cavity a high inversion population density is necessary to overcome the losses and bring about threshold. In the case of a semiconductor the carrier densities rise well above the normal levels expected for population inversion. If at some point the Q of the laser is switched from a low value to a high value the excess carrier density leads very rapidly to a very high photon density, much higher in fact than would normally be expected. This abnormally high photon density destroys the population inversion as the excess carriers are depleted so the output of the laser is limited in duration to a very short pulse. If the cavity is then switched back to a low Q value the process must begin again and repetitive switching therefore produces a train of short pulses at high frequency. Passive Q-switching requires a component within the optical cavity that self-switches, and the most obvious candidate is a saturable absorber. This is a device that has a high absorption at low photon density but a low absorption at high intensity. A simple two-level system with a finite number of states available for absorption is all that is required. At high optical intensities the populations of the two levels will be equal and the absorption will decrease to zero. Population inversion in such a system is not possible so the transition from absorption to gain does not occur. When the photon density is low after the output pulse the relaxation of the populations in the two-level system back to the thermal equilibrium distribution causes the absorption to rise again, and hence the process of passive Q-switching leads to the pulse train already described.

Self-pulsation entirely eliminates the need for expensive drive electronics, and furthermore is can be achieved with only a minor change to the laser design. A quantum well placed within the optical cavity and in close proximity to the quantum wells comprising the active region will act as a saturable absorber. There is a finite number of states available for absorption and carrier confinement ensures that carriers do not diffuse away from the absorbing region to depopulate the upper levels as they would in a bulk semiconductor. Pulse repetitions of the order of 100 MHz occur [49,50]. Upon switch-on the laser will of course exhibit self-pulsation due to the relaxation oscillations but these are usually transient. In the presence of the self-absorber, however, the oscillations should be sustained, but if the design is not optimised the self-pulsations die away [51]. The saturable absorber will of course be fed by the current and may itself achieve population inversion if it is too close to the active region. The absorber has to be below threshold itself so the quantum wells are designed so that most of the carriers are captured within the active layer.

The saturable absorber can be modelled using the standard rate equations [52]. As before, the carrier density in the quantum well is given by

$$\frac{dN_{qw}}{dt} = \frac{J_{inj} - J_l}{q} - SV_g \frac{dg}{dN}(N_{qw} - N_{0qw}) - \frac{N_{qw}}{\tau_{qw}} \tag{12}$$

where J_{inj} is the injected current, J_l the leakage current. The other terms have their usual meanings and represent respectively stimulated and spontaneous recombination. It is also necessary to construct a similar equation for the carrier density in the absorber.

$$\frac{dN_\alpha}{dt} = \frac{J_l}{q}\frac{V_{wq}}{V_\alpha} - SV_g \frac{d\alpha}{dN}(N_\alpha - N_{0\alpha}) - \frac{N_\alpha}{\tau_\alpha} \tag{13}$$

where α now represents the absorptivity. The photon density is given by

$$\frac{dS}{dt} = SV_g \frac{dg}{dN}\Gamma_{qw}(N_{qw} - N_{0qw}) + SV_g\Gamma_\alpha \frac{dg}{dN}(N_\alpha - N_{0\alpha}) - \frac{S}{\tau_s} + \beta BN_{qw}^2 \tag{14}$$

where the last two terms represent absorption and radiative recombination respectively. These equations show very clearly that there are two coupling mechanisms between the wells and the absorber. The first is the intended mechanism via the photon density and the second is a parasitic element via the carrier leakage. The leakage term was calculated by Summers and Rees [50] using a drift-diffusion model of carrier conduction. The carrier mobility in the AlGaInP cladding layers is low and a substantial electric field exists within the spacer between the wells and the absorber. The model described by Summers and Rees showed excellent agreement with experiment, from which it was concluded that above 60°C the leakage over the barrier is sufficiently large to begin to affect the recovery time of the absorber so that the oscillations are damped down.

10.8 Near infra-red lasers

A variety of wavelengths in the near infra-red ranging from about 700 nm to 1.5 μm are useful and laser sources can be found that cover most of this range. Sources emitting at 980 nm have already been described but particularly useful wavelengths include 740 nm and 808 nm. The first of these is finding increasing application in photodynamic therapy (PDT), a technique proving successful in the treatment of certain cancers and now being extended to other diseases. The technique utilises the characteristic absorption of photo-sensitive drugs injected into the patient. Until these are activated by light, they are harmless, and furthermore they tend to accumulate in cancers. Irradiating the patient with an intense source of light at the appropriate wavelength activates the drug but only in the region of the illumination. In this manner the conventional difficulty associated with chemo-therapy, namely its lack of specificity and its general toxicity, is overcome. The first chemicals absorbed at 630 nm and 660 nm, but now drugs are being developed with strong absorption above 740 nm. This is a far more convenient wavelength as the absorption in tissue is lower and the light can penetrate deeper into the body. Typically, structures in tensile strained quantum wells of GaAsP in AlGaAs cladding layers are used [53–55].

Sebastian *et al.* have also investigated Al-free lasers emitting at 808 nm, concentrating on the beam quality [56,57] and on optimising quantum well structures for high power operation [58]. 808 nm sources are required for pumping Nd-based solid state lasers, so very high power operation from multi-emitter bars is normally required. Long life-times and stability are essential. In fact, it is the diode lasers that are the principal limitation in the reliability of solid state lasers. Knauer *et al.* [59] have investigated the uptake of oxygen into the AlGaAs cladding layers of quantum well structures in GaAsP and shown that it very important for the reliability of sources at 730 nm but not so important for 800 nm sources. The reliability has been discussed by Diaz *et al.* [60] who also describe the production of lattice matched InGaAsP lasers at 808 nm by MOCVD [61]. Compared with AlGaAs lasers, which can cover a similar wavelength range corresponding to 1.42 eV to 1.9 eV, InGaAsP has a similar T_0 (~ 160 K) but a much higher intensity for COD corresponding to 6 MW cm^{-2} as opposed to 0.5–1 MW cm^{-2} for AlGaAs. AlGaAs tends to oxidise and DLD's are created relatively easily during degradation, with the result that the lifetime is ~ 10^5 hours, even if quantum well disordering is used to create non-absorbing mirrors. InGaAsP, on the other hand, has no Al to oxidise, except in the cladding layers where the intensity is relatively low and furthermore has the large In atom in its structure which effectively prevents dislocation motion. The surface recombination rate is an order of magnitude lower in this material than in AlGaAs, so the facet heating is lower. These three combine to give InGaAsP lasers a lifetime of over 10^7 hours.

10.9 Mid infra-red diode lasers

III–V lasers emitting in the wavelength range 2 μm–5 μm are very important for chemical detection. Many gases and organic compounds have characteristic absorption peaks within the mid-IR range of 3–5 μm, and optical sources that can match the absorption bands of particular chemicals are therefore required. In order to achieve these wavelengths III–V diode lasers are commonly based on antimony (Sb), and devices utilising (In,Ga)(As,Sb) have been investigated since the late 1970's, though high power devices did not start to appear until the mid- to late-1990s. These are still the subject of much research, and as with other lasers a bewildering array of techniques, ideas, and designs can be found within the literature. It is not possible, nor indeed even sensible, to attempt to summarise all this work, and so a synopsis of a few of the main ideas is given.

Drakin *et al.* [62] investigated DH lasers with $In_xGa_{1-x}Sb_{1-y}As_y$ quaternary active layers and $Al_{x1}Ga_{1-x1}Sb_{1-y1}As_{y1}$ confining layers deposited by LPE. Operating at 300 K in pulsed mode, these devices emitted at 2.2–2.4 μm. The substrates were *n*-type GaSb:Te or *p*-type GaSb:Si, both doped to $\sim 10^{18}$ cm^{-3}. The normalised threshold current density was 5.4 kA cm^{-2} at $\lambda = 2$ μm rising to 7.6 kA cm^{-2} at $\lambda = 2.4$ μm for devices of stripe width 12–14 μm with active regions more than 1 μm wide. Loural *et al.* [63] investigated the refractive index of very similar structures and found refractive indices for the active and cladding layers of 3.78 and 3.62 respectivey, and room temperature minimum threshold currents of 2 kA cm^{-2}. Martinelli and Zamerowski [64] investigated DH devices in InGaAs with InAsPSb cladding layers at $\lambda = 2.44$–2.52 μm. These were operated at 80–90 K with a threshold current density of 0.4 kA cm^{-2}. The output power saturated at 4 mW, despite a stripe width of 40 μm over a 200 μm cavity length, and above threshold kink in the power-current characteristic was observed. At 40 μm stripe width these were effectively broad area lasers and instability of the output characteristics is not surprising.

Low band-gap materials such as these suffer from problems of carrier confinement and Auger quenching of the luminescence. Auger quenching is simply an internal absorption of the photon at a transition of similar energy, in this case the split-off valence band which is at a similar energy to the band gap (figure 10.30). An extra term for photon loss would need to be included in the

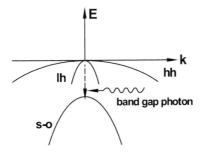

Figure 10.30. Auger quenching by intervalence band absorption.

rate equations as this photon is removed by the Auger absorption from the laser cavity. The carrier confinement is poor because the materials are themseves of low band gap and hence the barriers are correspondingly small. Lane et al. [65] have investigated carrier leakage in DH devices with 1 μm wide active layers of either InAsSb or InAs in InPAsSb cladding layers by means of the temperature dependent photoluminescence. A simple theory shows that for $T < 150$ K Auger recombination is not important and that carrier leakage is the dominant mechanism. Tsou et al. [66] conducted a theoretical study of carrier leakage in InAs active layer DH lasers and concluded that carrier leakage can be suppressed if the cladding layer band gap exceeds the active layer band gap by 0.4 eV.

It worth noting that in the field of mid-IR lasers III–V materials are not necessarily the preferred choice. IV–VI lead salts such as PbSe or PbSnTe have been available commercially for many years and in some respects are superior to III–V lasers. As discussed by Shi [67] lead salt lasers have lower Auger recombination rates than III–V materials, although Lane et al. showed in fact that some of the III–V materials have much lower Auger rates than previously thought, especially below 150–200 K. Nonetheless, Auger rates in III–V materials are significant and may limit the operation of these devices to low temperatures, and in this respect the IV–VI lead salts are superior. However, lead salts have other disadvantages compared with the III–V's; the traditional materials used for confinement do not offer good electrical confinement as the band offsets appear to be limited to ~ 100m eV. The electrical mobility decreases as band gap increases, setting an effective upper limit on the usable band gaps. Furthermore, the low thermal conductivity of the confinement and substrate materials limits the power dissipation in the devices. Shi therefore proposed in a theoretical study the use of GaSb cladding layers with PbSe as an active layer based on the idea that the total difference in band gap is ~ 420 meV; the refractive index change is large at λ around 4 μm (3.83 for GaSb and 5.0 for PbSe); the thermal conductivity of GaSb is higher than that of PbSe, so thermal dissipation is better; the high carrier mobility of GaSb will reduce the threshold current; and last, but by no means least, GaSb is available as a commercial substrate material. Theoretical studies such as this are very useful in elucidating the desirable properties of the materials required for good lasers, but such structures may not be realisable in practice. The formation of intermediate compounds such as Ga_2Se_3 at interfaces, or cross-diffusion of constituent elements, are practical issues that might well limit the usefulness of these material combinations. Nonetheless, hybrid IV–VI/III–V devices may well prove successful given the superiority of the III–V materials technology over IV–VI, particularly in respect of the availability of substrates. For this reason, this discussion of high power mid-IR lasers will be restricted to III–V devices and the reader is referred to Tacke [68] for a review of lead salt lasers.

Reports on high power mid-IR devices began to appear in the late 1990's. Yi *et al.* [69] compared the far-field pattern, perpendicular to the junction, of DH and MQW lasers in InAsSb/InPAsSb/InAs devices for operation at $\lambda = 3.2$–3.6 μm. Grown by MOCVD, the MQW lasers comprised ten compressively strained 10 nm wells of InAsSb separated by 40 nm barriers embedded in a 1 μm thick waveguide layer, and the DH lasers comprised a 1 μm thick InAsSb active layer ($E_g = 388$ meV) sandwiched between 1.5 μm thick InPAsSb cladding layers of band gap 520 meV. The DH lasers exhibited a far-field pattern that broadens very rapidly with temperature above 120 K (figure 10.31), whereas the MQW laser has a constant FWHM of 38° up to 150 K. Both lasers exhibited structure in the far-field pattern super-imposed on the Gaussian. The authors ascribed the cause to fluctuations of the refractive index in the active layer caused by compositional inhomogeneities and the consequent non-uniform carrier densities. The MQW lasers were more stable against this because of the smaller total volume of the active region. The properties of the beam in the plane of the junction were not described.

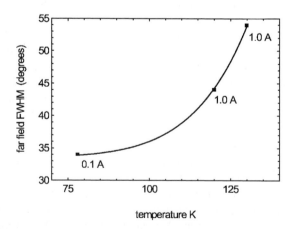

Figure 10.31. The variation of far-field pattern with temperature for InAsSb DH lasers. (After Yi *et al.* [69].)

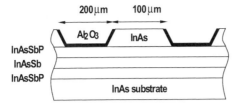

Figure 10.32. A high power InAsSb device. (After Rybaltowski *et al.* [71].)

High power operation at powers of 1 W and 3 W at 80 K was achieved in DH lasers of InAsSb sandwiched between InAsSbP cladding layers with total apertures of 100 μm and 300 μm respectively [70]. These devices exhibited excellent uniformity over their area, as demonstrated by the higher power devices. These consisted of three 100 μm wide stripes and the power simply scaled with the number of stripes, which suggests that they may be suitable for large scale production. The principal difficulty encountered with these lasers appeared to be their susceptibility to processing induced damage. To those familiar with silicon technology, and even GaAs technology to some extent, the lack of robustness of other materials can come as a surprise. InAs is easily damaged, even by the act of bonding a contact to the device, and this can lead to a lack of processing reproducibility. The surface passivating layer of Al_2O_3 between the stripes was used to good effect as an insulator that allowed bonding away from the stripe and helped to prevent damage (figure 10.32). The threshold current for these devices was a very low 40–50 A cm^{-2} with a characteristic temperature 30 K $< T_0 <$ 50 K. It is perhaps not surprising that these low band-gap devices are sensitive to temperature.

The maximum current that could be supplied was limited by the power supply to 30 A, and at this upper value the optical output power was approximately 3 W fom a three-stripe device. The estimated joule heating at this current was $\sim$ 23 W per stripe, demonstrating that carrier confinement is poor. This inevitably induces significant heating of the active region and wavelength shifts of over 50 nm were recorded in going from 0.15 A to 10 A per stripe. In fact, the output consisted of two well defined peaks themselves separated by about 20 nm at low current (3220 and 3200 nm) but which merged together to form a broad asymmetric peak at the higher currents (maximum intensity at $\sim$ 3275 nm). Temperature rises of $\sim$ 55 K in the active region were estimated from this wavelength shift. The far field pattern of these lasers was particularly good, with a divergence of $\sim$ 12°, a finding not easily explained by the authors based on their knowledge of the refractive index differences.

Such devices therefore hold out a lot of promise for high power operation, but clearly the three factors of mechanical hardness, low thermal conductivity, and high temperature sensitivity impose limitations. The last of these is due to carrier leakage at $T <$ 150 K, so in a subsequent development the cladding

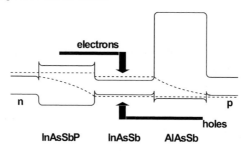

Figure 10.33. Electron confinement via an epitaxial layer of AlAsSb.

layers were optimised to give a better high power performance [71]. In particular, electron leakage was reduced by the use of AlAsAb ($E_g = 1.5$ eV), which has a very large conduction band offset with InAsSb (figure 10.33). This can only be used on one side of the laser as the electron injection would be severely affected, with some detriment also to the hole confinement. Therefore an asymmetric wave-guide exists, the consequences of which were not investigated. This laser also suffered from the poor thermal conductivity of these materials, with temperature rises of 70 K reported for the active region even during pulsed operation. In fact, the thermal conductivity of the AlAsSb is actually lower than the thermal conductivity of InAsSbP. The power reported in this mode was 6.7 W, and for CW operation output powers of 350 mW were reported for a single stripe 100 μm wide, with $T_0 = 54$ K.

Although the performance of these lasers is superior to that of the symmetrical structure to some extent, extending the structure to wavelengths beyond 4 μm is problematic. A heterojunction of InAsSbP/InAsSb still exists on the n-side and reduction of the band gap of the InAsSb active region to cover the longer wavelengths introduces such strain that misfit dislocations are generated that destroy the laser action through non-radiative recombination. One way of overcoming these difficulties is to employ strained layer superlattices, but here the intention is not to use the composition dependent energy levels to alter the wavelength, as in conventional multi-quantum well structures, or to use the strain to assist the high power operation through modifications to the valence band, as discussed previously. Rather, the intention is to fabricate a large number of thin layers all below the critical thickness so that devices can be tailored to give long wavelength emission without the deleterious effects of dislocations. The superlattice is intended to constitute the active region of the device so the barriers are chosen deliberately to have low conduction and valence band offsets to allow for uniform carrier densities over the length of the structure. Optical and additional electrical confinement is provided by wide band gap InAs$_{0.2}$Sb$_{0.28}$P$_{0.52}$ cladding layers on either side of the superlattice, lattice matched to the InAs substrate.

Of course, the well width dependence of the energy levels cannot be ignored, but this is in fact an advantage of these particular structures. The first report of these strained superlattice mid-IR lasers [72] showed that by fixing the composition, and hence the strain, of the wells and barriers, and further simplifying the design so that the wells have the same thickness as the barriers, simply varying the well width from 1 nm to 23 nm allows for operation between 3.3 μm and 4.5 μm. Thus an 80 period $InAs_{0.75}Sb_{0.25}/InAs_{0.95}P_{0.05}$ superlattice of 5.2 nm wells and barriers (as determined experimentally) with an average strain of 0.79% fabricated into 100 μm wide stripes demonstrated a maximum power per stripe of 546 mW from a cavity 1.8 mm in length at a wavelength of ~ 4.01 μm. The threshold current was 100 A cm^{-2} and $T_0 = 27$ K. The relatively poor thermal sensitivity characterised by the low T_0 is compensated for by the fact that the junction temperature in these lasers was estimated to be no more than 20 K above the heatsink, a significant improvement over the double heterostructure lasers operating around 3 μm. Variations in the structure to allow operation out to 4.8 μm are described in [73] and InAsSb strained quantum wells with InAlAsSb barriers operating at 3.9 μm have been described by Choi and Turner [74]. InAlAsSb has a larger conduction band offset with InAsSb than either InGaAs or InAlAs so these lasers could operate pulsed up to 165 K. At 80 K the threshold current was as low as 78 A cm^{-2} with powers of 30 mW/facet from a 100 μm wide stripe on a 1000 μm long cavity.

InAsSb is not the only III–V contender for mid-IR operation, however. GaInAsSb has also been investigated extensively, with MBE being the preferred method of gowth. The MBE of these materials is by no means straight forward, though, leading to poor reproducibility in the fabrication of laser structures. Mourad *et al.* [75] describe the difficulties of controlling the group V stoichiometry, as competition between the As and Sb depends on the fluxes of the two as well as the substrate temperature. These authors overcame this difficulty by the ingenious use of short-period superlattices grown by modulated MBE. These consist of alternate layers of GaInAs and GaInSb, achieved by alternate opening and closing of the group V shutters in sequence while the group III fluxes are kept constant throughout. The sequence of alternate layers of either a binary or ternary compound has the properties of a quaternary compound of the average composition. Mourad's devices were optically pumped by diode laser arrays emitting at 808 nm and emitted at 1.994 μm with output powers of 1.9 W/facet at a pump power of 8 W at 82 K. In view of the recognised, but imperfectly understood, role that electrical injection has on optical damage at facets it is by no means clear that similar powers could be achieved through electrical injection.

MBE growth of a different composition designed to emit at 2.63 μm was described by Cuminal *et al.* [76] in which direct control of the composition was achieved through a combination of appropriate flux and substrate temperature. The substrate temperature is in fact crucial, and an optimum exists around 400°C. Below this Sb clusters form on the surface and above this re-evaporation of the group V elements leads to the formation of point defects deleterious to laser operation. The lasers utilised a separate confinement structure in which five compressively strained quantum wells of GaInAsSb (7 nm) separated by four GaSb barriers (30 nm) were contained in 0.1 μm GaSb spacers which were themselves contained within 2.1 μm thick wider band gap AlGaAsSb optically confining cladding layers. These quantum wells are in fact of type II structure so there is no direct overlap of the electron and hole wavefunctions except at high injection. Both broad area and narrow stripe lasers were fabricated (100 μm and 8 μm respectively) which were operated pulsed at room temperature. For the broad area lasers the cavity lengths ranged from 240 μm to 1100 μm with power outputs of 60 mW/facet and 20 mW/facet respectively. The narrow stripe lasers exhibit red-shifted output indicative of junction heating (see Munoz *et al.* [77] for a discussion on the band gap shift in GaSb and GaInAsSb with temperature).

The importance of Auger processes in limiting the laser output in these materials was demonstrated by Turner *et al.* [78] who achieved 1 W CW at 2.05 μm from a 100 μm aperture structure fabricated from a single strained quantum well in GaInAsSb/AlGaAsSb. The current at this power was just less than 1 A and the device was operated at 10°C with a characteristic temperature $T_0 = 65$ K. The cavity length was 2.2 mm and the facets were coated to 1% and 99% reflectivities. At 3 mm cavity length the threshold current was as low as 50 A cm^{-2}. These authors modelled the output of their laser structure using band structure calculations to compute the gain but when Auger processes are included in the calculation the agreement between theory and experiment is excellent. The use of such long cavities implies a low gain in these devices. The same group [79] had previously investigated InAsSb strained quantum wells in InAlAsSb barriers for operation out to 3.9 μm and achieved 30 mW/facet. InAlAsSb provided a larger conduction band offset than either InGaAs or InAlAs, and in a similar vein Newell *et al.* [80] considered higher concentrations of Al in AlGaAsSb barriers to increase the valence band offset with GaInAsSb quantum wells. Strain can be used to deepen the heavy hole but in GaInAsSb the saturated gain in strained quantum wells is low [78] and the use of a higher Al concentration is therefore attractive. It brings with it problems of its own however, most noticeably the uneven pumping of the quantum wells caused by preferential recombination close to the barriers.

It is clear from the preceding discussion on mid-IR lasers that there is a bewildering amount of information in the literature. The field is not as mature as other areas of diode laser design and manufacture because the materials themselves present so many problems. The growth of active layers, achieving the right band offsets, the correct use of strain, tailoring the wave-guide, etc. are all demanding tasks in their own right. Not surprisingly this is still a very active field of research and development and as yet there is no clear picture as to the technology that will dominate the market place. Nonetheless high power arrays have been fabricated covering nearly the whole range of wavelengths in the mid-IR from 2 μm to 5 μm.

10.10 Diode pumped solid state lasers

Strictly, diode lasers are also solid state devices but the term "solid state laser" is reserved for the particular class of laser that uses intra-ionic transitions from impurity atoms within a dielectric host as the source of radiation. Solid state lasers require optical pumping sources to generate the population inversion and various diode lasers have already been mentioned in this regard. The most obvious question is; Why use a diode laser to pump a solid state laser? Why not simply use the diode itself? Although recent developments in IR lasers mean that many common, or at least similar, wavelengths are accessible from both sources, solid state and diode lasers have very different properties. The applications to which they may be put are therefore very different.

The most common solid state laser materials centre on the rare earth (RE) ions Nd and Er, but there are in fact many different possible emitting species in many different possible dielectric hosts. For diode pumped solid state lasers (DPSSL) Nd and Er are technologically favoured ions though others such as Yb, Th, Ho, and Dy, are used. Transition metals (TM) rarely feature in DPSSL, though Cr:LiSAF (LiSrAlF$_6$) has emerged as an important material for ultra-short pulses. The rare earth ions are used because the emission and absorption wavelength is relatively independent of the host material and also characterised by a narrow line width. Transition metal ions, such as Cr (ruby, alexandrite) and Ti (Ti:sapphire), on the other hand are heavily influenced by the crystalline environment and the absorption bands are broad. This makes diode pumping relatively less useful, as one of the key elements in the design of a solid state laser is matching the pump source to the absorption bands. In RE ions, where the absorption bands are sharp, diode pumping has the obvious advantage that the pumping source is monochromatic and that the wavelength can be tuned, by varying the temperature of the diode if necessary, to match the absorption band exactly. Where broad band sources, such as flash lamps, exist for pumping broad absorption bands in which a large amount of the pump light can be utilised there has to be a good reason for choosing monochromatic sources.

Conventional lamp-pumped solid state lasers require some sort of cavity to house the lamp and direct the radiation on to the laser rod, so the transfer of radiation from the pump to the laser is not tremendously efficient. Not only must the cavity be reflecting, but there will be emission at wavelengths not absorbed by the laser rod. There will also be a lot of heat generated, which must be conducted away, and cooling pipes can also absorb some of the pump radiation. Lasers, on the other hand, are efficient light sources in themselves, monochromatic, and highly directional. There is no need to build a cavity because whatever is incident on the laser rod is absorbed, apart from the small fraction reflected from the surface. In addition, the pump lasers can be mounted on coolers which do not interfere with the beam, all of which means that a diode-pumped solid state laser is typically 25% efficient whereas a lamp-pumped device has a typical efficiency of 2–3%. Diode pumping therefore means that solid state lasers can be made much smaller in size than conventional lasers, and high repetition rate, short-pulse systems capable of being frequency doubled or tripled by means of non-linear optical elements on the output are very common.

The advantages of the diode-pumped laser over lamp-pumped lasers is therefore clear. The advantage over the direct use of the pump laser itself can be summarised as follows:

- diode lasers have a much larger linewidth;
- energy can be stored in the upper levels of solid state lasers so Q-switching can release this energy to produce short pulses of very high power;
- diodes, on the other hand, are essentially CW devices, and pulsed operation essentially alters the available energy rather than the power;
- high power diode lasers cannot generally be switched quickly;
- solid state lasers have a much better beam quality than diode lasers.

The diodes can be arranged in relation to the laser crystal in a variety of formats, but perhaps the most useful is the end-pumped system using the expanded beam from a diode laser array (figure 10.34). However, other diode configurations for end pumping are possible [81]. There is a limit to how powerful such lasers can be because the length of the laser rod has to be less than the absorption depth of the pump beam, but it allows for small and very compact sources. For high power output it is best to direct the pump lasers on to the side of the rod.

The laser cavity can be integrated into the laser crystal, to give the so-called "micro-chip" laser [82]. This is a recent concept in solid state laser manufacture which has considerable advantages as a result of its particular configuration. The solid state material is fabricated in chip form using techniques developed for the semiconductor industry. A boule of laser material is diced and polished on both sides to form a wafer ~0.5 mm thick. A typical wafer may be 125 mm in diameter, and each wafer is diced to make a chip 1 mm × 1 mm. With something like 250 wafers per boule and 6000 chips per wafer, truely low cost, large scale manfucturing is possible. Each chip is mounted on to a pump laser and the whole device is mounted onto a cooler. The complete laser can be packaged in industry standard TO–3 cans, which gives an indication of their size. Microchip lasers have been fabricated in Cr:LiSAF, Yb:YAG, and Tm/Ho devices. The laser cavities are integrated onto the chip

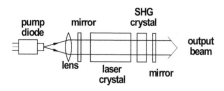

Figure 10.34. An end-pumped solid state laser using a diode pumping source and including second harmonic generation (SHG) to double the frequency.

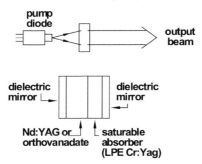

Figure 10.35. End-pumping in a micro-chip laser.

that reflect at 1.06 μm for the feedback but transmit at 808 nm for the pump wavelength. Figure 10.35 shows a typical device, including a saturable absorder of LPE-grown Cr:YA. The saturable absorber only allows light to be transmitted at high optical intensities so at low light intensities nothing is transmitted on to the mirror and there is no feedback. The population in the upper laser levels continues to build up, meanwhile, and eventually enough amplification of a spontaneous event will occur in a single pass to allow sufficient transmission onto the mirrors for feedback to begin. This is the process of passive Q-switching.

Micro-chip lasers are single mode devices. The short cavities allow for single mode operation even at very high pumping powers which are rarely, if ever, used in practise. Laterally the devices are also single mode in CW operation. Thermal changes to the refractive index of the laser chip form a waveguide which operates at stable fundamental mode. The output beam is also polarised. According to Zayhowski [82], when only one spatial mode operates the two orthogonal polarisation modes compete for the same gain so the first polarisation to achieve threshold lases at the expense of the other. Pulsed operation by both passive and active Q-switching are described in detail by Zayhowski. Passively Q-switched Yb:YAG has been described by Spuhler *et al.* [83]. Using a semiconductor saturable absorber, the pulse length at 1030 nm was 530 ps with 1.9 kW peak power at a repetition rate of 12 kHz.

Micro-chip lasers in Nd:YVO$_4$ are also common. Nd:YVO$_4$ is an excellent crystal for high power, stable and cost-effective diode-pumped solid-state lasers [84] and furthermore has larger absorption coefficient by a factor of five at a pump wavelength of 808 nm when compared with Nd:YAG, which makes it more efficient. It is also more tolerant to pump detuning. The pump absorption bandwidth of Nd:YVO$_4$ is much broader than that of Nd:YAG, extending from 802 to 820 nm, whereas YAG pumping requires 808 ± 3 nm. The latter requires strict temperature control of the diode to ensure that the pump source does not drift off the pump resonance. Micro-chip lasers, and vanadate lasers in particular, are high gain devices operating on the edge of stability and interesting effects such as self-Q-switching can be observed [85].

The intensity of the beam from micro-chip lasers makes them ideal for frequency doubling [86]. The laser sources are compact, reliables, and capable of delivering tens of milli-Watts in the visible. Huber *et al.* [87] have also described compact diode pumped CW devices emitting in the visible. Although the powers emitted may be of the order of a Watt or so, the beam quality is far superior to that of a diode laser. Solid state lasers such as the micro-chip design essentially convert the poor output from a diode laser to high quality diffraction limited output without the need for collimators, astygmatic lenses, cylindrical lenses, or any of the other paraphernalia associated with the beam quality of diodes. Moreover, Q-switched devices can deliver peak powers of 50 kW or so at the fundamental modes [88].

Other useful solid state ions include erbium, Tm, Tm/Ho, and Yb/Er. These ions, commonly sited in YAG, emit in the near infra-red at around 1.5 to 2.5 μm. Many of the applications for these wavelengths lie in the medical field, but Er, emitting at ~ 1.6 μ, is ideally matched to the requirements for long haul fibre communications systems. Er doped fibre amplifiers are one of the principal uses for this system. Er has been pumped mainly by 1480 nm diodes but 980 nm is now more common. A mode-locked Er fibre laser capable of delivering pulses of duration 50 ps to 200 ps was described by Okhotnikov *et al.* [89] and Laporta *et al.* [90] described Erbium-Ytterbium microlasers, especially Er-Yb doped phosphate glasses which can be fabricated into fibres. Lasers that emit between 1520 nm–1560 nm are also eye-safe and therefore have potential application in range-finding systems and other optical instruments.

Perhaps the most exciting development in diode pumped solid state lasers lies in the field of femto-second devices. Pulse duration and bandwidth are correlated via the Fourier transform and for very short pulses of the order of femto-seconds, frequency components extending over a very wide range corresponding to something like 300 nm in wavelength range are required. For this reason Ti:sapphire is commonly used as the basis for femto-second sources, but this requires pumping in the green. Cr:LiSAF (Cr:LiSrAlF$_6$), on the other hand absorbs at 650 nm, and as described, high power AlGaInP/GaInP diodes exist for this region. Cr:LiSAF also has a broad band emission extending over 400 nm and is therefore suited to the fabrication of diode pumped femto-second optical sources. Transform-limited pulses as short as 113 fs and modelocked output powers up to 20 mW are possible for less than 110 mW of laser-diode pump power [91]. Sorokin *et al.* [92] provides a review of other materials for these ultra-short pulse devices.

10.11 Summary of materials and trends

High power diode lasers are devices that emit more than a few tens of milli-Watts of optical power. The main limitations of the power output are degradation mechanisms leading to catastrophic optical damage, and thermal roll-over. The latter is caused by the inability to conduct the heat out of the active regions at very high drive currents which leads to temperature rises in the active region and the loss of electrical confinement, and in the case of wavelength selective feedback a shifting of the gain spectrum outside the feedback range. The power can be scaled to tens or even hundreds of W simply by increasing the total emitting aperture, but broad area lasers, effectively anything wider than ~ 15 μm, suffer from filamentation and poor output beam quality. The effect is caused principally by phase-amplitude coupling which can lead to self-focusing effects and spatial hole-burning, otherwise understood as stimulated emission induced carrier depletion within a localised volume. High quality single mode devices can be fabricated using ridge wave-guide structures, tapered gain regions, spatial filtering, or low-gain long cavity devices, or combinations of these.

Techniques that lower the optical intensity at the facet are especially favoured as the total output power can be increased in materials subject to COD. Tapered ridge waveguides and quantum wells lasers with large optical cavities, and hence low confinement, fall into this category. Techniques to coat or modify the facets can be used, such as quantum well intermixing to produce non-absorbing mirrors, but these are expensive. Double heterostructure lasers are still produced but in many materials there is a trend towards the use of quantum wells. These have a smaller active region and the gain can be tailored more easily to control the phase-amplitude coupling coefficient. Strained quantum wells have a modified valence band structure that is beneficial to the laser operation, and the use of Al-free materials enhances the lifetime. In particular, In based devices are less susceptible to degradation because of the size of the In atom, which acts as a dislocation block. Al-free lasers in the near infra-red have very long lifetimes and are used extensively to pump solid state sources. In the visible, AlGaInP is preferred over AlGaAs because it has a higher COD intensity and longer lifetime. The growth of ternary and quaternary compounds other than AlGaAs has presented a considerable technical challenge but high quality materials are now available that will cover the entire range from 650 nm to 5 μm. In the visible, GaInP quantum wells feature prominently, and in the near infra-red InGaAs is a common material. In the mid-IR antimonide lasers are proving very interesting, but as yet are low temperature devices. The main structures in this wavelength range are double heterostructure, possibly asymmetric in order to maximise the electron confinement.

The uses of the lasers are as varied as the designs available, but the principal application is still as pumping sources for solid state materials. Diode lasers themselves are finding increasing application in materials processing and medical applications, but the compact nature of the diode pumped systems and their relatively high efficiency over conventional solid state lasers makes these attractive competitors. Furthermore, diode pumped solid state lasers provide advantages not available directly from diodes of comparable wavelength. In particular, beam quality, pulse duration, and output power can be superior in DPSSL for some applications. In general DPSSL utilise small crystals in the end pumped or side pumped geometry, and the directional nature of the diode pump eliminates the need for an optical cavity. Indeed, diode pumping has made possible the mass production of solid state lasers in the form of microchip lasers. These utilise technology originally developed for the semiconductor industry that make it possible to fabricate thousands of devices from a single boule of laser material.

10.12 References

[1] Treusch H G, Ovtchinnikov A, He X, Kansar M, Mott J and Yand S 2000 *IEEE J Selected Topics in Quantum Electronics* **6** 601–604
[2] Hess O and Kuhn T 1996 *Progress in Quantum Electronics* **20** 85–179
[3] Thompson G H B 1980 *Physics of Semiconductor Laser Devices* (John Wiley) p 380
[4] Yariv A 1989 *Quantum Electronics* 3rd Edition (New York: John Wiley & Sons) pp 577–599
[5] Marciante J R and Agrawal G P 1998 *IEEE Photonics Technology Letters* **10** 54–57
[6] Marciante J R and Agrawal G P 1997 *IEEE J. Quant. Elect.* **33** 1174–1179
[7] Marciante J R and Agrawal G P 1996 *Appl. Phys. Lett.* **69** 593–595
[8] Szymanski M, Kubica J M, Szczepanski P and Mroziewicz B 1997 *J. Phys. D: Appl. Phys.* **30** 1181–1189
[9] Wolff S, Messerschmidt D and Fouckhardt H 1999 *Optics Express* **5** 32–37
[10] Wolff S and Fouckhardt H 2000 *Optics Express* **7** 222–227
[11] Lang R J, Dzurko K, Hardy A A, Demarrs S, Schoenfelder A and Welch D F 1998 *IEEE J. Quant. Elect.* **34** 2196–2210
[12] Kano F, Yamanaka T, Yamamoto N, Mawatari H, Tohmori Y and Yoshikuni Y 1994 *IEEE J. Quant. Elect.* **30** 533–537
[13] Tan G L and Xu J M 1998 *IEEE Photonics Tech. Lett.* **10** 1386–1388
[14] Eliseev, P G 1996 *Prog. Quant. Electr.* **20**(1) 1–82
[15] Lee H H 1993 *IEEE J. Quant. Electr.* **29** 2619–2624
[16] Henry C H, Petroff P M, Logan R A and Merritt F R 1979 *J. Appl. Phys.* **50** 3271–3732
[17] Guo S, Miwa S, Ohno H, Fan J F and Tokumoto H 1995 *Materials Science Forum* **196–201** 1949–1954
[18] Ng S L, Ooi B S, Lam Y L, Chan Y C, Zhou Y and Buddhudu S 2000 *Surface and Interface Analysis* **29** 33–37
[19] Okayasu M, Fukuda M, Takeshita T, Uehara S and Kurumada K 1991 *J. Appl. Phys.* **69** 8346–8351
[20] Taguchi A and Kageshima H 1999 *Physical Review B* **60** 5383–5391
[21] Rechenberg I, Richter U, Klein A, Hoppner W, Maege J, Beister G and Weyers M 1997 *Microscopy of Semiconducting Materials 1997, Institute of Physics Conference Series* **157** 557–560

[22] Kamajima T and Yonezu H 1980 *Jpn. J. Appl. Phys.* **19**(Supplement 19-1) 425–429
[23] Hull D 1975 *Introduction To Dislocations* 2nd Edition (Oxford: Pergamon Press)
[24] Buda M, van de Roer T G, Kaufmann L M F, Iordache G, Cengher D, Diaconescu D, Petrescu-Prahova I B, Haverkort J E M, van der Vleuten W and Wolter J H 1997 *IEEE J. Selected Topics in Quantum Electronics* **3** 173–179
[25] Iordache G, Buda M, Acket G A, van de Roer T G, Kaufmann L M F, Karouta F, Jagadish C and Tan H H 1999 *Electronics Letters* **35** 148–149
[26] Buda M, van der Vleuten W C, Iordache G, Acket G A, van de Roer T G, van Es C M, van Roy B H and Smalbrugge E 1999 *IEEE Photonics Technology Letters* **11** 161–163
[27] Mawst L J, Battacharya A, Lopez J, Botez D, Garbuzov D Z, DeMarco L, Connolly J C, Jansen M, Fang F and Nabiev R F 1996 *Appl. Phys. Lett.* **69** 1532–1534
[28] Ebert G, Beister G, Hulsewede R, Knauer A, Pitroff W, Sebastian J, Wenzel H, Weyers M and Trankle G 2001 *IEEE. J. Selected Topics in Quantum Electronics* **7** 143–148
[29] Pessa M, Napi J, Savlainen P, Toivonen M, Murison R, Ovtchinnikov A and Asonen H 1996 *J. Lightwave. Technol.* **14** 2356–2361
[30] Fukuda M, Okayasu M, Temmyo J and Nakano J 1994 *IEEE J. Quant. Electr.* **30** 471–476
[31] Hiramoto K, Sizugawa M, Kikawa T and Tsuji S 1999 *IEEE J. Selected Topics in Quantum Electronics* **5** 817–821
[32] Horie H, Ohta H and Fujimori T 1999 *IEEE J. Selected Topics in Quantum Electronics* **5** 833–838
[33] Jaeckel H, Bona G-L, Buchmann P, Meier H P, Vettiger P, Kozlovsky W J and Lenth W 1991 *IEEE J. Quant. Electr.* **27** 1560–1567
[34] Coldren L A and Corzine S W 1995 *Diode Lasers and Photonic Integrated Circuits* (John Wiley) p 325
[35] Williams K A, Plenty R V, White I H, Robbins D J, Wilon F J, Lewandowski J J and Nayar B K 1999 *IEEE J. Selected Topics in Quantum Electronics* **5** 822–831
[36] Moerman I, van Daele P P and Demeester P M 1997 *IEEE J. Selected Topics in Quantum Electronics* **3** 1308–1320
[37] Walpole J N 1996 *Optical and Quantum Electronics* **28** 623–645
[38] Bhattacharya A, Zorn M, Oster A, Nasarek M, Wenzel H, Sebastian J, Weyers M, Trankle G 2000 *J. Cryst. Growth* **221** 663–667
[39] Haberland K, Bhattacharya A, Zorn M, Weyers M, Zettler J-T and Richter W 2000 *J. Electr. Mater.* **29** 94–98
[40] Nadja S P and Kean A H 2000 *J. Cryst. Growth* **217** 345–348
[41] Pietzonka I, Sass T, Franzheld R, Wagner G and Gottsvalk V 1998 *J. Crystal Growth* **195** 21–27
[42] Jen H R, Cheng M J and Stringfellow G B 1996 *Appl. Phys. lett.* **48** 1603–1605
[43] Nadja S P, Kean A H, Streater R W and Springthorpe A J 2000 *J. Cryst. Growth* **222** 226–230
[44] Smowton P M, Thomson J D, Yin M, Dewar S V, Blood P, Bryce A C, Marsh J H, Hamilton C J and Button C C 2001 *Semicond. Sci. Technol.* **16** L72–L75

[45] Smowton P M, Thomson J D, Yin M, Dewar S V, Blood P, Bryce A C, Marsh J H, Hamilton C J and Button C C 2002 *IEEE J. Selected Topics in Quantum Electronics* **38** 285–290
[46] Lichtenstein N, Winterhoff R, Scholz F, Schweizer H, Weiss S, Hutter M and Reichl H 2000 *IEEE J. Selected Topics in Quantum Electronics* **6** 564–569
[47] Hiroyama R, Inoue D, Nomura Y, Shono M and Sawada M 2002 *Jpn. J. Appl. Phys.* **41**(pt 1) 1154–1157
[48] Hiroyama R, Inoue D, Nomura Y, Ueda Y, Shono M and Sawada M 2002 *Jpn. J. Appl. Phys.* **41**(pt 1) 2559–2562
[49] Hoskens R C P, van de Roer T G, van der Poel C J and Ambrosius H P M 1995 *Appl. Phys. Lett.* **67** 1343–1345
[50] Summers H D and Rees P 1998 *IEEE Photonic Tech. Lett.* **10** 1217–1219
[51] Adachi H, Kamiyama S, Kidoguchi I and Uenoyama T 1995 *IEEE Photonic Tech. Lett.* **7** 1406–1408
[52] Summers H D and Rees P 1997 *Appl. Phys. Lett.* **71** 2665–2667
[53] Sumpf B, Beister G, Erbert G, Fricke J, Knauer A, Pittroff W, Ressel P, Sebastian J, Wenzel H and Trankle G 2001 *Electr. Lett.* **37** 351–353
[54] Sumpf B, Beister G, Erbert G, Fricke J, Knauer A, Pittroff W, Ressel P, Sebastian J, Wenzel H and Trankle G 2001 *IEEE Photonic Tech Lett.* **13** 7–9
[55] Erbert G, Bugge F, Knauer A, Sebastian J, Thies A, Wenzel H, Weyers M and Trankle G 1999 *IEEE Journal Of Selected Topics In Quantum Electronics* **5** 780–784
[56] Sebastian J, Beister G, Bugge F, Buhrandt F, Erbert G, Hansel H G, Hulsewede R, Knauer A, Pittroff W, Staske R, Schroder M, Wenzel H, Weyers M and Trankle G 2001 *IEEE Journal Of Selected Topics In Quantum Electronics* **7** 334–339
[57] Hulsewede R, Sebastian J, Wenzel H, Beister G, Knauer A and Erbert G 2001 *Opt. Commun.* **192** 69–75
[58] Knauer A, Bugge F, Erbert G, Wenzel H, Vogel K, Zeimer U and Weyers M 2000 *Journal Of Electronic Materials* **29** 53–56
[59] Knauer A, Wenzel H, Erbert G, Sumpf B and Weyers M 2001 *Journal Of Electronic Materials* **30** 1421–1424
[60] Diaz J, Yi H, Jelen C, Kim S, Slivken S, Eliashevich I, Erdtmann M, Wu D, Lukas G and Razeghi M 1996 *Inst. Phys. Conf. Ser.* **145** 1041–1046
[61] Razeghi M, Eliashevich I, Diaz J, Yi H J, Kim S, Erdtmann M, Wu D and Wang L J 1995 *Mat Sci Eng B – Solid State Materials for Advanced Technology* **B35** 34–41
[62] Drakin A E, Eliseev P G, Sverdlov B N, Bochkarev A E, Dolginov L M and Druzhinina L V 1987 *IEEE J. Quant. Electr.* **QE-23** 1089–1094
[63] Loural M S S, Morosini M B Z, Herrera-Perez J L, Von Zuben A A G, da Silveira A C F and Patel N B 1993 *Electronics Letters* **29** 1240–1241
[64] Martinelli R U and Zamerowski T J 1990 *Appl. Phys. Lett.* **56** 125–127
[65] Lane B, Wu D, Yi H J, Diaz J, Rybaltowski A, Kim S, Erdtmann M, Jeon H and Razeghi M 1997 *Appl. Phys. Lett* **70** 1447–1449
[66] Tsou Y, Ichii A and Garmire, E M 1992 *IEEE J. Quant. Electr.* **28** 1261–1268
[67] Shi Z 1998 *Appl. Phys. Lett.* **72** 1272–1274
[68] Tacke M 2001 *Philosophical Transactions of the Royal Society of London Series A-mathematical Physical and Engineering Sciences* **359** 547–566

[69] Yi H, Rybaltowski A, Diaz J, Wu D, Lane B, Xiao Y and Razeghi M 1997 *Appl. Phys. Lett* **70** 3236–3239
[70] Rybaltowski A, Xiao Y, Wu D, Lane B, Yi H, Feng H, Diaz J and Razeghi M (1997) *Appl. Phys. Lett* **71** 2430–2432
[71] Wu D, Lane B, Mohseni H, Diaz J and Razeghi M 1999 *Appl. Phys. Lett* **74** 1194–1196
[72] Lane B, Wu Z, Stein A, Diaz J and Razeghi M 1999 *Appl. Phys. Lett* **74** 3438–3440
[73] Lane B, Tong S, Diaz J, Wu Z and Razeghi M 2000 *Materials Science and Engineering* **B74** 52–55
[74] Choi H K and Turner G W 1995 *Appl. Phys. Lett.* **67** 332–334
[75] Mourad C, Gianardi D, Malloy K J and Kaspi R 2000 *J. Appl. Phys.* **88** 5543–5546
[76] Cuminal Y, Baranov A N, Bec D, Grach P, Garcia M, Boisser G, Joullie A, Glastre G and Blondeau R 1999 *Semicond. Sci. & Technol.* **14** 283–288
[77] Munoz M, Pollak F H, Zakia M B, Patel N B and Herrera-Perez J L 2000 *Phys. Rev. B* **62** 16600–16604
[78] Turner G W, Choi H K and Manfra M J 1998 *Appl. Phys. Lett.* **72** 876–878
[79] Choi H K and Turner G W 1995 *Appl. Phys. Lett.* **67** 332–334
[80] Newell T, Wu X, Gray A L, Dorato S, Lee H and Lester L F 1999 *IEEE Photonics Technology Letters* **11** 30–33
[81] Wyss C P, Luthy W, Weber H P, Brovelli L, Harder C and Meier H P 1999 *Optical and Quantum Electronics* **31** 173–181
[82] Zayhowski J J 1999 *Optical Materials* **11** 255–267
[83] Spuhler G J, Paschotta R, Kullberg M P, Graf M, Moser M, Mix E, Huber G, Harder C and Keller U 2001 *Applied Physics B* **72** 285–287
[84] Chen Y F, Tsai S W and Wang S C 2001 *J. Chem. Appl. Phys. B* **73** 115–118
[85] Conroy R S, Lake T, Friel G J, Kemp A J and Sinclair, B D 1998 *Optics Letters* **23** 457–459
[86] Sinclair B D 1999 *Optical Materials* **11** 217–233
[87] Huber G, Kellner T, Kretschmann H M, Sandrock T and Scheife H 1999 *Optical Materials* **11** 205–216
[88] Molva E 1999 *Optical Materials* **11** 289–299
[89] Okhotnikov O G, Araujo F M and Salcedo J R 1994 *IEEE Photonics Technology Letters* **6** 933–935
[90] Laporta P, Taccheo S, Longhi S, Svelto O and Svelto C 1999 *Optical Materials* **11** 269–288
[91] Hopkins J M, Valentine G J, Agate B, Kemp A J, Keller U and Sibbett W 2002 *IEEE Journal Of Quantum Electronics* **38** 360–368
[92] Sorokin E, Sorokina I T and Wintner E 2001 *Appl. Phys. B* **72** 3–14

Problems

1. A $Ga_{0.1}Al_{0.9}As$-$Ga_{0.4}Al_{0.6}As$ DH high power laser, cavity length 500 μm and stripe width 50 μm, has an internal optical power density on the facet of 200 MW cm^{-2}. Ignoring spontaneous recombination, estimate the total

current density and the carrier density within the active region. The active region, of refractive index 3.518, is 1 μm thick, as are the barrier layers. As before, assume the diffusion coefficient for electrons in AlGaAs is 100 cm^2 s^{-1} and that the conduction band offset is 40% of the total.
2. Calculate the leakage current for the laser in (1) above for a barrier with 30% aluminium and compare this with the stimulated emission current.
3. Estimate the output power density for the laser in (1).
4. The formulation of an effective time constant for stimulated emission is somewhat unrealistic. It satisfies our intuition that as the photon density increases the time constant should decrease, but at high photon densities as in (1) above the time constant is only 3 ps. Gain compression will cause the time constant to settle at some limiting value which would probably be treated as an adjustable parameter in a model. However, a reasonable estimate will be about 10^{-10} s. Assuming this to be the case, calculate the carrier densities at time constants of 1 ns and 0.1 ns. Hence estimate the leakage currents for the two barriers.
5. The temperature of the laser in (1) rises to 50°C above room temperature. Calculate the change in wavelength and estimate the change in leakage current for a barrier of $x = 0.3$ and a stimulated emission time constant of 1 ns.

Chapter 11

Blue lasers and quantum dots

The blue laser is a subject in its own right. Its development is not much a story of the development of a new type of laser but of a materials system with the right bandgap and optical properties that is robust enough for the extreme operating conditions of the diode laser. It is also the story of one particular man's efforts. Shuji Nakamura was a researcher at Nichia corporation in Japan and worked on gallium nitride despite being told not to. He developed the nitride blue LED which led to the laser, and in a recent summing up of a court action over patent rights, the judge described the invention as a "totally rare example of a world class invention achieved by the inventor's individual ability and unique ideas in a poor research environment in a small company" [1]. Strong words, but there is no doubt about Nakamura's influence on the development of the blue laser.

The gallium-aluminium-indium nitride alloy system has overtaken the II–VI materials ZnSe and ZnS, which in some respects were the natural choice for blue emission. These are naturally occurring compounds that have been known for a long time in the semiconductor community. ZnSe has a band gap of 2.7 eV at room temperature (460 nm) and is closely lattice matched to GaAs, and the bandgap of ZnS, which is lattice matched to silicon and GaP, is 3.7 eV (335 nm). The epitaxial growth of both the (Zn,Mg)(S,Se) and (Zn,Cd)(S,Se) systems by MBE and MOVPE has been investigated extensively over the years. The addition of magnesium widens the band gap for waveguide cladding layers and the addition of Cd narrows the bandgap for quantum wells (see figure 11.1).

One of the major problems with wide band gap materials has been the development of stable p-type doping. By contrast, n-type doping is easy in ZnSe and ZnS. As grown, the materials are slightly n-type, particularly ZnSe on GaAs. Gallium out-diffusion during growth can lead to n-type doping with the Ga sitting substitutionally for the Zn. Similarly indium is a common dopant that can be introduced intentionally, but both are fairly mobile and not especially stable for use in devices. However, chlorine, from group VII, is an effective n-type dopant that substitutes for the chalcogen (S or Se) and can yield relatively high densities of free electrons [3], but achieving high

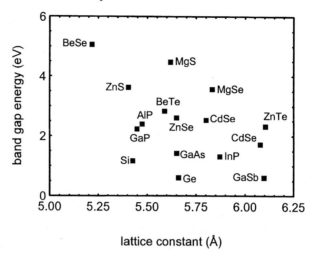

Figure 11.1. Band gaps and lattice constants of the important II–VI materials and some of the possible substrates. (After Nurmikko and Gunshor [2].)

concentrations of stable dopants has presented some challenges [4]. However, these materials resisted p-type doping. A group V atom (e.g. Sb or P) substituting for the chalcogen would be expected to lead to an excess hole concentration, but self-compensation through the formation of deep levels or associated vacancies cancelled out the acceptors. Group I elements such as Li could be made to substitute for the Zn atom and render the material p-type but the small atom size and high mobility under the influence of an electric field made this type of doping ultimately unstable. Stable p-type doping was eventually achieved using nitrogen as an acceptor. The development of the RF excited nitrogen plasma source for MBE enabled a small fraction of a nitrogen molecular beam (up to 8% in some cases) [5] to be dissociated into highly reactive monatomic nitrogen which doped the epitaxial layers p-type without the need for further processing.

Even so, the technology was not without its problems. Nitrogen is less soluble in ZnSe than in ZnTe [6] so the maximum concentration of acceptors will lie around 10^{17} cm^{-3}, though in the ZnMgSSe system concentrations as high as 10^{20} cm^{-3} have been reported [3]. Moreover, the presence of nitrogen tends to give rise to deep levels, and compensating defects can also occur, especially at high doping concentrations. Hauksson et al. [7] reported a deep compensating donor at 44 meV which is believed to be a selenium vacancy complexed to a substitutional nitrogen and a zinc atom (designated VSe-Zn-NSe), while others have reported strong donor-acceptor pair emissions [8–10]. There is also evidence of bandgap shrinkage [11]. These difficulties notwithstanding, p–n junction LEDs and lasers have been made [12–14], including quantum well structures [15] operating under pulsed conditions at room temperature as well as CW diodes operating at room temperature [16].

The growth and device fabrication technology of II–VI materials was fairly well advanced. Even the contacts to the *p*-type material, which can be difficult because the valence band lies at a deep energy within the solid and it is not easy to find a material to align to it, have been developed to the point where CW diodes can operate at room temperature for several hundred hours. Several reviews of II–VI blue-green lasers emitting in the range from 461 nm to 514 nm have been published in addition to those referenced above [17]. Reviewing blue and green lasers in 1997, Nurmikko and Gunshor [2] report that in fact II–VI technology was well in advance of III-nitride technology, principally because of the ease of growth and the suitability of the substrates. Although both types of laser had been developed sufficiently by that time to possess similar lifetimes of a few hundred hours, the differences in the quality of material between the two was stark. The first II–VI lasers were demonstrated in 1991 where as the nitride laser was not demonstrated until some five years later. Moreover, the II–VI lasers had a much higher quantum efficiency than their nitride counterparts, had lower threshold electron-hole pair densities by at least an order of magnitude, and the optical cavities were much better behaved. Nitride lasers appeared to show evidence of "sub-cavities", possibly cracks in the material which provided optical feedback from within the material itself.

The idea that a laser could contain cracks like this and still operate was something of a surprise, but in fact it is the reason for the success of the nitrides. Although there were, and still are, many problems associated with the development of both II–VI and III-nitride materials, the nitrides appeared to be very robust materials that tolerated amounts of damage and lattice disruption that simply could not be tolerated in other materials. There are no closely matched suitable substrates for GaN and its alloys with indium and aluminium, but epitaxial growth on sapphire was still achieved despite a massive 14% lattice mismatch. The material is stuffed full of defects (of the order of 10^{10} cm^{-2}) but this does not prevent the emission of light. Contrast this with GaAs where a defect density of 10^3 cm^{-2} begins to affect the light output and by 10^6 cm^{-2} light emission ceases totally. Moreover, the defects in GaN have a very low mobility (the velocity is estimated to be lower than in GaAs by a factor of 10^{20} [18]) so once present their location will remain fixed, unlike in GaAs where the defects can migrate towards the active regions and kill the luminescence.

Not only is GaN stable against large concentrations of defects, the defects themselves are also stable against multiplication. This is in stark contrast to the II–VI materials that have relatively low densities of misfit dislocations by virtue of their similar lattice constants to the substrates, but these dislocations propagate into the active layer and can multiply under the extreme injection conditions of the diode laser. It was due to improvements in the control of these defects in the II–VIs that led to the attainment of lifetimes of several hundreds of hours, but that is not enough for a commercial device in whch lifetimes of 10^4 hours are required. The chief difficulty is that II–VI compounds are much softer than nitrides, and are susceptible to degradation by defect multiplication. Landwehr et al. [19] have commented in detail on this aspect and identified the use of BeSe and BeTe as possible components to increase the covalency of the solid.

The "strength" of a semiconductor is influenced by the covalency. One only has to think of silicon and diamond as examples of a very strong covalent bond. The shear constant for these two are 951 GPa and 102 GPa respectively, compared with 65 GPa for GaAs and 32.2 GPa for ZnSe [20]. By this measure GaAs is twice as strong as ZnSe. The covalent bond is highly directional, which is why it is so robust. In silicon, for example, there are four valence electrons in the s and p orbitals. The s electrons have a spherical symmetry where as the p electrons reside in orbits that are mutually perpendicular to each other. The s and the p electrons hybridise to form four equivalent bonds oriented tetragonally (figure 11.2). The angle between the bonds is maintained not only by coulombic repulsion but also by exchange forces, so called because quantum mechanics forbids the electrons to exchange places by occupying the same physical space. These forces are extremely strong and hold the atom in place against shear forces acting to break the bond. Dislocations involve shear forces and broken bonds so covalent solids are therefore robust and stable against their multiplication. Quite simply, it is difficult to displace and dislocate the atoms in the structure, whereas for an ionic bond the force between ions is predominantly coulombic so there is no real directionality. It becomes easier, therefore, to displace atoms by shear forces and to break the bonds.

Of course, a truly ionic bond is somewhat of an idealisation. In semiconducting materials such as GaAs and ZnSe the bond is partially ionic and partially covalent. Rather than a complete transfer of charge from one atom to another there is simply a shift in the charge distribution away from the centre so each atom is partially ionised and some polarity in the bond exists. The polarity, and its complementary property, the hybrid covalency, can be defined in terms of the properties of the bond and the constituent atoms, as described in detail by Harrison [21]. The hybrid energy is the expectation energy of the bond, i.e. a weighted average of both the s and p eigen energies, and for sp^3

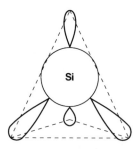

Figure 11.2. The tetragonal orientation of hybrid covalent bonds in silicon.

hybrids typical of tetragonal materials (the superscript "3" simply means that the probability of finding the electron in a p state is thrice that of finding it in a s state), the hybrid energy is simply

$$\varepsilon_h = \frac{(\varepsilon_s + 3\varepsilon_p)}{4}. \qquad (1)$$

The hybrid polar energy is simply half the difference between the hybrid energies of the two materials, i.e.

$$V_3^h = \frac{(\varepsilon_h^1 \varepsilon_h^2)}{2} \qquad (2)$$

where the notation V_3 is the same as that used by Harrison. For an elemental semiconductor $V_3 = 0$, i.e. there is no polarity in the bond whatsoever. The hybrid covalent energy is defined simply in terms of the interatomic spacing d,

$$V_2 = \frac{2.16\hbar^2}{md^2} \qquad (3)$$

where, again, Harrison's terminology is retained and $\hbar^2/m = 7.62$. The polarity is then

$$\alpha_p = \frac{V_3}{|V_{2,3}|} \qquad (4)$$

and the hybrid covalency is

$$\alpha_c = \frac{V_2}{|V_{2,3}|} \qquad (5)$$

where

$$|V_{2,3}| = (V_2^2 + V_3^2)^{1/2}. \qquad (6)$$

The energies V_2 and V_3 define the *bond energies* associated with these elements. In the non–polar solid the bonding and anti-bonding states, which respectively form the valence band and the conduction band, are separated by $2V_2$ whereas in polar solids the bonding and anti-bonding states are separated by $2|V_{2,3}|$.

Intuitively we would expect the covalency α_c to imply something quantitative about the distribution of charge between the atoms. Pseudopotential theory will allow such calculations but they are beyond the scope of this book. Rather, the most important aspect of the covalency relevant here is its relationship to the shear constant, which is defined for a zinc blende structure in terms of the elastic constants as $c_{11} - c_{12}$ (see appendix IV for further details on the elastic constants). Harrison [22] shows that

$$c_{11} - c_{12} \propto V_2 \alpha_c^3. \qquad (7)$$

The experimentally measured shear constants for GaAs and ZnSe have already been given above, and can be compared with calculated values of 114.8 GPa

Table 11.1. Lattice parameters, elastic constants, bond energies, and covalency for a number of III–V material [24,25].

Material	Lattice constant (Å)	C_{11} ($\times 10$ GPa)	C_{12} ($\times 10$ GPa)	Shear constant ($\times 10$ GPa)	Bond length (Å)	V_2 (eV)	V_3 (eV)	Co-valency α_c
BN	3.618	82	19	63	1.57	6.68	2.42	0.94
BP	4.5383	31.5	10	21.5	1.97	4.24	0.85	0.98
GaP	5.451	14.05	6.203	7.847	2.36	2.92	1.72	0.86
InP	5.873	10.11	5.61	4.5	2.54	2.55	1.82	0.81
AlAs	5.661	11.99	5.75	6.24	2.43	2.79	1.53	0.88
GaAs	5.653	11.9	5.38	6.52	2.45	2.74	1.51	0.88
InAs	6.058	8.329	4.526	3.803	2.61	2.42	1.61	0.83
AlSb	6.1355	8.769	4.341	4.428	2.66	2.33	1.19	0.89
GaSb	6.094	8.834	4.023	4.811	2.65	2.34	1.17	0.89
InSb	6.479	6.669	3.645	3.024	2.81	2.08	1.28	0.74

for Si, 68.6 GPa for GaAs, and 36.5 GPa for ZnSe [23]. Although the agreement is not exact it is nonetheless good, and it is clear that the more polar the semiconductor the smaller the shear constant.

The lattice parameter is also an important factor in the crystal strength. Table 11.1 shows the variation of the elastic constants [24] with the lattice parameter and bond length as well as the hybrid polar and covalent energies [25] and the covalency. Boron nitride is included here because it is a naturally occurring tetragonal nitride and therefore belongs in this group. The other group-III nitrides listed in table 11.2 naturally form in the hexagonal phase but can be grown epitaxially in a cubic form [26]. The general trend is for the shear constant to reduce as the lattice parameter increases, though there is the odd compound out of sequence. Similarly the covalent energy V_2 also decreases, but because of the general decrease in V_3 there is no clear trend in the covalency. However, the decrease in V_2 is enough to explain the decrease in shear constant from equation (7).

The role of the lattice constant is further emphasized in figure 11.3 where the reduced lattice constant is plotted as a function of lattice parameter for the materials in table 11.1 [24]. When the elastic constants are reduced by the factor e^2/b^4, where e is the electronic charge and b is the nearest neighbour distance, there is a very clear dependence on the lattice constant a according to equations (8).

$$C^*_{11} = C_{11} \cdot \frac{e^2}{b^4} = 1.333 + \frac{2.998}{a}$$
$$C^*_{12} = C_{12} \cdot \frac{e^2}{b^4} = 0.1521a \tag{8}$$

Table 11.2. Lattice parameters, elastic constants, bond energies, and covalency for the II–VIs and III-nitrides. The lattice constants for the Be-containing compounds are taken from [28] whilst the shear constants of Be-containing compounds are taken from [29]. The elastic constants of ZnS and are ZnSe taken from [30]. Note that the elastic constants of the nitrides are calculated values rather than experimental, as there is a paucity of reliable experimental data as yet [26].

Material	Lattice constant (Å)	C_{11} (×10 GPa)	C_{12} (×10 GPa)	Shear constant (×10 GPa)	Bond length (Å)	V_2 (eV)	V_3 (eV)	Co-valency α_c
BeS	4.865			11.98	2.10	3.73	3.04	0.78
BeSe	5.139			8.52	2.20	3.40	2.67	0.79
ZnS	5.4102	10.40	6.50	3.9	2.34	3.01	3.45	0.66
BeTe	5.626			5.42	2.40	2.86	2.20	0.79
ZnSe	5.6676	8.59	5.06	3.53	2.45	2.74	3.08	0.66
AlN*	4.37	34.8	16.8	18	1.89	4.61	3.31	0.81
GaN*	4.50	29.6	15.4	14.2	1.94	4.37	3.29	0.8
InN*	4.98	18.4	11.6	6.8	2.15	3.56	3.39	0.72

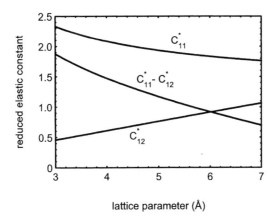

Figure 11.3. The reduced elastic constants [24] as a function of lattice parameter.

These trends imply the following about semiconducting materials; both the covalency and the lattice parameter indicate the strength of the material. Intuitively we would expect a small lattice parameter to indicate a strong bond, and hence a greater resistance to lattice distortion, but where materials have a similar lattice parameter, such as GaAs and ZnSe, the covalency is a good indicator of the difference between the two. GaAs and ZnSe are both formed from elements from the same period of the table, but Ga and As are separated by only one element, Ge, whereas Zn and Se are separated by three so we would expect GaAs to be more covalent than ZnSe. This is reflected in the polar energy, and through it the covalency, and although the change in covalency is not that large (0.88 to 0.66) the effect of taking the cube for the shear strength leads to a factor of 2.9.

Similar trends can be seen in table 11.2. BeSe and BeS both have smaller lattices than ZnS and ZnSe and are therefore stronger. In fact, ZnS and ZnSe have significantly smaller covalencies than any of the III–V materials listed. BeTe bucks the II–VI trend somewhat, just as some compounds in table 11.1, but as Be is a very small atom the covalency is high. In general, though not always, the lighter the metal, the higher the covalency. In the nitrides, for example, the covalency decreases steadily from 0.94 for BN to 0.72 for InN. Put simply, a smaller metal atom is less likely to lose electrons in a partially ionic bond. This is why the beryllium compounds have a high shear constant comparable to that of GaAs and silicon, and why the nitrides of aluminium and gallium, which have much smaller lattices, are even stronger still. It should be noted that the elastic constants for the cubic nitrides have been calculated owing to the scarcity of experimental data, but the agreement between experiment [27] and calculation for the hexagonal phase of InN suggest that these values can be taken as a pretty reliable estimation. AlN and GaN are seen to be very strong and InN is very similar to BeSe.

The use of the Be-chalcogenides should therefore lead to more stable devices if the growth technology can be perfected. However, the beryllium containing chalcogenides essentially represented a new class of materials for which the growth and doping had not been perfected in the late 1990s when these materials were being written about as a possible way out of the difficulties that II–VI lasers faced [31]. Given the already significant achievements in nitride technology that had occurred by this time, II–VI materials were effectively unable to compete and have now slipped way behind the III-nitrides as the focus of commercial and industrial interest. II–VI materials have some interesting properties that make them model systems for research so there are still a significant number of papers being published on their growth and characterisation, and occasionally papers appear on laser devices using Be-containing materials [32]. At this stage, however, it seems that the nitrides will remain the material of interest for the semiconductor community, not only for lasers but for high temperature electronic devices. The developments in this area in recent years ensure that nitrides have a strong future and the compatibility between lasers and other devices within a single material system is an important economic driver.

For these reasons the remainder of this chapter on blue lasers will concentrate on nitride materials. This will lead on naturally to a discussion of quantum dots. Nurmikko and Gunshor write in their review of blue-green lasers in 1997 [2] that research had indicated that topological and compositional fluctuations in the In composition within a quantum well lead to isolated regions of high indium concentration that may be just a few nanometers across. This leads to quantum confinement in three dimensions; a so-called "quantum dot". Quantum dots have also been fabricated deliberately in a wide variety of materials, including II–VIs and conventional III–Vs, and lasers based on the concept are close to commercial reality, but the III-nitrides seem particularly well-suited to this concept. Indeed, it will become clear that a considerable effort has been expended in establishing the 2-D growth regime necessary to the formation of good quality quantum wells, whereas quantum dots exploit the 3-D growth that was a common feature of these materials. It may well be that quantum dots represent the best way of extending the useful range of these materials into the green and yellow emitters.

11.1 Nitride growth

The growth of epitaxial nitrides of aluminium, gallium, and indium, has been described in a number of reviews [18,33–35], so only a brief resumé will be given here. The reader is referred to these sources and the references therein for a detailed history of the growth techniques as well as an up-to-date exposition of available methods.

Attempts have been made to grow GaN in bulk form since the 1930s by passing ammonia over the hot metal. More recently single crystal growth of GaN has been achieved using a nitrogen pressure of 20 kbar [36]. Thermodynamic calculations suggest that the equilibrium pressure of N_2 over GaN is over 45 kbar, but this sort of pressure is not accessible to present day growth technology. The growth mechanism appears to be dissolution of nitrogen into liquid gallium followed by diffusion of nitrogen to a cold zone where GaN is precipitated. The nitrogen molecule dissociates on adsorption onto the liquid metal surface and is therefore easily absorbed into the metal. However, at 20 kbar the solubility of nitrogen in Ga is only about 1% at about 1900 K and at lower temperatures the solubility is even lower. Establishing a temperature gradient allows control over the growth process because adsorption and dissolution occurs at the higher temperature but super-saturation occurs in the growth region at the lower temperature and solid GaN is precipitated. It appears that the diffusion of the nitrogen to the colder growth zone is the limiting process in the growth rather than adsorption and dissolution.

These developments in bulk single crystal growth are relatively recent, and although the crystals have a very low dislocation density of 100 cm^{-2} they are still small at only 1 cm or so in size. They are not suitable for commercial device fabrication and the most significant achievements in GaN devices to date have been realised in epitaxial material grown on either sapphire or SiC. However, it took several years to refine the growth from the first large area CVD on sapphire, which occurred in 1969. Figure 11.4 shows the lattice constants and band gaps of both the cubic and hexagonal phases of the nitrides with the lattice parameters of SiC and sapphire also shown as dashed lines. Note that GaN has a minimum lattice mismatch with sapphire of 13.9%, depending on the orientation. Growth on different planes of sapphire has been described extensively by Ambacher [33] but it is not absolutely essential to an

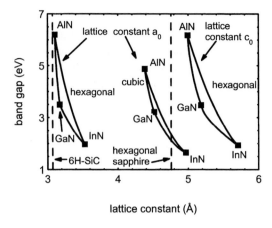

Figure 11.4. Lattice constants of III-nitrides (after Ambacher [33]) with the lattice constants of common substrates 6H-Sic and sapphire also shown.

understanding of light emitting devices and will not be discussed in any detail here.

One of the key steps to producing high quality films has been the growth of a buffer layer of AlN at a fairly low temperature ($\sim 500°C$) followed by the subsequent growth at a high temperature ($\sim 1000°C$). This is often called the "two step method", as the growth is interrupted so that the main epitaxial layer nucleates on the buffer layer instead of on the sapphire. Whether CVD or MBE is preferred, the use of a low-temperature grown buffer appears to be essential to achieving high quality material. The buffer layer growth is columnar [33,37] because the lattice mismatch is so large that growth is three dimensional and occurs in domains. The domains may be slightly mis-oriented with respect to each other, but even if they had the same orientation the mismatch is so large that they would not join perfectly when meeting. This results in a distinctly columnar structure within the film, with each column being ~ 0.1 μm in diameter. Jain *et al.* [34] discuss the buffer layer in more detail and describe the use of both GaN buffers and InN buffers only 30 nm or so thick. The latter resulted in a defect density in an epitaxial overlayer of GaN seven times smaller than exists in a GaN layer on a GaN buffer.

Control of the defect density in GaN has been quite difficult to achieve. These sort of lattice mismatches will inevitably result in a very short critical thickness. In fact it is about 3 nm [38]. However, most of the misfit dislocations are confined to the interface [33] but still, there are about 10^{10} dislocations/cm^2 in the GaN. Epitaxial lateral overgrowth (ELOG) results in much lower densities of defects. ELOG is achieved by depositing 0.1 μm of silicon dioxide on top of approximately 0.2 μm of GaN and patterning the oxide into windows roughly 4 μm wide separated by stripes roughly 7 μm wide (figure 11.5).

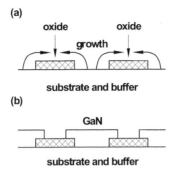

Figure 11.5. Epitaxial lateral overgrowth of GaN.

Precise dimensions vary from laboratory to laboratory. Deposition of GaN onto this system causes growth laterally over the oxide and at a thickness of about 10 μm the stripes coalesce into a single film. The defect density over the oxide is much lower than the defect density in the window by about three orders of magnitude, which is the desired effect but has the unfortunate consequence of requiring precise alignment of devices with the underlying stripe pattern. A second lateral epitaxial overgowth onto stripes aligned with the first windows results in a GaN substrate with a low defect density across the whole of the wafer. ELOG has been described in detail by Beaumont *et al.* [39]. A related process is pendeo-epitaxy, where a seed layer of GaN on SiC is masked and etched down into the substrate (figure 11.6). CVD growth on the top of the GaN is prevented by the mask and nucleation occurs instead on the etched GaN side

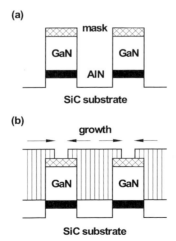

Figure 11.6. Pendeo-epitaxy of GaN.

walls. Eventually the growing surfaces meet and growth then continues upward and laterally over the masks to produce a low defect density GaN substrate. Both ELOG and pendeo-epitaxial films grown thick enough can serve as substrates for device quality thin films grown by either MBE or MOVPE.

Although other forms of CVD have been used to grow GaN, metalloorganic technology achieved maturity in the mid-1990s and dominates today. The trimethyl compounds of aluminium, indium, and gallium are the common precursors for the metal and ammonia is used for the nitrogen. Control of the carrier concentration has been one of the biggest problems. As grown, GaN is always n-type with a background carrier concentration of $\sim 10^{16}$ cm^{-3}, but it is not clear whether nitrogen vacancies or residual impurities such as oxygen are responsible [34]. However, GaN can be doped n-type during growth by both Si and Ge to densities of about 10^{19} cm^{-3}. Mg is an acceptor that gives hole concentrations in excess of 10^{18} cm^{-3}, but in as-deposited Mg-doped material the hole concentration is very low because residual hydrogen from the source gases passivates the acceptors. It was a chance observation that low energy electrons incident on GaN activated the Mg atoms that led to the development of the the thermal annealing processes now routinely used to activate these acceptors in as-deposited GaN. As-deposited InN is still very heavily n-type and attempts to dope it p-type have so far been unsuccessful [35].

Annealing processes have proven particularly difficult to control in these materials. As reviewed by Ambacher, the pressure of N_2 over InN, GaN or AlN rises dramatically at 630°C, 850°C and 1040°C respectively as the material decomposes and nitrogen is driven off. These are effectively the upper working temperatures of these materials as the equilibrium vapour pressure is so high that it is not possible to put a high enough overpressure on the surface to prevent dissociation. Therefore growth by MBE, which will be described in more detail later, dopant activation, and contact formation, must all occur below these temperatures. Mg-doped material is typically annealed between 500–700°C in nitrogen to activate the dopants, and contacts to n-type material are made by annealing a 25 nm Ti-/100 nm Al bi-layer at 500°C. Note that the Ti separates the Al from the nitride. This structure has a specific resistance lower than 10^{-3} Ω cm^2, but rapid thermal annealing at 900°C for 30 s will reduce it to below 10^{-5} Ω cm^2. Contacts to p-GaN are usually made from thin layers of Ni and Au deposited onto the GaN, with the Au layer adjacent to the GaN. Annealing in an oxygen ambient at >400°C causes the nickel to oxidise. Ga is dissolved in the gold and is gettered by the oxide, leaving behing Ga vacancies at the surface of the GaN [40]. These vancancies enhance the hole concentration at the surface and reduce the contact resistance.

Growth by VPE is done at higher temperatures because there is a trade-off between decomposition of the deposited material and decomposition of the reacting gases. The optimum growth temperature for GaN is therefore 1050°C. There are several methods of VPE, and the hydride method (HVPE) in which liquid Ga reacts with HCl to form GaCl and hydrogen, is very common.

$$2Ga(l) + 2HCl(g) \rightleftharpoons 2GaCl(g) + H_2(g). \tag{9}$$

Gallium chloride reacts in the gas phase with ammonia (NH_3) to form solid GaN, hydrogen chloride and yet more hydrogen.

$$GaCl + NH_3 \rightleftharpoons GaN(s) + HCl(g) + H_2(g). \tag{10}$$

For deposition temperatures above 800°C reactions at the surface also have to be considered. An activated complex of Ga, Cl, and H forms on the surface which then decomposes to give a "free" Ga atom adsorbed but not bonded chemically onto the surface, as well as hydrogen chloride, and monatomic hydrogen. The adsorbed Ga then absorbs the nitrogen from the decomposing ammonia and GaN is formed provided the ratio of V:III is high enough. Usual values lie between 500 and 700.

This is an attractive growth method for thick films because the growth rate can be several tens, sometimes hundreds, of micrometres per hour. Very thick, fully relaxed quasi-bulk GaN films can be grown which can then be separated from the substrate and used as substrates themselves. AlN can be grown in a similar way, and also the Al-Ga-N alloys, and both $GaCl_3$ and $AlCl_3$ can be used as alternative metal sources. This technique is not suitable for quantum wells, however, as it takes less than a second to grow several nanometres of material. MOCVD is therefore preferred. The chemistry of MOCVD deposition of AlN has been discussed by Ambacher [33] but there appears to be a paucity of similar information related to the deposition of GaN. The chemistry is quite complex and involves the formation of intermediate compounds, as well as much higher V:III ratios. Typically GaN and AlN are deposited above 900°C but InN deposition occurs at a much lower temperature around 550°C.

In fact, there are two identifiable growth regimes for InN. The first occurs at a temperature around 500°C [35] and leads to the best quality films. At temperatures above this significant decomposition of the film occurs, but lower than this and metallic droplets can form on the surface of the film. In addition, a much higher V:III ratio is required, of the order of 10^5, if the formation of In droplets is to be avoided even at 500°C. The role of ammonia, or how the ammonia decomposes within the growth chamber, is not entirely clear. The second regime occurs at temperatures greater than 650°C and here a high V:III ratio impedes the growth. It is thought that the decomposition of ammonia is much enhanced reducing the need for such a high input ratio, but further, leading to a significant increase in the amount of hydrogen present. Hydrogen depresses the growth but the mechanism is not clear [35]. Lower growth temperatures can be realised with alternative sources of nitrogen, such as hydrazine (N_2H_4), di-methyl hydrazine, and other organo-nitrides, but the films tend to be contaminated with carbon. InN films are strongly *n*-type, irrespective of the dopants in the gas mixture. GaInN alloys also have to be grown at lower temperatures than GaN in order to avoid loss of In from the growing surface. This means that for heterostructures it is necessary to grow a small amount of the subsequent barrier at a lower temperature than is ideal for this material, or risk losing In from the surface as the growth is interrupted and the temperature raised to the optimum.

Growth by MBE has taken a long time to develop and lags behind MOCVD. The essential problem is the nitrogen source. One of the principal advantages that MBE has over MOCVD is the lower growth temperature, but this fact itself makes it impossible to use ammonia in its molecular form. It simply will not react with free Ga at the surface below 450°C [34]. Even so, the incorporation efficiency is only 0.5% at 500°C, reaching 3.8% at 700°C, and remaining pretty much constant thereafter. This is probably because nitrogen is lost from the surface. Dimethylhydrazine [$(CH_3)_2NNH_2$] decomposes at a lower temperature but introduces carbon into the layer. These problems are effectively overcome using a reactive source of nitrogen. An RF plasma generated using a frequency of 13.56 MHZ, a standard in sputtering systems, contains a small fraction of monatomic nitrogen and combining plasma generated nitrogen beams with metals from conventional effusion cells is effective for growing these films. However, RF generated plasmas contain high energy particles and the impact of these on the growing surface is detrimental to the optical and electrical properties of the films. Electron cyclotron resonance (ECR) at 2.45 GHz is more effective because the higher frequency leads to a greater fraction of reactive species and the geometry of the ECR allows low energy particles to be extracted, thus avoiding impact damage in the film. Growth rates are typically around 1 μm h^{-1} or slightly less, which is slow enough for control of quantum wells but high enough to grow relatively thick barrier layers.

In common with MOCVD, a two-step process is necessary for heteroepitaxial growth on sapphire, with the growth being interrupted so that the buffer layer can be annealed. Control of alloy composition is quite simple in principle in III–Vs because if the films are grown under V-rich conditions the total growth rate is governed by the total arrival rate of the metal atoms and the composition is governed by their relative arrival rates. However, in practise loss of the more volatile metal in the alloy as well as surface segregation might occur in the nitrides, leading perhaps to spatial non-uniformity of the composition. In addition, growth is usually 3-dimensional, which is probably caused by the low growth temperature relative to the melting point (typically 600–700°C compared with ~2500°C for GaN) because the mobility of the atoms at the surface is so low in GaN compared with, say, GaAs. Two-dimensional growth, in which the whole layer essentially grows together, has been the key to the growth of high quality layers in other material systems and the lack of it in these materials may be another reason why the material quality is relatively poor. It is easy to imagine also that if the growing surface is contoured then depending on the dimensions of the contours relative to the quantum well thickness, the quantum well will either be distorted but still contiguous or disrupted entirely in the plane of the film so that isolated regions of low band gap material exist.

11.2 Optical and electronic properties of (Al,Ga,In)N

The accepted view of the range of band gaps covered by these alloys is 1.89 to 6.2 eV, as shown in figure 11.4, but recently some controversy has arisen over the band gap of InN. Undoubtedly the early work on sputtered polycrystalline films suggests a band gap of this magnitude, but larger band gaps have been observed and this was thought to arise from partial oxidation of the InN. The value of 1.89 eV was a limiting value approached in conditions where oxidation had been eliminated. However, single crystal InN films have been grown with band gaps ranging from 0.7 to 2.0 eV [41]. The cause is attributed to the Moss-Burstein shift; essentially a band-filling effect in which the effective band gap is different from the density of states band gap. The reason why polycrystalline films appeared to have such a high band gap is simply because the carrier density is so high. Even now, intrinsic or p-type InN has not been grown, and as-grown material always has a background carrier density approaching 10^{20} cm^{-3}.

Such large variations in band gap with carrier density can actually be observed in light emitting devices. Nakamura, in a review of InGaN-based violet laser diodes [42] showed the output wavelength of single quantum well LEDs varying from 680 nm to 630 nm as the forward bias current is increased by a factor of four. This was interpreted in terms of band-filling. However, the longest of these wavelengths lies below the then accepted value of the band gap for InN, which was explained in terms of the quantum confined Stark effect. These materials are piezo-electric in the hexagonal form so the lattice distortion due to strain in the quantum well results in an internal electric field which separates the electron and hole wavefunctions and lowers the emission energy below the band gap (figure 11.7). Now that the energy gap has been shifted downwards, it seems much more likely that this observation is due entirely to

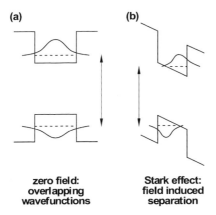

Figure 11.7. Reduced wavefunction overlap in the quantum confined Stark effect.

band-filling. For the purposes of this chapter, however, the bandgap will be assumed to follow figure 11.4, as most of the published work has assumed such a variation and a comprehensive re-interpretation of available experimental data has not yet been performed.

In fact it is remarkable how little is actually known about these materials. The In content of a given device has often been estimated from the output wavelength, but given the doubts over the band gap of the material past estimates may well have been wrong. Even theoretical calculations to confirm the experimental observation of a smaller band gap have thrown up answers ranging from 0.01 to 1.79 eV, but there are seemingly reliable calculations that confirm the band gap to be 0.85 ± 0.01 eV [43]. Similar calulations on a number of III–V materials were performed so that precise chemical trends could be derived, and the accuracy of these suggests that the band gap of InN is reproduced by these calculations. Similarly, confusion has also existed over the band offsets. Calculations performed on the InN/GaN system in 1996 gave a valence band offset of 0.48 eV as opposed to a measured value of 1.05 eV [44] but this offset is not likely to be affected much by the controversy over the InN band gap as most of the additional offset is expected to appear in the conduction band. Vurgaftman and Meyer [45] have reviewed the band parameters for the nitride system and recommend 0.5 eV for the InN/GaN valence band offset and 0.85 eV for the GaN/AlN valence band offset.

The electron effective mass is generally taken to be 0.20 m_0 [45]. Reports of the hole effective mass of GaN vary from 0.3 to 2.2. Šantić [46] recommends a density of states value (ie. averaged over all the valence band maxima) of 1.25 by averaging over all the various published reports. Hexagonal GaN has three valence bands but unlike GaAs and the more common cubic III–V materials they are none of them degenerate at $k=0$ because of the non-cubic field [44]. The so-called "A" band has the lowest energy and the heaviest mass and is therefore the band that forms the hole quantised states within a quantum well. Šantić's average value is 1.58 but reported values range between 1.00 and 2.07. In fact the effective mass is k-dependent as the bands are not parabolic and varies linearly with k from 1.2 to 1.8 [47] but right at the band edge it may be as low as 0.3 [48]. Similar values have been assumed in calculations of quantum confined states in alloys containing small amounts of indium. Shi [49] calculated the exciton states of $In_{0.15}Ga_{0.85}N$ quantum dots and used a hole mass of 1.1 m_0.

InGaN alloys are far from homogeneous. Alloy fluctuations, i.e. isolated patches of low band gap material, are formed by phase segregation of In. The idea has been put forward that these patches form quantum dots, which are semiconductor structures in which quantum confinement in all three dimensions occurs. They have been deliberately grown in a number of materials systems [50] using the so-called "Stranski-Krastanov" growth mode, which results in the spontaneous ordering of semiconductor heterojunctions into coherent islands. Initially, the deposited layer grows as a two-dimensional wetting layer covering the entire surface for at least one monolayer. Then, three-dimensional islands form on the wetting layer to relieve the stress elastically. If more material were to be deposited these islands would eventually coalesce to produce a single crystal epitaxial film, in which the stress would be relieved plastically if the thickness exceeds the critical limit. However, if the growth is stopped well short of this point these islands remain isolated and growth of an epitaxial wide band gap layer on top then results in a layer containing islands of narrow band gap material confined in three dimensions by material of a wider band gap. In essence these are solid state crystals with an atom-like electronic structure.

These self-assembled quantum dots are probably quite different from the structures observed in InGaN. Studies of spatially resolved cathodoluminescence revealed luminescent patches some 60 nm wide [51], but this is the emitting area rather than the confinement area. The quantum wells seem to be atomically smooth so alloy fluctuations would appear to be responsible rather than confinement in isolated patches of quantum well material contained within barrier material. If the latter were to occur, and such structures could conceivably form if irregular growth onto a rough surface caused the well material to break up, then the emitting area and the confinement area would be the same. However, if alloy fluctuations are responsible carriers can diffuse out of the potential minimum to some extent, making the emitting area larger than the confinement area. The carrier diffusion length has been estimated at 60 nm or so [34] but quite what is the confinement dimension is another matter. If it is larger than about 20 nm lateral confinement effects might not be significant as the Bohr radius for the electron is ≈ 5 nm. It will then be more appropriate to think of these areas as discs in which 1-D confinement occurs, but if the confinement area is smaller quantum effects may be significant.

Alloy fluctuations give rise to several characteristic effects, as summarised by Davidson *et al.* [52] and Lin *et al.* [53]:

- PL lines are broad and are increasingly so the wider the well. This is in contrast to most III–V quantum well systems where the linewidths increase with decreasing well width as interfacial fluctuations become relatively more important.
- PL lines are red shifted from absorption edges and also photo-conductivity edges by up to nearly 200 meV. The shift increases linearly with increasing In composition.
- PL lines also appear at sub-band gap energies, even in bulk InGaN films grown on GaN buffer layers.
- PL energies shift with increasing concentrations of free carriers.
- The integrated PL intensity varies with temperature according to

$$I = \frac{I_0}{[1 + A \cdot e^{T/T_0}]} \qquad (11)$$

which is usually found in amorphous semiconductors, whereas the energy of the photo-conductivity maximum varies with temperature according to the well-known Varshni relationship (equation chapter 10, 5) appropriate to band edge phenomena
- The time-dependence of the photo-current shows a "stretched exponential" behaviour

$$I(t) = I(0) \exp\left[-\left(\frac{t}{\tau}\right)^\beta\right] \quad (12)$$

which has been observed in disordered systems.

The commonly accepted explanation for these observations is that the band structure resembles that of an amorphous semiconductor, in which the valence and conduction band densities of states do not rise sharply at a well-defined energy but decrease exponentially with energy into the band. Carriers trapped in these so-called "band tails" are strongly localised and therefore have a very low mobility compared with carriers in the extended states at higher energy. The band gap is defined in terms of a mobility edge rather than in terms of the density of states. Similarly, the optical band gap is not as well defined as it is in an ordered crystalline solid where the optical absorption rises sharply from zero to typically $\sim 10^4$ cm^{-1} at the band edge. Optical absorption in the band tails is manifested as a slow decline in absorption with photon energy called an Urbach tail. The principal difference between a disordered solid and alloy fluctuations lies in the origin, and hence the nature, of the states comprising the band tails. In a disordered solid the band tail states arise from microscopic defects which can act as non-radiative centres through close coupling with other near-neighbour defects. In an alloy the localised states arise from compositional fluctuations and are therefore spatially more extended, with the consequence that they can act as efficient centres for carrier collection and recombination, which explains the characteristic PL and photoconductivity behaviour described above: the photoconductivity is measuring the higher energy extended states and the PL is measuring the localised, sub-band gap states. If indeed quantum effects occur as a result of this confinement the radiative efficiency will be enhanced via an increased oscillator strength. It should also be borne in mind that the quantum confined Stark effect will also contribute to an increased linewidth, and this undoubtedly occurs in addition to localisation effects, but heavy doping of the barrier layers can help screen out the piezo-electric field.

It is worth mentioning here the bowing parameter. The energy gap of an alloy of A and B is assumed to follow a simple quadratic form [45]

$$E_g(A_{1-x}B_x) = (1-x)E_g(A) + xE_g(B) - x(1-x)C \quad (13)$$

where C is known as the bowing parameter. If $C=0$ the band gap of the alloy varies linearly over the compositional range, but if $C>0$ the band gap at intermediate compositions can actually be lower than the band gaps at either end. It has always been assumed that the bowing parameter is large in this alloy system so that a localised increase in In content leads to a non-linear decrease in the band gap and hence strong localisation. With the discovery that the band gap of InN is nowhere near as large as originally thought the bowing parameter needs to be re-assessed, but this has not yet been done yet. It is not clear exactly how the band gap varies with In composition, nor is it clear how the In composition fluctuates within the alloy. It is safer for the present simply to accept that localisation occurs with identifiable energy differences of around 200 meV without worrying about the actual composition.

As In has to be added to bring the band gap into the visible, some segregation seems unavoidable, but, as will be shown, it can be reduced to manageable levels for devices in the wavelength range 400–420 nm. GaN on its own will emit in the UV (~360 nm), which will be useful for many applications such as data storage and chemical spectroscopy, but of course quite inappropriate for the human eye. However, it was recognised fairly early on that GaN itself is a poor luminescent emitter and even small amounts of In (>1%) significantly improved the luminescent output (figure 11.8). In fact, it was not possible initially to observe stimulated emission in GaN alone [54], partly, it is believed, because a considerable density of states exists within the band gap. These states give rise to a long exponential absorption tail extending down to 1.5 eV and act as non-radiative recombination centres. The states are probably physically located at threading dislocations. More recent work comparing the luminescent efficiency of GaN/AlGaN quantum wells grown homo-epitaxially on GaN with identical hetero-epitaxial structures grown on sapphire has indicated an increase in the brightness by a factor of 20 due to the

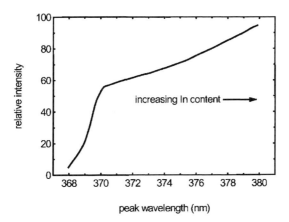

Figure 11.8. Variation of the luminescent output with band gap energy in InGaN. (After Nakamura [42].)

reduction in the density of threading dislocations [55]. In all other respects, such as the temperature dependence of the wavelength, these structures behaved identically apart from the intensity of the emission.

The same improvements in substrate technology that have allowed useful emission from GaN are also available to InGaN but, somewhat oddly, there are conflicting reports about the magnitude of the improvement it brings. In the light emitting diodes reported by Nakamura [42] the intensity of light output does not significantly improve with a reduction in the density of dislocations from over 10^{10} cm^{-2} to less than 10^6 cm^{-2} (which is an average over all the substrate including defect free regions) except at relatively low forward currents, but in those reported by Mukai et al. [56] significant differences in light intensity as a function of current were reported. It is fairly well established that the main effect of threading dislocations in InGaN is to reduce the area of material over which radiative recombination occurs [56], thereby reducing the intensity, but otherwise the light output is not significantly affected. Dislocations act as recombination centres but as the carrier diffusion length is estimated to be around 50–60 nm from spatially resolved cathodoluminescence, a dislocation density of at least 10^{15} cm^{-2} is required to affect the recombination rate significantly, and strong luminescence is observed even in highly defective material. In fact the presence of alloy fluctuations suppresses non-radiative recombination because the carriers are confined to the fluctuation and recombine radiatively long before they reach the dislocations.

Alloy fluctuations would therefore seem to be beneficial for LEDs, but the same is not necessarily true for laser diodes. It is important to emphasise that the intensity is affected via the emitting area rather than through the PL lifetime, which has been found to be insensitive to dislocation density but strongly dependent on the details of the interfacial quality or material purity [57]. In contrast, the size of the alloy potential fluctuations has been shown to depend directly on the material quality via the growth mode, with homo-epitaxial material showing fluctuations as low 3 meV compared with 170 meV in hetero-epitaxial material [58]. Hetero-epitaxial material can grow in spirals around dislocations that break through to the surface of the film whereas homo-epitaxial growth tends to be 2-D [59]. It appears then that the alloy fluctuations are associated with dislocations but are located sufficiently far way for confined carriers to recombine radiatively. Dislocations notwithstanding, large fluctuations are apparently beneficial for laser operation at low powers because population inversion occurs within the fluctuations [60]. This leads to low threshold currents and higher gain for low powers, but at high powers the maximum modal gain is reduced due to saturation. Population inversion can only be maintained by the capture of more carriers into the potential fluctuation. A small alloy potential fluctuation, on the other hand, leads to a relatively large threshold current as the carriers are spread over the volume of the crystal but also a narrower gain spectrum and a higher modal gain at high power output [58], not only because of reduced saturation but also because of

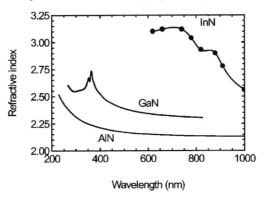

Figure 11.9. The wavelength dependent refractive index of GaN [61], InN [62], and AlN [63]. The data for InN exists at specific wavelengths only and the line is simply to aid the eye.

the higher density of states at a given photon energy. That said, the increase in laser operating lifetime has been correlated with the decrease in defect density arising from the use of bulk GaN substrates, so all round it is better for laser fabrication to have a high quality material with correspondingly low alloy fluctuations.

In order to shift the band gap to higher energies AlGaN alloys are necessary. These materials find use in electrical and optical confinement layers for GaN quantum wells and potentially as components in Bragg reflectors. Figure 11.9 shows the refractive indices [61–63]. Resonant cavity LEDs have already been fabricated using this technology [64] but as yet the vertical cavity surface emitting laser seems somewhat far off. There are a number of problems associated with this material system. First, the lattice mismatch could be as high as 2.5% and high Al content films tend to crack upon growth. It seems to be well established that cracking occurs during growth [64] but this can in fact be used to advantage [65] through lateral overgrowth across the cracks to produce a high quality, crack-free continuous layer. Second, there is some evidence that this material disorders naturally in much the same way as InGaN [66] but quite possibly the consequences are not so serious. As yet the use of $Al_xGa_{1-x}N$ as an active layer within a device has not been reported so compositional fluctuations will affect devices indirectly, either through fluctuations in the optical properties or through fluctuations in the barrier potential. Third, the resistivity increases dramatically for $x > 0.2$ for as yet unknown reasons, but in general the aluminium content of barriers through which a current is required to flow is kept low. The bowing parameters for the different valleys are small; 0.7 eV for the Γ valley, 0.61 eV for the X valley, and 0.8 eV for the L valley [45]. The recommended value for GaN/AlN valence band offset is 0.85 eV, i.e. 31%.

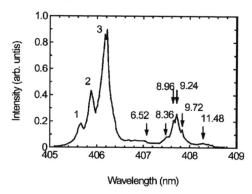

Figure 11.10. The output spectrum of an early InGaN multiquantum well laser with the integer and fractional mode indices of the peaks. (After Nakamura *et al.* [68].)

11.3 Laser diodes

The first InGaN laser diodes operated at room temperature lasted only a few hours, but this was rapidly extended to several hundred hours, and then to several thousand through improvements in the substrates and the film quality [67]. Even so, the output characteristics have not always been as expected, and some appreciation of the problems can be gained from looking at the output spectrum from an early (1996) MQW laser on sapphire [68] at about 20 mA above threshold (figure 11.10). Unlike the spectrum just above threshold, where a single peak at 406.7 nm was observed, multiple peaks are present. If the first three peaks correspond to well defined modes with a spacing of 0.255 nm, then not only is the mode spacing too large for the size of cavity, but the longer wavelength peaks have fractional mode indices and therefore don't fit into the mode structure. In fact, they appear to fit reasonably well a mode structure with a spacing of 0.11 nm.

One possible explanation, already mentioned, is the possibility that structural defects, perhaps cracks, effectively sub-divide the cavity. Alternatively, the large mode spacing could be a result of large carrier induced changes to the refractive index, as proposed by Jiang *et al.* [69]. Taking into account changes in the refractive index, then

$$\lambda \Delta m + m \Delta \lambda = 2L \cdot \Delta n = 2L(\Delta n_\lambda + \Delta n_n) \qquad (14)$$

where Δn_λ and Δn_n are respectively the material dispersion and the carrier-induced refractive index changes. Writing the dispersion as

$$\Delta n_\lambda = \frac{\partial n}{\partial \lambda} \Delta \lambda \qquad (15)$$

and taking $\Delta m = 1$ because we are interested in the mode spacing, then, after some manipulation,

$$\Delta\lambda = (\Delta\lambda)_0 \left[1 - \frac{2L}{\lambda} \Delta n_n \right] \tag{16}$$

where

$$(\Delta\lambda)_0 = \frac{\lambda}{\left[2L \frac{\partial n}{\partial \lambda} - m \right]} \tag{17}$$

is the mode spacing due to dispersion alone.

Applying this to the the mode spacing in figure 11.10, where $L = 6.7 \times 10^5$ nm, and taking the refractive index for GaN from figure 11.10 to be $n = 2.52$, with $\partial n/\partial \lambda \approx -0.0024$ (nm)$^{-1}$, then $(\Delta\lambda)_0 \approx 0.035$ nm. This estimate of the refractive index is admittedly uncertain. The quoted composition is 20% In in the well, which will increase the refractive index above that of GaN, but by how much is unclear. Tyagai's data [62] does not extend out this far and though other data for the refractive index has been presented [70] it is often lower than in figure 11.9 and sometimes lower even than the refractive index of GaN. Therefore it is not consistent with the trend from large band gap, low refractive index AlN through to low band gap, high refractive index InN that figure 11.9 clearly shows. A refractive index of 2.52 is therefore a good estimate in the absence of reliable data, but in any case a small change in either direction will not significantly affect the mode spacing. Taking into account the carrier induced refractive index change,

$$\Delta n = -\left[\frac{2\pi q^2}{n_0 m^* \omega^2} \right] \cdot n_e \approx -3.2 \times 10^{-22} n_e \tag{18}$$

with an effective mass of $0.2m_0$, gives a carrier density of around 6×10^{18} cm^{-1}, which is lower than the quoted threshold carrier density of 1.3×10^{19} cm^{-1}. However, this last value has been estimated from the threshold current, and if these laser diodes act like the light emitting diodes reported by Mukai *et al.* [56] the threshold carrier density may well be much smaller than that calculated. Certainly, the calculations by Yamaguchi *et al.* [58] indicate that in InGaN with large alloy fluctuations gain is possible for carrier densities below 10^{19} cm^{-1}, but the emission is below the average band gap, whereas for material with small alloy fluctuations a higher carrier density is needed and the emission occurs above the average band gap.

Whilst this explanation is certainly possible it doesn't fit all the facts. The carrier densities don't quite match, nor is it clear why additional peaks at a lower mode spacing and longer wavelength should appear at a relatively high current above threshold. The two emissions seem to be completely independent, which is of course possible if carrier localisation at fluctuations is occuring with no communication between the fluctuations. If this were the case, however, the longest wavelength fluctuation would be expected to emit first. Fluctuations that are independent cannot be described by a Fermi distribution function [71], so in effect all the fluctuations experience the same population independent of the particular energy levels, which will vary slightly with both the composition and the size. It is only as the carrier density increases and carriers can escape from the fluctuations that it becomes meaningful to describe the population by a Fermi function, so under these conditions the higher energy emissions would occur last. The model of carrier-induced refractive index change implies the contrary; a smaller mode spacing has a smaller carrier density and should therefore occur at a lower current. Moreover, the effect of carrier induced changes would be expected to apply generally to InGaN, but the majority of diode lasers now behave rather well in comparison. Figure 11.11 shows one of the first multiquantum well devices deposited in a commercial reactor [72]. It is deposited on sapphire and operated under pulsed conditions, but all the observable emission lines, and even two barely noticeable, but distinct features marked "?", fit a well-defined mode separation of 0.043 nm, which in the absence of dispersion gives a refractive index of ∼2.52 for an 800 μm long cavity. There are some missing modes and possibly the intensity distribution of the modes is not what might be expected for a smooth gain function, but clearly the modes are all propagating in the same cavity structure, unlike the modes in figure 11.10.

Mode spacings ranging from ≈0.03–0.05 nm, depending on the cavity length, have been reported in a number of papers on laser diodes emitting in the range 400 nm–410 nm [73–75]. These are consistent with an effective refractive index of ≈3.5, which may be a little high according to figure 11.10, but takes no account of the high dispersion around 400 nm. Incidentally, the paper by Miyajima *et al.* [75] demonstrates convincingly the importance of the lateral overgrowth for improving the emission properties. Spatially resolved photoluminescence over the surface of the InGaN film shows a dramatic

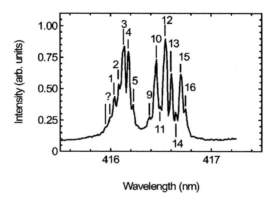

Figure 11.11. The output spectrum of the first InGaN multiquantum well diode grown in a commercial reactor. (After Park *et al.* [72].)

reduction in intensity over both the seed region and where the overgrown regions meet. Miyajima *et al.* also used cleaved facets, which are not common in early hetero-epitaxial nitride lasers. Whilst the substrates may cleave smoothly the film on top does not behave so well [76], with the result that a lot of effort has been put into developing a wet-etching technology for the facets [77]. The development of thick quasi-bulk GaN substrates has the added advantage that formation of the laser cavity by cleaving is made easier.

At present quantum well laser diodes emitting in the wavelength range $400 \leq \lambda \leq 420$ nm are available commercially. Not only do these devices have relatively low threshold current densities (≈ 1500 A cm^{-2}) compared with longer wavelength lasers (5000 A cm^{-2} at 450 nm) but they also have long lifetimes, reported in 2001 to be around 6000 hours compared with 3000 hrs at 430 nm and just a few hundred hours at 450 nm [78]. However, extending the wavelength outside this range in either direction is problematic. InGaN LEDs show the largest external quantum efficiency between 400 and 450 nm but dropping dramatically at both shorter and longer wavelengths [78]. At longer wavelengths the In alloy fluctuations become a serious source of inefficiency as the band tail states require filling before lasing action occurs among the extended states within the band. This leads to very high threshold currents, which were so high for $\lambda > 455$ nm that room temperature CW operation was impossible even in 2001 [79]. Although the InGaN will allow in principle emission out to 650 nm, which is the effective wavelength limit of the conventional III–V DH or quantum well lasers described in chapters 5 and 6, in practice it will be very difficult to achieve emission in the green and yellow using quantum well nitrides unless similar improvements in the material quality that have been made at low In concentrations can be extended to higher concentrations. The development of GaN substrates, for example, has reduced the alloy fluctuations for In concentrations below 10% to negligibly low levels so that devices exhibit a distributed loss of ≈ 50 cm^{-1} and characteristic gain of 70 cm^{-1} [80]. Whether phase segregation will continue to be a problem at higher In compositions is not yet clear.

At the other end of the spectrum, progress has been made down to $\lambda \approx 360$ nm, close to the band gap of GaN. Kneissl *et al.* have investigated InGaN quantum well laser diodes with In compositions from 0.2% to 2.7% [81] and AlGaN quantum well laser diodes with active regions containing 2% Al [82]. Both of these compositions are very close to the GaN band gap and of course require optical and electrical confinement layers containing Al. This presents a significant challenge, because the lattice mismatch between AlGaN and GaN puts an upper limit on the Al content before cracking of the AlGaN layer occurs during growth. Kneissl *et al.* overcame the problem to some extent by using short-period superlattice confinement layers with an average Al content of 12% and 0.8 μm thick for the InGaN diodes and 16% Al and 0.6 μm thick for the AlGaN diodes, with an additional 0.1 μm of lower Al content AlGaN separating the superlattice from the quantum wells. These devices were grown on sapphire substrates, and therefore might not be expected to be as efficient as devices grown on GaN, but even so the InGaN devices had a constant threshold current of 5 kA cm^{-2} over the range $378 \leq \lambda \leq 368$ nm rising very sharply to 17 kA cm^{-2} at 363 nm. The AlGaN laser diodes had threshold currents of 23 kA cm^{-2}. In similar laser diodes fabricated on GaN substrates, but with the addition of ZrO_2/SiO_2 dielectric mirror deposited over the facets to increase the reflectivity, Masui et al. [83] reported threshold currents of around 3 kA cm^{-2} at wavelengths down to 364 nm but rising sharply to over 5 kA cm^{-2} at 363 nm. The increase in threshold current was ascribed to absorption in the cladding layers, and with the limitations on the amount of Al that can be incorporated into the layers this would seem at present to be a fundamental problem. Nonetheless, these diodes exhibited lifetimes of 2000 hours, which is good enough for some applications.

11.4 Quantum dot lasers

As discussed, the idea of alloy fluctuations in the plane of the well naturally leads on to the idea of quantum dots in these materials. Potential fluctuations in alloys with less than 10% In seem to have been largely eliminated by the use of high quality GaN substrates, but InGaN alloys would seem to lend themselves to the formation of zero-dimensional systems because of the large lattice mismatch and the tendency to 3-D growth under certain ratios of N:In flux. Zero-dimensional systems have been described at length by Yoffe [84] where three regimes of confinement have been identified. Strong confinement corresponds to systems that are smaller than the Bohr radius of both carriers as well as the exciton. The electron and hole can be thought of as separate confined particles and excitons are not formed within the micro-crystallite. The other extreme is weak confinement, in which the microcrystallite dimensions are larger than the characteristic Bohr radii. An exciton forms and the exciton motion is quantised. Intermediate between these two, the electron motion is quantised but the hole, which usually has a much higher effective mass and therefore smaller Bohr radius, is not and interacts with the electron via the Coulomb potential. Which of these three conditions obtains depends not just upon the size but also on the material itself through the effective masses. In GaN the electron Bohr radius is about 5 nm whereas the hole radius is ≈ 7 Å. Yoffe has considered mainly II–VI micro-crystallites embedded in glass matrices, but his paper also covers the general theoretical background to the confined states.

In general, numerical methods have to be used to determine the energy states. As in the case of a quantum well, analytical solutions in the effective mass approximation exist, but in a real solid where some penetration of the wavefunction into the barrier occurs, only a numerical solution will yield the quantised states. In relation to quantum dot lasers, Grundmann [71] has considered self-organised structures of the Stranski-Krastanow type. Self-organised means that the structures are not defined by external means such as lithography. These can be fabricated by etching through a quantum well to form a disc perhaps 100 nm or more in diameter, and despite the relatively large size in the plane, they can still exhibit quantum dot like behaviour [85]. For a laser, however, there is the added complexity of interrupting the growth to etch down through the well. Self-organised systems are far better, as the quantum dots are formed during growth and are naturally covered by the barrier materials so that conventional diode laser fabrication technology can be used for contacting and cleaving. Pseudopotential methods and the 8-band k · p method both yield very similar results for structures less than ≈ 10 nm across. The energy separation is much larger than in a quantum well, and the oscillator strength is very large. The conservation of momentum that's so important in other lasers utilising extended states, and quantum wells are included in this because the states are extended in the plane of the well, simply does not apply here, even though the electron and hole states might not have the same wave vectors. The localisation of the electron and hole wavefunctions leads to an uncertainty in the carrier momentum that is at least one order of magnitude greater than the photon momentum.

One might suppose, in view of the quantisation, that the linewidth would be very narrow but fluctuations in the sizes of the micro-crystallites leads to imhomogeneous broadening of several meV. Nonetheless, in the conventional III–V materials examined to date high material gains of the order of 10^5 cm^{-1} are possible, but in practise modal gains of 10 cm^{-1} are common. This means that the cavities have to be very long in order to achieve threshold, but threshold currents as low as 12 A cm^{-2} at low temperature have been achieved [see 71 and references therein]. These structures are therefore very promising as laser materials. Broad area, high power devices can be achieved that will emit up to the limit for catastrophic optical damage. For example, Zhukov *et al.* [86] have reported six layers of stacked InAs/InAlAs quantum dots with a total area density of 6×10^{11} cm^{-2} fabricated into 100 μm wide emitters giving 3.5 W from a single emitter 920 μm long. These are clearly attractive attributes for diode lasers, and quantum dot materials have been used as the active medium of every type of laser discussed within this book, with the exception of the DH device. However, quantum dots are not without their problems. The bottleneck for these devices is the carrier capture time into the quantum dot. As each dot is independent of each other population inversion can only be re-established when a new carrier is captured into the dot, which leads to a strong saturation of the gain. The independence of the dots has the advantage, however, that carriers will not diffuse towards non-radiative centres.

Key to Stranski-Krastanow growth is a lattice mis-match between film and substrate. There has to be a strain to be relieved by the formation of hillocks on top of the very thin 2-D layer, which makes InGaN an ideal material at higher In compositions. Indeed, many of the quantum dot attributes mentioned above are recognised as already operating in these materials, but these effects have arisen naturally and are therefore not optimised. With the band gap of InN having been pushed down to ≈ 0.9 eV, the deliberate growth of InGaN quantum dots could cover the entire visible spectrum. Jiawei *et al.* [87], in a review of GaN-based quantum dots, reports on structures containing a fixed 16% In (±2%) at 1.5, 3, and 5 mono-layer thicknessess that luminesce at 451 nm, 532 nm, and 656 nm respectively. The efficiency of these structures relative to the inefficient quantum well LEDs described earlier has not been mentioned, but the principle has been demonstrated and the growth requirements are more in keeping with what seems to be the natural trends in these materials. Moreover, it is not strictly necessary to add In to the active material. GaN quantum dots have been grown on AlN and these too luminesce from 2.13 eV to 3.18 eV, probably because the internal electric fields lower the emission energy.

11.5 Summary

The development of blue laser diodes has been described as a triumph in the materials engineering of the III-nitrides. These materials are much more robust than II–VI compound semiconductors, as explained by their higher covalency. Several key steps were necessary to realise laser diodes; the development of MOCVD; the two-step growth method; thermal annealing of Mg-doped material; and the growth of thick, quasi-bulk GaN as a substrate. Early laser diodes demonstrated anomalous modal behaviour consistent with the formation of cracks within the cavity, but rapid improvment in the quality of the substrates and the epitaxial layers on top soon resulted in devices that exhibit a well-behaved mode structure. The quantum efficiency of the devices is highest when small amounts of In are added to the active region, corresponding to emission between 400 nm and 420 nm, though devices emitting down at 360 nm, corresponding to the band gap of GaN, have been improved recently. The large lattice mis-match in these materials makes them prone to alloy fluctuations, but for In concentrations < 10% these appear to be minimal. However, at 16% In quantum dots can be grown that will emit over the entire visible part of the spectrum, and these material would seem to be ideally suited to development of this type of structure.

11.6 References

[1] Whitaker T 2004 *Compound Semiconductor* **10** 18
[2] Nurmikko A and Gunshor R L 1997 *Semicond. Sci. Technol.* **12** 1337–1347
[3] Faschinger W 1995 *J. Cryst. Growth* **146** 80–86
[4] Desnica U V 1998 *Prog. Cryst. Growth Charact. Mater.* **36** 291–357

[5] Kurtz E, Einfeldt S, Nurnberger J, Zerlauth S, Hommel D and Landwehr G 1995 *Phys. Status Solidi B – Basic Res.* **187** 393–399
[6] Fan Y, Han J, Gunshor R L, Walker J, Johnson N M and Nurmikko A V 1995 *J. Electron. Mater.* **24** 131–135
[7] Hauksson I S, Simpson J, Wang S Y, Prior K A and Cavenett B C 1992 *Appl. Phys. Lett.* **61** 2208–2210
[8] Kothandaraman C, Neumark G F and Park R M 1995 *Appl. Phys. Lett.* **67** 3307–3309
[9] Kamata A and Moriyama T 1995 *Appl. Phys. Lett.* **67** 1751–1753
[10] Fujita S, Tojyo T, Yoshizawa T and Fujita S 1995 *J. Electron. Mater.* **24** 137–141
[11] Zhu Z Q, Takebayashi K, Yao T and Okada Y 1995 *J. Cryst. Growth* **150** 797–802
[12] Lee M K, Yeh M Y, Huang H D and Hong C W 1995 *Jpn. J. Appl. Phys. Part 1 – Regul. Pap. Short Notes Rev. Pap* **34** 3543–3545
[13] Han J, Gunshor R L and Nurmikko A V 1995 *J. Electron. Mater.* **24** 151–154
[14] Prete P and Lovergine N 2002 *Prog. Cryst. Growth Charact. Mater.* **44** 1–44
[15] Ohno T, Kawaguchi Y, Ohki A and Matsuoka T 1994 *Jpn. J. Appl. Phys. Part 1 – Regul. Pap. Short Notes Rev. Pap.* **33** 5766–5773
[16] Han J, Gunshor R L and Nurmikko A V 1995 *Phys. Status Solidi B – Basic Res.* **187** 285–290
[17] Kolodziejski L A, Gunshor R L and Nurmikko A V 1995 *Annu. Rev. Mater. Sci.* **25** 711–753
[18] Orton J W and Foxon C T 1998 *Rep. Prog. Phys.* **61** 1–75
[19] Landwher G, Waag A, Fischer F, Lugauer H-J and Schüll K 1998 *Physica E* **3** 158–168
[20] Harrison W A 1980 *Electronic Structure and the Properties of Solids; the Physics of the Chemical Bond* (San Francisco: W H Freeman & Co.) p 189
[21] Harrison W A 1980 *Electronic Structure and the Properties of Solids; the Physics of the Chemical Bond* (San Francisco: W H Freeman & Co.) p 67
[22] Harrison W A 1980 *Electronic Structure and the Properties of Solids; the Physics of the Chemical Bond* (San Francisco: W H Freeman & Co.) p 189
[23] Harrison W A 1980 *Electronic Structure and the Properties of Solids; the Physics of the Chemical Bond* (San Francisco: W H Freeman & Co.) p 188
[24] Azuhatay T, Sotayz T and Suzuki K 1996 *J. Phys.: Condens. Matter* **8** 3111–3119
[25] Harrison W A 1980 *Electronic Structure and the Properties of Solids; the Physics of the Chemical Bond* (San Francisco: W H Freeman & Co.) p 115
[26] Kim K, Lambrecht W R L and Segall B 1996 *Phys. Rev. B* **53** 16310–16326
[27] Wang K and Reeber R R 2001 *Applied Physics Letters* **79** 1602–1604
[28] Lide D R ed. 1913–1995 *CRC Handbook of Chemistry and Physics* 75th Edition. (Boca Raton: CRC Press)
[29] Vériè C 1998 *Journal of Crystal Growth* **184/185** 1061–1066
[30] Martin R M 1970 *Phys. Rev. B* **10** 4005–4011
[31] Vériè C 1998 *J. Electron. Mater.* **27** 782–787
[32] Nekrutkina O V, Sorokin S V, Kaigorodov V A, Sitnikova A A, Shubina T V, Toropov A A, Ivanov S V, Kop'ev P S, Reuscher G, Wagner V, Geurts J, Waag A and Landwehr G 2001 *Semiconductors* **35** 520–524
[33] Ambacher O 1998 *J. Phys. D: Appl. Phys.* **31** 2653–2710

[34] Jain S C, Willander M, Narayan J and Van Overstreaten R 2000 *J. Appl. Phys* **87** 965–1006
[35] Bhuiyan A G, Hashimoto A and Yamamoto A 2003 *J. Appl. Phys* **94** 2779–2808
[36] Czernetzki R, Leszczynski M, Grzegory I, Perlin P, Prystawko P, Skierbiszewski C, Krysko M, Sarzynski M, Wisniewski P, Nowak G, Libura A, Grzanka S, Suski T, Dmowski L, Litwin-Staszewska E, Bockowski M and Porowski S 2003 *Physica Status Solidi A – Applied Research* **200** 9–12
[37] Akasaki I 2000 *J. Cryst. Growth.* **221** 231–239
[38] Kim C, Robinson I K, Myoun J, Shim K, Yoo M-C and Kim K 1996 *Appl. Phys. Lett.* **69** 2360–2362
[39] Beaumont B, Vennéguès Ph. and Gibart P 2001 *Phys. Stat. Sol. (b)* **227** 1–43
[40] Jang H W, Kim S Y and Lee J L 2003 *J. Appl. Phys.* **94** 1748–1752
[41] Bhuiyan A G, Sugita K, Kasashima K, Hashimoto A, Yamamoto A and Davydov V Y 2003 *Appl. Phys. Lett.* **83** 4788–4790
[42] Nakamura S 1999 *Semicond. Sci. Technol.* **14** R27–R40
[43] Wei S-H, Nie X, Batyrev I G and Zhang S B 2003 *Phys. Rev. B* **67** 165–209
[44] Wei S H and Zunger A 1996 *Appl. Phys. Lett.* **69** 2719–2721
[45] Vurgaftman I and Meyer J 2003 *J. Appl. Phys.* **94** 3675–3795
[46] Šantić B 2003 *Semicond. Sci. Technol* **18** 219–224
[47] Shields P A, Nicholas R J, Peeters F M, Beaumont B and Gibart P 2001 *Phys. Rev. B* **64** art. no. 081203
[48] Shields P A, Nicholas R J, Takashina K, Grandjean N and Massies J 2002 *Phys. Rev. B* **65** 195–320
[49] Shi J-J 2002 *Solid State Communications* **124** 341–345
[50] Bennett B R, Shanabrook B V, Thibado P M, Whitman L J and Magno R 1997 *J. Cryst Growth* **175** 888–893
[51] Chichibu S, Wada K and Nakamura S 1997 *Appl. Phys. Lett.* **71** 2346–2348
[52] Davidson J A, Dawson P, Wang T, Sugahara T, Orton J W and Sakai S 2000 *Semicond. Sci. Technol.* **15** 497–505
[53] Lin T Y, Fan J C and Chen Y F 1999 *Semicond. Sci. Technol.* **14** 406–411
[54] Pankove J I 1997 *MRS Internet J. Nitride Semicond. Res.* **2** art. 19
[55] Grandjean N, Massies J, Grzegory I and Porowski S 2001 *Semicond. Sci. Technol.* **16** 358–361
[56] Mukai T, Nagahama S, Iwasa N, Senoh M and Yamada T 2001 *J. Phys.: Condens. Matter* **13** 7089–7098
[57] Chichibu S F, Marchand H, Minsky M S, Keller S, Fini P T, Ibbetson J P, Fleischer S B, Speck J S, Bowers J E, Hu E, Mishra U K, DenBaars S P, Deguchi T, Soto T and Nakamura S 1999 *Appl. Phys. Lett.* **74** 1460–1462
[58] Yamaguchi A A, Kuramoto M, Nido M and Mizuta M 2001 *Semicond. Sci. Technol.* **16** 763–769
[59] Heying B, Tarsa E J, Elsass C R, Fini P, DenBaars S P and Speck J S 1999 *J. Appl. Phys.* **85** 6471–6476
[60] Chichibu S F, Abare A C, Mack M P, Minsky M S, Deguchi T, Cohen D, Kozodoy P, Fleischer S B, Keller S, Speck J S, Bowers J E, Hu E, Mishra U K, Coldren L A, DenBaars S P, Wada K, Sota T and Nakamura S 1999 *Materials Science and Engineering B – Solid State Materials for Advanced Technology* **59** 298–306

[61] Yu G, Wang G, Ishikawa H, Umeno M, Soga T, Egawa T, Watanabe J and Jimbo T 1997 *Appl. Phys. Lett.* **70** 3209–3211
[62] Tyagai V A, Evstigneev A M, Krasiko A N, Andreeva A F and Malakhov V Ya 1977 *Sov. Phys. Semicond.* **11** 1257–1259
[63] Geidur S A and Yaskov A D 1980 *Opt. Spectrosc. (USSR)* **48** 618–622
[64] Natali F, Byrne D, Dussaigne A, Grandjean N and Massies J 2003 *Appl. Phys. Lett.* **82** 499–501
[65] Bethoux J-M, Vennéguès P, Natali F, Feltin E, Tottereau O, Nataf G, De Mierry P and Semond F 2003 *J. Appl. Phys.* **94** 6499–5607
[66] Laügt M, Bellet-Amalric E, Ruterana P and Omnès F 2003 *J. Phys. Chem. Sol.* **64** 1653–1656
[67] Nakamura S 1999 *Thin Solid Films* **343–344** 345–349
[68] Nakamura S, Senoh M, Nagahama S, Iwasa N, Yamada T, Matsushita T, Sugimoto Y and Kiyoku H 1996 *Appl. Phys. Lett.* **69** 1568–1570
[69] Jiang H X and Lin J Y 1999 *Appl. Phys. Lett.* **74** 1066–1068
[70] Yang H F, Shen W Z, Qian Z G, Pang Q J, Ogawa H and Guo Q X 2002 *J. Appl. Phys.* **91** 9803–9808
[71] Grundmann M 2000 *Physica E* **5** 167–184
[72] Park Y, Kim B J, Lee J W, Nam O H, Sone C, Park H, Oh E, Shin H, Chae S, Cho J, Kim I-H, Khim J S, Cho S and Kim T I 1999 *MRS Internet J. Nitride Semicond. Res.* **4** 1–4
[73] Nakamura S 1999 *Materials Science and Engineering B* **59** 370–375
[74] Kneissl M, Bour D P, Van de Walle C G, Romano L T, Northrup J E, Wood R M, Teepe M and Johnson N M 1999 *Appl. Phys. Lett.* **75** 581–583
[75] Miyajima T, Tojyo T, Asano T, Yanashima K, Kijima S, Hino T, Takeya M, Uchida S, Tomiya S, Funato K, Asatsuma T, Kobayashi T and Ikeda M 2001 *J. Phys.: Condens. Matter* **13** 709–714
[76] Abare A C, Mack M P, Hansen M, Sink R K, Kozodoy P, Keller S, Speck J S, Bowers J E, Mishra U K, Coldren L A and DenBaars S P 1998 *IEEE J. Select. Topics in Quant. Electr.* **4** 505–509
[77] Stocker D A, Schubert E F, Boutros K S and Redwing J M 1999 *MRS Internet J. Nitride Semicond. Res.* 4S1, G7.5
[78] Mukai T, Nagahama S, Yanamoto T and Sano M 2002 *Phys. Stat. Sol. (a)* **192** 261–268
[79] Nagahama S, Iwasa N, Senoh M, Matsushita T, Sugimoto Y, Kiyoku H, Kozaki T, Sano M, Matsumura H, Umemoto H, Chocho K, Yanamoto T and Mukai T 2001 *Phys. Stat. Sol. (a)* **188** 1–7
[80] Kneissl M, Van de Walle C G, Bour D P, Romano L T, Goddard L L, Master C P, Northrup J E and Johnson N M 2000 *J. Lumines.* **87–89** 135–139
[81] Kneissl M, Treat D W, Teepe M, Miyashita N and Johnson N M 2003 *Appl. Phys. Lett.* **82** 2386–2388
[82] Kneissl M, Treat D W, Teepe M, Miyashita N and Johnson N M 2003 *Appl. Phys. Lett.* **82** 4441–4443
[83] Masui S, Matsuyama Y, Yanamoto T, Kozaki T, Nagahama S and Mukai T 2003 *Jpn. J. Appl. Phys.* **42** L1318–L1320
[84] Yoffe A D 2002 *Advances in Physics* **51** 799–890
[85] Tarucha S, Honda T, Austing D G, Tokura Y, Muraki K, Oosterkamp T H, Janssen J W and Kouwenhoven L P 1998 *Physica E* **3** 112–120

[86] Zhukov A E, Kovsh A R, Ustinov V M, Livshits D A, Kop'ev P S, Alferov Z I, Ledentsov N N and Bimberg D 2000 *Materials Science and Engineering B* **74** 70–74

[87] Jiawei L, Zhizhen Y and Nasser N M 2003 *Physica E* **16** 244–252

Problems

1. Consider the laser in figure 11.10. This has a cavity length of 670 μm. Calculate the expected mode spacing assuming a refractive index of 2.5 and a dispersion of -0.0025 nm^{-1} at 406 nm and -0.0024 nm^{-1} at 408 nm. Calculate the effective cavity length corresponding to the two emissions.
2. Consider Miyajima's laser [75] emitting at 404 nm with a cavity length of 600 μm and a mode spacing of 0.038 nm. Calculate the effective refractive index assuming first, no dispersion, and second, a dispersion of -0.0025 nm^{-1}.
3. Calculate the power reflectivity of a typical GaN facet at 400 nm and the mirror loss for a 500 μm long cavity. What cavity length is needed to reduce the mirror loss to that of a conventional III–V laser with a similar cavity length?

Chapter 12

Quantum cascade lasers

The quantum cascade laser is a triumph of solid state engineering. It exists only as a consequence of developments in so-called "band-gap engineering" and unlike a conventional diode laser does not rely on band-to-band recombination. This idea was pioneered by Federico Capasso, then at Bell Labs, who, along with others, started to develop complex quantum devices based on the idea of epitaxy with abrupt interfaces using molecular beams. Several applications of epitaxial growth have already been described in this book, such as double heterostructures, quantum wells, and distributed Bragg reflectors, but of these only the quantum well represents any sort of band gap engineering. In the DBR the electronic properties are not of primary importance beyond achieving the necessary conductivity to allow the mirrors to act as contacts. In the DH laser the width of the active region is still large enough for the material to be considered bulk-like, and the band gap of the active region is identical to the band gap of bulk material. In the quantum well, however, the lowest electron energy level effectively becomes the conduction band, and so the band gap is altered over the bulk material by the zero-point energy of both the electrons and the holes. In a material containing many quantum wells, sometimes with barriers thin enough for the wells to interact with each other, the effective band gap can be made to change throughout the structure simply by altering the well width, even though the well and barrier compositions remain fixed.

The quantum cascade laser uses this type of band gap engineering and consists of a series of quantum wells and barrier which together make up both active regions and injector regions. Usually, but not always, active regions comprise the layers in which the radiative transitions occur among the sub-bands of quantum wells. Injector regions, on the other hand, comprise a series of interacting quantum wells – so-called "superlattices" – which give rise to a band of states, called a "mini band" – such that electron transport perpendicular to the wells can occur. A distinction is made here between mini-bands and sub-bands, which are the discrete states localised in quantum wells whereas mini-bands are the states associated with a superlatice. They arise from a succession of interacting wells, that is to say, wells in which the barriers are sufficiently thin that the wavefunctions are not confined simply to one well but penetrate through the barriers and extend over the whole of the superlattice. The bands also extend over a limited range in energy and the number of states in a band is equal to the number of wells that make up the superlattice.

The design of the active and injector regions is not straightforward and cannot be reduced to a few simple equations. It will become clear in this chapter that numerical computation is the only tool that the designer has to ensure that a particular choice of material composition and a particular growth sequence will result not only in the right energy levels, but also rates of population and depopulation from the levels that will allow population inversion. Transport through the mini-bands also has to be modelled numerically using Monte Carlo techniques, so what follows in this chapter, unlike the descriptions of other lasers in this book, is not a set of concepts or equations that can be used as design principles, but an exploration of some of the ideas and structures that have so far been reported in the literature. To do anything more requires a level of understanding beyond anything that can be given here.

A quantum cascade laser will contain several injector-active pairs so electrons injected into the system cascade through and are re-used in order to generate multiple photons per electron, as illustrated in figure 12.1 [1]. Consider an electron injected into the upper laser level (ULL) from the right. If the lower laser level (LLL) is empty then by definition population inversion

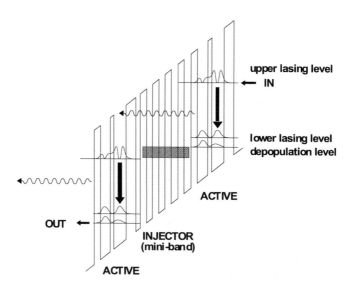

Figure 12.1. A schematic of a quantum cascade laser (after Gmaichl *et al.* [1]) showing two active regions at either end of a superlattice injector. The possibility of more than one photon per electron is clearly indicated, but it should be noted that the photons are emitted in the plane of the layers rather than in the growth direction as indicated above.

between these two levels exists and stimulated emission is possible. If the assumption is made that electrons can only reach the LLL via the ULL, and in an ideal device this will apply, then clearly if the exit from ULL to the depopulation level (DPL) and then to the miniband is slower than the total time from taken to enter the ULL and to make the transition to the LLL, charge will build up within the lower states of the quantum well and population inversion will be lost. For laser action to occur, therefore, it is necessary for the dwell time in the ULL to exceed the relaxation time from the LLL. In an ideal case electrons entering the LLL would be swept out instantaneously, thereby presenting an empty level to the electron in the ULL.

Population inversion in these systems is mathematically very simple. The rate of change of the carrier concentration in the LLL, n_L, can be expressed as a function of the excess concentration in the upper and lower levels, i.e.

$$\frac{dn_L}{dt} = \frac{n_U}{\tau_U} - \frac{n_L}{\tau_L} \qquad (1)$$

where $\tau_{U,L}$ are the lifetimes of the respective levels. These are global lifetimes that represent all the possible paths out of the state, including relaxation by phonon emission which is, in fact, the dominant relaxation mechanism. Clearly, the presence of multiple pathways will reduce the level lifetime. In steady state, therefore,

$$\frac{dn_L}{dt} = 0 \qquad (2)$$

and

$$\frac{n_U}{n_L} = \frac{\tau_U}{\tau_L}. \qquad (3)$$

As argued, the lower level lifetime has to be shorter than the upper level lifetime.

There is a natural upper limit to the photon energy that can be obtained from a cascade laser. Quite clearly, the energy cannot be greater than the conduction band offset, so even for the most favourable material system, AlGaAs/GaAs, energies will be limited to a few tenths of an electron-volt. However, one of the features of the quantum cascade laser should be apparent by now; the emission wavelength is independent of the band gap of the constituent materials but is dependent instead on the width of the quantum wells in the active region, making this type of device structure very versatile. A very wide range of energies out into the far infra-red are possible. In mid-IR quantum cascade lasers, corresponding to the wavelength range from about 5–20 μm, of which the 8–14 μm atmospheric transmission window is the most important, rapid relaxation from the LLL to the DPL is achieved by making the LLL-DPL energy separation the same as, or very close to, the longitudinal optical phonon energy so that the probability of relaxation by single phonon emission is very high. In GaAs this is ~36 meV.

From this perspective the LO phonon energy represents a seemingly natural limit on the energy of the laser emission. It follows that if resonant phonon emission can be used to depopulate the LLL then it can also depopulate the ULL. If the emission energy of the laser is lowered by widening the quantum well so that the ULL-LLL energy separation approaches phonon resonance the LLL-DPL separation will naturally move off the resonance, leaving a very short upper lifetime and a relatively longer lower lifetime. The possibility of population inversion is thereby destroyed. It should be emphasised that this is a problem in scattering rates rather than energy levels. Long wavelength luminescence is possible simply by adjusting the well width to achieve the correct energy level separation, and luminescence at 70–80 μm was reported by several groups throughout the late 1990's, but for a laser the additional requirement on the scattering times out of the states is a very severe limitation. If the ULL-LLL transition energy is reduced below the phonon energy then very careful engineering of the LLL-DPL energy level is necessary to ensure the scattering rate will allow population inversion. Other relaxation mechanisms, such as electron-electron scattering, become important at such low subband separations. Even so phonon scattering can still be used with the correct design of laser, but the problem wasn't solved until 2001, since when laser emission wavelengths out to ~ 100 μm have been achieved. The quantum cascade laser is thus a very powerful tool for long wavelength applications in both communication and medicine.

This use of subbands is a significant departure from other diode lasers, but it's not the whole story. It is only the "quantum" but not the "cascade", and this is the second, and perhaps most radical, departure from conventional laser technology, for the electron is re-used in other laser transitions. An electron leaving the DPL enters the miniband states of the next injector region and is transported through the superlattice into the ULL at the other end. Minibands in superlattices are described in detail on page 390. It is sufficient for the present to realise that minibands arise from the interacting states of a series of closely spaced quantum wells. If all the wells and barriers are identical the minibands lie at a common energy throughout the structure, and the bottom of the lowest miniband effectively becomes the conduction band edge within the structure. However, under an applied field this miniband will not be flat, as in figure 12.1, but will in fact break up into a series of discrete states called a Wannier-Stark ladder. In the quantum cascade laser it is necessary to design the injector region so that the electron states of the interacting quantum wells align under the action of an electric field to produce a mini band that is essentially flat. This is achieved by grading the thicknesses of the layers and also the composition within the layers, so that under zero applied bias the band offsets resemble more of a saw-tooth structure than a square well. As the electron is transported through the mini band, its energy relative to the bottom of the well, which is essentially the bulk conduction band edge, increases, and the electron emerges from the mini band at the next ULL. In this way electrons are re-used in the lasing process, unlike the conventional diode laser where the band-edge recombination removes the electron from the lasing process.

The full extent of the numerical computation necessary to design a quantum cascade laser should be apparent by now. Not only is it necessary to solve the Schrödinger equation to calculate the wavefunctions of the states in the wells, especially if, as in figure 12.1, the states at either end of the laser transition extend over two or three wells, but it is also necessary to simulate the transport and scattering processes. The variation of the energy of the states with applied voltage has to be considered as part of the design, and normally will involve different states of the different wells being brought into resonance at a particular applied field. In addition, of course, the superlattice states also have to be designed to allow transport at the same electric field as the laser transitions occur. The only effective way to do this is to design a structure and to simulate its operation using whatever simplifications and approximations are appropriate. It has been reported, for example, that full details of the transport will not be revealed without a full stochastic Monte Carlo simulation, as details of the electron distributions in the states and the minibands, as well as the effective electron temperature, are all but impossible to determine analytically [2]. Calculations of the electron-electron and electron-phonon scattering suggest, however, that the electrons in the various states reach thermal equilibrium among themselves, but not necessarily with the lattice, so they can be regarded as having the same temperature even if its absolute value is uncertain [3]

Even the full details of the electron-phonon scattering are hard to compute, but the calculations are probably accurate enough for the electron energies involved. Certainly, such calculations are beyond the scope of this book, but the underlying principles can be described. In essence, the calculation proceeds by defining a deformation potential for the formation of the phonon, i.e. the effect on the lattice potential of displacing an ion. For small displacements this potential has the form of simple harmonic motion, which results from a Taylor expansion of the potential to second order and putting the linear term to zero because the ion rests at a potential minimum. It is assumed that this potential interacts weakly with the electron so that Fermi's Golden Rule can be applied, which of course calculates transition rates in terms of a matrix element. The matrix element can only be calculated accurately for real systems by numerical integration, as the wavefunction itself has to be calculated numerically by solution of Schrödinger's equation. The inverse of the scattering rate gives the lifetime.

Similarly, electron-electron scattering rates can also be calculated [4–6], but self-consistent calculations are very difficult. That is, scattering rates usually depend on the electron density but if scattering changes the population of a state then strictly the scattering rate should also change in response. However, static populations are normally used to calculate the scattering rates. Electron-electron scattering can be inter-subband in nature, such that electrons are transferred from one subband to another, or intra-subband, such that the electrons involved remain within the same subbands, but exchange energy and momentum. In particular, Harrison [3] has identified a "bi-intrasubband" mechanism in which two electrons, which may be in different sublevels or in the same sublevel, interact with scattering rates of the order of 10^{13}–10^{14} s^{-1} for all realistic temperatures but the final sublevels do not change. The populations do not change as a result of such scattering but the exchange of energy and momentum between the electrons means that the electrons in all the subbands can be considered to have the same uniform temperature. Quite what this temperature will be is difficult to say but it will, in all likelihood, exceed the lattice temperature. So, whilst the inter-subband electron-electron scattering rates are generally much lower than the intra-subband scattering rates, inter-subband electron-electron scattering can dominate over phonon scattering both at low temperatures where the phonon density is low and at small intersubband energy separations corresponding to far infra-red emissions.

In summary, quantum cascade lasers differ from conventional diode lasers in several ways:

- the devices are unipolar; that is to say, only one carrier type is involved;
- population inversion is achieved between sub-bands of a quantum well system rather than between electron and hole states;
- the dependence of the lasing wavelength on band gap is therefore broken, and a very wide range of wavelengths can be engineered in well-behaved and well-understood materials systems such as GaAs/AlGaAs;
- the wavelengths extend from the near infra-red to the very far infra-red close to 100 μm;
- the localised and discrete nature of the lasing energy levels means that the system appears to be almost atomic-like, with the exception that of course in-plane dispersion exists
- the in-plane dispersion in each sub-band is of the same sign, unlike bipolar devices, so even if transitions occur between states at in-plane $k \neq 0$ the wavelength is very similar to transitions at $k = 0$ (see figure 12.2), and the gain spectrum is correspondingly narrow;
- charge neutrality cannot exist in the same way as in the bipolar device, in which equal concentrations of electrons and holes are injected;
- space charges will exist therefore within the quantum well, which can only be offset by background doping;
- the linewidth is expected to be almost Schawlow-Townes – as like the refractive index in this wavelength region is determined principally by electronic transitions at much higher photon energies, leading to a very small linewidth enhancement factor.

In some respects, however, the devices are similar:

- Light emission occurs perpendicular to the direction of current flow.
- Optical confinement is achieved through the use of wave guiding structures.

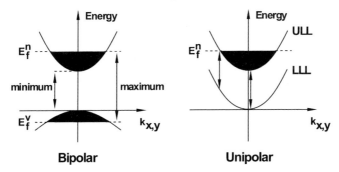

Figure 12.2. The contrast between the gain spectrum of a unipolar quantum cascade structure and a conventional bipolar device. The similar dispersion for the upper and lower laser levels of the unipolar device results in a similar transition energy irrespective of the in-plane wave vector whereas the bipolar device has widely separated maximum and minimum transition energies caused by the separation of the Fermi levels.

- The cavity reflectors can be formed by cleaving or through the fabrication of distributed Bragg reflectors.

Not only is the detailed theory of quantum cascade lasers too complex to describe simply, but it is less than 10 years since the first report of a practical quantum cascade laser appeared in the press and the field is still rapidly developing. Much theoretical and experimental work on gain [7–9] and transport [10–12] has already been done in addition to that already referenced. It is left to the interested reader to seek out these works; what follows in this chapter are some of the simplified theoretical considerations by way of explanation of the differences summarised above as well as some of the significant developments that have taken place in recent years.

12.1 Quantum cascade structures

The unipolar nature of the quantum cascade laser offers a degree of flexibility to the device designer that bipolar devices do not. The wavelength range of the bipolar devices is limited by the quality of materials available, and even though low band gap materials exist that will allow laser operation in the 2 μm–5 μm wavelength range (see chapter 10), extending the wavelength beyond this poses real difficulties in terms of epitaxial growth and carrier confinement. The quantum cascade laser overcomes these difficulties by relying not on fundamental material properties such as the band gap but on engineered properties such as the subband spacing. The choice of materials is irrelevant to some extent, provided of course that the conduction band offsets are large enough for effective confinement at the lattice matched composition and the material growth is well enough controlled to allow abrupt interfaces in well defined layer sequences. Lattice matching is essential because of the total thickness of a typical device. Any sort of strain built up over the layer sequence will result in relaxation unless the strain can be compensated, and whilst such systems have been demonstrated in the laboratory [13] they are far from commercialisation.

Quantum cascade lasers have employed several different schemes for the active regions. The first laser [13] utilised a diagonal transition rather than the vertical transition shown in figure 12.1. In a diagonal scheme the final state of the transition is centred on a different well from the initial state so the electron must move sideways in real space. Such a requirement increases the upper state lifetime, which is essentially dominated by optical phonon scattering. Central to the diagonal transition is the notion of "anti-crossing". The confined states move in energy with the application of an electric field but of course the motion is relative to the point of reference. Some states will move up and others will move down, but when they coincide in energy they will anti-cross, as in figure 12.3. The states are resonant at this point and coupling between the two occurs. The subbands of the active region are designed to anti-cross at the operational electric field of the laser.

Lasers based on inter-subband transitions have been reported at various wavelengths in the near infra-red, at various powers amd temperatures of operation. A sense of the progress can be gleaned from Table 12.1, in which some of the experimental devices that have appeared in the literature are

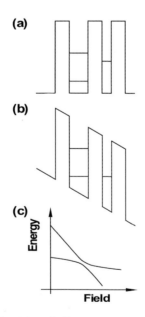

Figure 12.3. Anti-crossing in states of adjacent quantum wells.

Table 12.1. Progess in quantum cascade structures.

λ (μm)	Power (W)	Reflector	T (K)	Material	Reference
4.2	0.008 pulsed	cleaved facet	90	InGaAs/InAlAs	14
4.3	0.004 pulsed	cleaved facet	125	InGaAs/InAlAs	15
4.3	0.03 pulsed	cleaved facet	125	InGaAs/InAlAs	16
5	0.2 pulsed	cleaved facet	320	InGaAs/InAlAs	17
5.3	1.15 pulsed	DFB	420	InGaAs/InAlAs	18
5.9	0.67 pulsed	coated facet	400	InGaAs/InAlAs	19
8.5	0.06 pulsed	DFB	300	InGaAs/InAlAs	20
8.5	0.01 pulsed	cleaved facet	305	InAs-GaAs-AlAs	21
10.16	0.23 pulsed	DFB	85	InGaAs/InAlAs	22
	0.08 pulsed		300		

described. Most noticeably, the peak powers and operating temperatures have improved steadily with time, though the extension of the quantum cascade concept to the GaAs material system took some time, the earliest devices all exploiting the InGaAs-InAlAs system. The list of devices and achievements given in table 12.1 is by no means comprehensive and the structures themselves will not be described as they differ mainly in the layer sequence rather than the composition. The distributed feedback reflectors described by Hofstetter et al. are chemically wet-etched into the top waveguide layer from the surface down to a depth of 100 nm for the lasers at 5.3 μm [18] (grating period = 825 nm) and 400 nm in the lasers emitting at 10.16 μm [22] (period 1.59 μm). The semiconductor/air interface is thus a vital part of the waveguide structure, which is of course a potential weakness of the device. However, these structures are a long way from commercialisation. Gmachl et al. [20] preferred to regrow a thick InP layer over the grating, but the problem with all-semiconductor waveguides, and one neatly avoided by Hofstetter et al., is free-carrier absorption, which was shown in chapter 4 to depend on the square of the wavelength. The top part of the waveguide must also serve as the contact layer and some absorption of the laser radiation is inevitable.

In addition to the discrete subbands in quantum wells some lasers have been designed with a superlattice active region so that optical transitions occur from the bottom of one miniband to the top of another. Minibands are less susceptible to imperfections in the growth as there exists a range of energies into which electrons can be injected. A typical such device is a 9 μm $Ga_{0.47}In_{0.53}As/Al_{0.48}In_{0.52}As$ superlattice structure grown by gas source MBE at the Center for Quantum Devices at Northwestern University, Illinois, and described by Razeghi and Slivken [23]. The active region comprises six well and barrier pairs and the injector comprises seven well and barrier pairs as detailed in table 12.2. Several such active/injector stages will be included in a

Table 12.2. Layer sequences used by Razeghi and Slivken [23].

Active		Injector	
Well	Barrier	Well	Barrier
5.6	0.9	2.8	2.5
5.2	0.9	2.8	2.5
4.8	0.9	2.7	2.6
4.5	1.0	2.5	2.6
4.1	1.1	2.4	2.7
3.8	2.5	2.4	2.9
		2.3	3.5

single device, so the necessity to produce 26 layers per stage of an accuracy much greater than 0.1 nm illustrates the demands of the growth technology. Furthermore, in mid-IR ranges the dominant optical interaction in matter is free carrier absorption so the background doping has to be very carefully controlled. If it is too high the absorption losses will be large but if it is too small it will not counteract the space charge induced by high electron injection, which will subsequently distort the internal electric field and disrupt the laser operation.

After growth the devices are patterned into ridge or double channel waveguides by photolithography and chemical or plasma etching. The waveguides are typically 15–25 μm wide (which would normally be considered as a broad area in high power double heterostructures but because of the longer wavelength will still support a single lateral mode) and then cleaved into 3 mm long cavtities. At room temperature these lasers are capable of delivering peak powers of 7 W at 3.26 A with a slope efficiency of 4.4 W/A, which equates to 32 emitted photons per injected electron. Operating temperatures up to 470 K have been achieved.

12.2 Minibands in superlattices

Minibands can be best understood through the Kronig–Penney model, which was first published in 1931 [24] as an explanation of the band structure of solid state materials in terms of the propagation of an electron inside the periodic potential due to the atoms in the lattice. The form of the model successfully predicts the existence of bands inside solids but the potential is assumed to be square, and therefore not representative of real bulk materials. For this reason it has never really found much use in the calculation of real band structures, although it has not been entirely forgotten [25]. However, with the advent of the

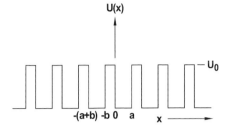

Figure 12.4. The square potential Kronig–Penney model.

superlattice the Kronig–Penney model has found renewed favour because the square potential is nothing less than an idealisation of the abrupt band structure of the heterostructure superlattice. Indeed, Leo Esaki, the founder of modern semiconductor quantum engineering, reports that in making the first superlattice he was in fact trying to make a practical Kronig–Penney structure [26]. It is not surprising, therefore, that the Kronig–Penney model should apply so well to these structures.

The Kronig–Penney potential is shown in figure 12.4. A potential barrier of width b and height U_0 alternates with a potential well of width a. There are at least three methods of calculating the properties of an electron propagating in such a structure; the boundary matching method, the transfer matrix method, and the Green's function method. The transfer matrix method is a standard formulation for treating wave propagation in periodic structures and relates the wave properties, i.e. amplitude and phase, in one layer to the wave properties in the preceding layer, but it is easy to lose sight of the essential physics in the process. The Green's function method is inherently mathematical and will not be described further, so the boundary matching method will be used here. In essence, this requires the electron wavefunction and its derivative to be continuous across each interface in the structure, thereby matching the wavefunction and its derivative at each boundary.

We start, as before, with Schrödinger's equation.

$$\left[\frac{p^2}{2m_0} + V(r)\right]\psi(r) = E\psi(r) \tag{4}$$

where

$$\frac{p^2}{2m_0} = \frac{-\hbar^2}{2m_0}\nabla^2 \tag{5}$$

and $V(r)$ is the spatially varying potential. All other symbols have their usual meaning. In the Kronig–Penney model as first envisaged the square potential is intended as an approximation to the lattice potential itself, but here the square potential respresents the superlattice conduction band which is super-imposed onto the lattice potential. This is an important difference between the two cases. In the presence of the lattice potential only the wave function is given by the Bloch function (see Appendix III)

$$\psi_{nk} = u_{nk}(r) \exp(jkr) \qquad (6)$$

but in the superlattice we resolve the potential into its lattice (L) and superlattice (SL) components

$$V(r) = V_L(r) + U_{SL}(r) \qquad (7)$$

and write the wave function as

$$\psi(r) = F(r) u_{n0}(r) \qquad (8)$$

where $u_{n0}(r)$ is the Bloch function of the nth band at $k=0$, as explained in appendix III, and $F(r)$ is a slowly varying envelope. Strictly, for this formalism to apply the potential $U_{SL}(r)$ also has to vary slowly, which clearly it does not. It is for this reason that the superlattice is divided into its constituent parts of well and barrier with an effective wavefunction defined in each, which, together with the derivative, are matched at the boundaries. Strictly, it is the envelope function rather than the Bloch function that is matched at the boundary, because the Bloch function varies on the scale of the underlying lattice. Moreover, because the superlattice wells and barriers are lattice matched the Bloch function retains its periodicity across the boundary, and as with the treatment of the confined states in a quantum well, the Bloch function can be discarded. Therefore, the superlattice envelope function is replaced with a wavefunction

$$\psi(z+d) = e^{ikd}\psi(z) \qquad (9)$$

which is itself a Bloch function but periodic in the superlattice period $d = a + b$.

The wave functions in both the well (A) and the barrier (B) can be defined. It follows immediately that

$$\psi_A(0) = e^{-ikd}\psi_B(d) \qquad (10)$$

so the four boundary conditions of interest can be defined as

$$\psi_A(a) = \psi_B(a) \qquad (11)$$

$$\frac{1}{m_A^*}\psi_A'(a) = \frac{1}{m_B^*}\psi_B'(a) \qquad (12)$$

$$\psi_A(0) = e^{-ikd}\psi_B(d) \qquad (13)$$

and

$$\frac{1}{m_A^*}\psi_A'(0) = \frac{1}{m_B^*}e^{-ikd}\psi_B'(d) \qquad (14)$$

The wavefunction can be constructed from a mixture of left and right propagating plane wave functions. Within the well, $0 < z \leq a$

$$\psi_A(z) = C_A e^{ik_A z} + D_A e^{-ik_A z} \tag{15}$$

where C_A and D_A are amplitude coefficients, and within the barrier, $a < z \leq d$

$$\psi_B(z) = C_B e^{ik_B z} + D_B e^{-ik_B z} \tag{16}$$

where the wavevectors are

$$k_A = \frac{1}{\hbar}\sqrt{2m_A^* E} \tag{17}$$

and

$$k_B = \frac{1}{\hbar}\sqrt{2m_B^*(E - U_0)}. \tag{18}$$

For an electron energy $E < U_0$ the wavevector k_B is imaginary, and the transformation can be made

$$k_B = i\mathrm{K}_B \tag{19}$$

where K_B is real and corresponds to $E > U_0$. Substituing the well and barrier wavefunctions into the four boundary conditions leads to a set of simultaneous equations for which the determinant must vanish. The solution to this equation for $E < U_0$ can be written [26]

$$\cos(kd) = \cos(k_A a)\cosh(\mathrm{K}_B b) - \frac{1}{2}\left[\mathrm{K} - \frac{1}{\mathrm{K}}\right]\sin(k_A a)\sinh(\mathrm{K}_B b) \tag{20}$$

where

$$\mathrm{K} = \frac{k_A m_B^*}{k_B m_A^*}. \tag{21}$$

A similar expression exists for the case $E > U_0$.

In solid state physics texts dealing with the atomic Kronig–Penney model this expression is often simplified by assuming that the barrier thickness tends to zero in the limit as U_0 tends to infinity and also that the effective masses are both unity [27]. This is clearly not appropriate for a superlattice and which the barrier width and height are fixed by the details of the growth, but it is intructive nonetheless to make the assumptions. Then

$$\frac{P}{k_A a}\sin(k_A a) + \cos(k_A a) = F(k_A a) = \cos(ka) \tag{22}$$

where P is a positive finite number. This function is plotted in figure 12.5 for $P = 2\pi$. The straight lines at $F(k_A a) = \pm 1$ represent the limits of the cosine on

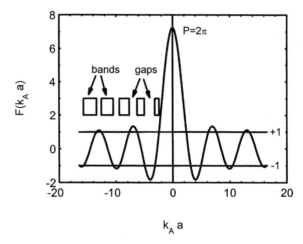

Figure 12.5. Solution of the Kronig–Penney model.

the right hand side. Values of $F(k_A a)$ lying outside this range are, by definition, invalid, and these define the energy gaps by way of forbidden values of the wavevector. The boxes represent the energy bands spanning the range $|F(k_A a)| < 1$. At low $k_A a$ the bands are narrow and widely spaced but as $k_A a$ increases the bands become wider and the gaps narrower. Mathematically, the first term on the left hand side dominates for low values of $k_A a$ but this rapidly decays and gives way to the cosine term on the left hand side for large $k_A a$. The function $F(k_A a)$ therefore has no regions of invalidity at large $k_A a$ and the energy gaps disappear as the bands touch. For smaller values of P the bands will merge at lower values of $k_A a$ but there will always be an energy gap at low energies.

The same general results apply to superlattices, but the limitation of a definite barrier width and a finite barrier height means that above barrier states will exist. These will also exhibit band gaps for electron energies just above the barrier but at high electron energies the bands will merge to form a continuum. These states are not of primary interest in the quantum cascade laser, so figure 12.6 shows the confined states of a GaAs/AlGaAs Kronig–Penney superlattice as a function of both well width and barrier width [26]. "Confined" in this context means that the electron energy lies below the barrier height, but of course it is the nature of a superlattice that, unlike a single quantum well with wide barriers, electrons can propagate along the superlattice. For the well-width dependence a fixed barrier width of 5 nm was assumed and similarly for the barrier width dependence a fixed well width of 5 nm was assumed. The conduction band offset, i.e. barrier height, occurs at ~190 meV so for a well thickness of zero, corresponding to bulk AlGaAs there are no states in the well. It is easy to imagine the top of the first band continuing up in energy and

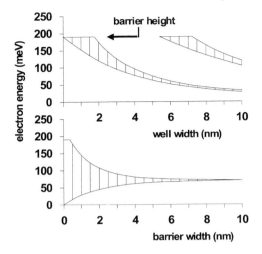

Figure 12.6. Superlattice states in a GaAs/AlGaAs periodic structure. (After Steslicka [26].)

meeting the y-axis at ~ 370 meV, where in fact it coincides with the bottom of the second band. There are therefore no band gaps at zero well-width, which is what would be expected on intuitive grounds.

The band states behave very much like the confined states of an isolated well, with the obvious exception of the width of the band, which is a direct result of the communication with neighbouring wells. Thus, the first band starts to descend into the well immediately the well thickness exceeds zero but doesn't become fully confined until a thickness of 1.7 nm, at which point the top of the band descends into the well. The band energy decreases with well width, as does the width of the band. In effect the electrons spend more time in the well than in the barriers and there is correspondingly a much smaller probability, though still non-zero, of communication with neighbouring wells. Eventually, as the well width increases above 5.3 nm, the second band descends into the well. In this particular configuration no third band is present within the well, but in fact a third and fourth band do exist above the barrier height. These bands exist because even though the wavevector is real at every point within the superlattice the discontinuous change in potential at the interfaces leads to discontinuous changes in the wavevector, and because the system is periodic stop bands will occur for any wavevector satisfying the Bragg condition (see chapter 9). At any given well width the band gap narrows as the electron energy increases and the bands become wider, in accordance with figure 12.5, so eventually a continuum of states forms above the barrier. As in the bulk, the bands can be described with reference to a mini-Brillouin zone corresponding to momenta $\pm m\pi/d$, where d is the superlattice period.

The effect of increasing the barrier width can also be understood by similar intuitive reasoning. At zero barrier width the band structure should conform to the GaAs band structure, and in figure 12.6 the band extends over the whole energy range of the well. This well is deep enough for one confined state only at a width of 5nm and as the barrier width increases the band becomes progressively narrower until eventually it becomes a discrete state at ~ 70 meV. In effect all communication with neighbouring wells has ceased and the electron is isolated. Above the barrier, bands exist but the bands are in fact widest for narrow barriers. For wide barriers the band widths decrease but so also do the band gaps, because the bands must tend to the continuum of states.

The superlattice structures described above are uniform and therefore the minibands are also uniform, but the electronic structure is modified under the presence of an electric field. In a bulk material the conduction band will simply have the potential associated with the electric field super-imposed upon it, but in a superlattice the effect of the electric field is to separate in energy the individual states of the quantum wells so that they are no longer in resonance and no longer interact. The superlattice breaks up into a series of discrete states at well defined energy intervals, a so-called Wannier–Stark ladder [28]. In a quantum cascade laser the layer thicknesses and compositions are graded so that at zero field states do not align but under the action of an electric field these states are brought into resonance and a miniband is formed. This is somewhat different from the idealised superlattices described above but the principle remains the same. The miniband states arise from interacting states of a series of quantum wells, and the number of states within a mini band corresponds to the number of interacting wells. In addition, the mini band extrema correspond to the mini Brillouin zone boundaries.

12.3 Intersubband transitions

The selection rules for interband optical transitions were derived in chapter 6 using the envelope functions in the growth direction whilst retaining the periodic Bloch functions in the plane of the well. The principal restriction on heavy hole valence band-to-conduction band transitions was shown to arise from the magnitude of the envelope function overlap integral. For states with the same sub-band index the overlap integral is the highest, but where the sub-band indices differ (odd-even or even-odd) transitions are forbidden. For intersubband transitions, however, the selection rules are reversed; odd-to-odd or even-to-even transitions are generally forbidden, but even-to-odd transitions, and vice versa, are allowed, as the momentum operator is a differential and flips the parity of the final state (figure 12.7). As the intersubband states are derived from the same basis set – the conduction band for electron states and the valence band for hole states – flipping the parity has the effect of making the overlap integral for odd-to-even transitions non-zero. In figure 12.1, for example, the lower state of the transition has almost opposite parity from the upper state.

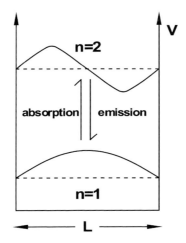

Figure 12.7. Intersubband transitions between states of different parity.

Intersubband transitions in quantum wells can also exhibit very large oscillator strengths. Following West *et al.* [29] the electric dipole approximation will be used assuming the states of an ideal infinite square well potential. This replaces the electromagnetic interaction Hamiltonian $-q/m \cdot A \cdot p$ with the dipole operator $-q \cdot E \cdot r$, where q is the electronic charge, r is the position vector, A is the scalar potential, and E is the electric field (see Appendix V). Physically this corresponds to the classical idea of an electric field E exerting a force qE which acts over, i.e. displaces the charge, a distance z. The negative sign implies that energy is gained from the electric field and that work is done on the charge in moving it against the field to a higher potential. The essence of the dipole approximation is that the wavelength is large compared with the dimension of the quantum system under consideration. It is often used in atomic systems where the atom is of the order of an Ångström or so and the wavelength of visible light is at least 300 nm, and in the quantum wells considered here the intersubband transition energies correspond to wavelengths several micrometres long whereas the well is just a few nanometres thick. The momentum matrix element is dispensed with and the integral over the product of states can be derived quite easily.

Assume now an idealised infinite square well potential with wave functions [29]

$$\psi_n = A_n \sin(nk_q z) \exp(ik_t r_t) \qquad (23)$$

where k_t is the transverse (in-plane) wave vector and $k_q = \pi/L_w$ is the wave vector in the growth direction z, with L_w being the width of the well and n is the quantum number of the state. The average dipole moment is

$$\mu = q \int \psi_n(z) z \psi_t(z) dz = q \int A_n^* \sin(nk_q z) z A_m \sin(mk_q z) dz. \qquad (24)$$

Using the identity

$$\sin(A)\sin(B) = \frac{1}{2}[\cos(A-B) + \cos(A+B)] \qquad (25)$$

and integrating over the width of the well only (the wavefunctions go to zero at the well boundary), then the dipole matrix element becomes

$$e(z) = \frac{qL_w}{\pi^2} \frac{8nm}{(m^2 - n^2)^2} \qquad (26)$$

where use has also been made of the normalisation of the constants

$$A_n^* A_m = \frac{2}{L_w}. \qquad (27)$$

The oscillator strength is given by

$$f \equiv \frac{2m_e \omega}{\hbar} \langle z \rangle^2 = \frac{m_e}{m^*} \frac{64(nm)^2}{\pi^2(m^2 - n^2)^3}. \qquad (28)$$

For a finite well these oscillator strengths are modified somewhat by the penetration of the wavefunction into the barriers, and for coupled wells, such as those illustrated in figure 12.1, where the wavefunction extends across more than one well, the oscillator strength is given by the numerical integration of the product of the two wavefunctions of the states at either end of the transition. If the well design is asymmetric the selection rules are not so stringent because the wavefunctions themselves are not simple harmonic functions and do not have well defined parity. Indeed, it is even possible that transitions will occur diagonally from one well to an adjacent well, as already discussed.

The polarisation properties of the laser are not affected by the symmetry of the wells. The fact of a dipole operator acting on envelope functions in the z-direction implies an electric vector in the z-direction, so quantum cascade lasers tend to be polarised with the magnetic vector in the plane of the well, say in the y-direction, and the light propagating also in the plane of the well but in the x-direction.

12.4 Intersubband linewidth

The assumption of similar effective masses for the different intersubband states is not alway realised in practise, and of course a different mass means a different in-plane dispersion. This causes the linewidth to broaden and affects the shape of the gain spectrum [30]. Within the Kane model the effective mass of the nth conduction subband is given by

$$m_n = m_0\left[1 + \frac{2E_n(0)}{E_G}\right] \qquad (29)$$

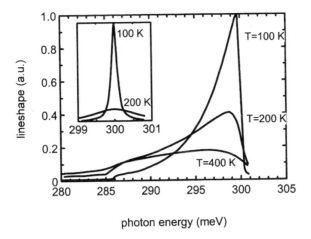

Figure 12.8. The lineshape for inter-subband transitions corresponding to different in-plane effective masses at different lattice temperatures with the same calculation for equal masses shown inset. (After Gelmont et al.)

where $E_n(0)$ is the energy of the nth band at $k_{x,y} = 0$ and E_G is the band gap energy. For two states $E_2(0) - E_1(0) = 0.3$ eV in InGaAs the effective mass ratio m_2/m_1 is large at 1.5, which leads to a linewidth as shown in figure 12.8. The linewidth is highly asymmetric with a tail extending to the long wavelength side. The inset shows the linewidth calculated for equivalent effective masses, i.e. identical dispersion. The shape is Lorentzian with a very narrow width (FWHM ~ 0.2 meV) and a strong temperature dependence. At a temperature of 200 K the line has broadened considerably and the height has been reduced by a factor of ~ 9. The corresponding calculations for T = 300 K and T = 400 K are not shown as the magnitude is too small to be visible.

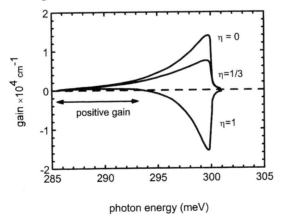

Figure 12.9. Gain spectra calculated from the lineshapes shown in figure 12.8 at different population ratios. (After Gelmont et al.)

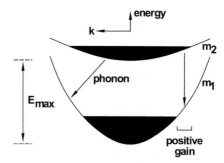

Figure 12.10. Intersubband transitions between subbands with different effective masses. The intersubband phonon scattering is also illustrated.

For subbands with different dispersion, however, the linewidth becomes much broader, extending over a range of ~ 15 meV. The shape arises from the transitions at higher energy corresponding to in-plane $k \neq 0$. The calculations considered in-plane phase relaxation by intra-subband scattering only, as intersubband scattering at such high energies relative to the phonon energy is considerably slower than the intra-subband scattering. Moreover, the important temperature identified is that of the electrons rather than the lattice, as fixing the lattice temperature at 100 K and varying the electron temperature produced almost identical results. Substituting this linewidth into a standard expression for gain – in this respect the quantum cascade laser resembles conventional edge-emitting devices – yields the gain spectra shown in figure 12.9 with the ratio $\eta = n_1/n_2$ as shown. For $\eta = 1$ population inversion does not exist and the gain is negative corresponding to absorption. Over a small energy range, however, positive gain exists due to the effective population inversion caused by the differences in dispersion between the two subbands (figure 12.10).

12.5 Miniband cascade lasers

There are several advantages if the transitions take place between minibands rather than discrete subbands, and several lasers emitting over a wide range of wavelengths have been demonstrated. The wavelength is determined principally by the energy separation between the minibands, for the oscillator strength is maximised at the mini-zone boundary. Optical transitions in minibands have been discussed extensively by Helm [10] in terms of a modified sum rule. In atomic systems the oscillator strength obeys the so-called f-sum rule

$$\sum_j f_{ij} = 1$$

which is modified in a superlattice by the presence of the effective mass m_{SL}, defined as

$$m_{SL}^{-1} = \frac{1}{\hbar^2} \frac{\partial^2 E_i(k_z)}{\partial k_z^2} \qquad (30)$$

to be

$$\sum_j f_{ij} = 1 - \frac{m}{m_{SL}}. \qquad (31)$$

Transitions from the bottom one miniband to the top of the miniband immediately lower in energy therefore have a higher oscillator strength than transitions from the top of the miniband to the bottom of the miniband, as shown in figure 12.11. An electron at $k = 0$ can also make intra-band transitions, which, from the sum rule, must reduce the oscillator strength for the inter-band transition. At $k = \pi/d$, however, the intra-band transitions are of an opposite sign representing the absorption of energy, so the oscillator strength for the downward transition is increased. It doesn't matter that the transitions occur at the zone boundary, for that is simply a consequence of the band index. For transitions occurring between the third and second minibands, for example, the transitions would occur at the zone centre.

Population inversion is easier to achieve in a superlattice active region provided the electron temperature is low enough for the bottom miniband to remain largely empty and provided the width of the miniband is larger than the optical phonon energy. If the electron temperature is too high then all the states in the miniband will be full and scattering between states will not occur. However, in an empty miniband the intra-miniband scattering times are usually much shorter than inter-miniband scattering times so that depopulation of the lower laser level is readily achieved. Similarly the doping level must be such that the Fermi level lies well below the top of the lower miniband. However, there is an additional requirement on the doping level. It should be optimised to ensure that the electric field does not penetrate significantly into the

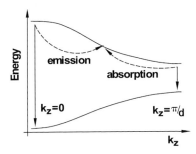

Figure 12.11. Enhanced oscillator strength at the miniband zone boundary.

superlattice so that the field is dropped almost entirely across the injector region.

These ideas were first demonstrated at $\lambda = 8\ \mu$m in an eight period superlattice with 1 nm thick barriers of $Al_{0.48}In_{0.52}As$ and 4.3 nm thick $Ga_{0.47}In_{0.53}As$ wells [31]. The fact of eight periods means that each mini-band comprises 8 states with an electroluminescence spectrum 30 meV wide compared with 10 meV for similar quantum cascade structures. This increases the threshold current somewhat despite the high oscillator strength of approximately 60, and an intra-miniband relaxation time of ~0.1 ps compared with the optical mode phonon scattering time of ~10 ps from the bottom of the second miniband to the top of the first miniband. An intrinsic superlattice has also been proposed [32], the main difference being that the dopants are placed in the injection and relaxation regions so that the dopants are separated from the extrinsic electrons. The electric field generated by this mechanism exists primarily in the active regions and is then cancelled by the applied field to give a similar profile as shown in figure 12.12.

As well as inter-miniband transitions, structures can be designed so that transitions occur between a bound state and a miniband [33,34] (figure 12.13). The essential idea is that a chirped superlattice gives rise to minibands under the influence of an electric field. These bands are never truly flat, but extend reasonably uniformly over a limited region of space until they break up to form the next miniband. In this scheme there is no separation of the active and the injection regions, as occurs in other devices, as the lower laser level and the injector are one and the same miniband. However, a single narrow well placed at strategic points within the superlattice gives rise to a discrete state which acts as the upper laser level. The upper miniband is effectively redundant. One of the potential disadvantages of this design is the low oscillator strength, as understood from the sum rule, and the relatively large linewidth. Transitions from the upper laser level to various points within the miniband are possible,

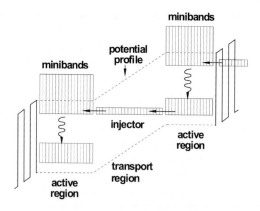

Figure 12.12. A schematic of a superlattice cascade laser.

Terahertz emitters

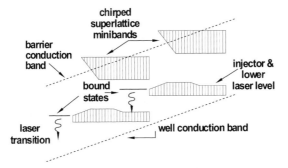

Figure 12.13. A bound-continuum cascade laser. (After Faist *et al.* [33].)

leading to a linewidth of some 20 meV or so [34]. However, this is offset by the realisation that room temperature operation will in any case cause a thermal broadening of this magnitude, and the advantage of the fast depopulation rate of the lower laser level makes this an attractive scheme for high temperature operation.

12.6 Terahertz emitters

The Terahertz frequency range is a part of the electromagnetic spectrum where radiation sources have been difficult to fabricate. Quantum structures can be designed with energy levels corresponding to this frequency range, and reports of electroluminescence at wavelengths out to 100 μm are common in a variety of materials systems. Converting this electroluminescence to coherent radiation is a difficult problem that has only recently been overcome. The essential difficulty is to design a structure with at least three levels such that there is a narrow energy separation between two upper levels with the lifetime in the upper level being longer than the lifetime in the lower level. For laser wavelengths greater than 35 μm, corresponding to the optical phonon energy in GaAs, scattering from the upper level by optical mode phonons can be discounted, but if the energy separation between the depopulation level and the lowest laser level is also less than the phonon energy, then phonon scattering out of this state can also be discounted. Therefore similar mechanisms of depopulation will apply to both the upper and lower laser levels and population inversion will be very difficult to achieve. The key to terahertz laser operation is therefore to design a system as illustrated in figure 12.14 in which the upper two levels have a low energy separation and the lower two levels correspond to the optical mode phonon energy for rapid depopulation.

Such a complicated energy level scheme cannot be designed using a single symmetric well. Stepped asymmetric quantum wells have been proposed [35], as well as double and triple quantum wells [35–37]. The asymmetric well system is not easy to understand intuitively, so let's concentrate intead on the

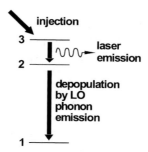

Figure 12.14. The basic energy level scheme for THz emission.

double or triple well system. A wide well is used to provide the upper and lower laser levels corresponding to $n = 1, 2$ of this well, and the adjacent narrow wells are used to provide intermediate levels corresponding to their $n = 1$ levels (figure 12.15). The well width essentially determines the energy level and the barrier width determines the dipole matrix element via the overlap integral of the states. For very narrow barriers the penetration of the wavefunction into the adjacent wells becomes large and the overlap integral is correspondingly increased, whereas in the limit of very wide barriers the states would remain isolated and transitions from the states originating in one well to the states originating in another would be forbidden.

The three-well system of figure 12.15 was designed with optical pumping by a CW CO_2 laser in mind. Such systems were extensively investigated before electrical pumping was realised because of the relative simplicity of the optically pumped structure [37]. Optical pumping has the advantage that the electron injector regions can be ignored, and provides a system by which model calculations of transition rates and the like can be tested experimentally. The design of FIR semiconductor lasers depends crucially on the optimisation of the relative transition rates among the various levels, including electron-phonon and electron-electron scattering. These can be calculated from Fermi's Golden

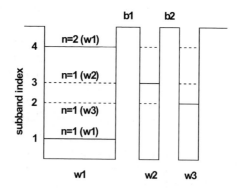

Figure 12.15. Energy levels from adjacent wells provide the key to THz emission.

Rule using appropriate deformation potentials, phonon modes, and electron densities, but self-consistent calculation is difficult. A full Monte Carlo technique, where the trajectory of several independant electrons is computed according to standard statistical methods, should really be employed in order to understand such structures [38,39], but the very fact that such calculations are necessary means of course that a succinct description of the principles behind the designs [40] is beyond the scope of this book. For electrical injection into a cascade structure, it is necessary to have only three levels from two quantum wells with injection directly into the upper laser level [41]. Alternatively, bound-to-continuum transitions might be used [42], or superlattice lasers [43], possibly chirped [44], to give relatively low threshold, long-wavelength (~ 66 μm) devices.

12.7 Waveguides in quantum cascade structures

Waveguiding structures in quantum cascade lasers present a serious technological challenge to the laser manufacturer. In waveguides based on the conventional refractive index difference the optical penetration into the cladding layers is proportional to the wavelength, which can be very long in the case of THz devices. If the cladding layer is not thick enough to contain the entire optical field some irreversible leakage, i.e. loss, out of the guide will occur. Of course, if the core of the guide is itself optically thick then for the fundamental mode the optical field at the core-cladding interface is small compared with the field at the centre of the guide so the loss from this mode will be small. However, power will also propagate in the higher order modes and these will be more lossy. If conventional waveguides are to be built then structures several times thicker than the wavelength need to be grown. Even in the GaAs/AlGaAs system difficulties occur for thicknesses greater than ~ 1.5 μm due to residual strain in the AlGaAs, so alternative approaches are necessary.

Sirtori *et al.* [45] proposed the use of heavily doped GaAs cladding layers for mid-IR wavelengths in the range 5–20 μm, and demonstrated the principle on a laser emitting at 8.92 μm. Heavily doped GaAs ($n \approx 5 \times 10^{18}$ cm^{-3}) has a plasma frequency around 11 μm, and around this wavelength the real part of the refractive index decreases. Heavily doped GaAs cladding layers will therefore provide a large refractive index difference at the wavelength of interest. The active region itself was quite small and consisted of only 36 periods, totalling just over 1.5 μm thickness, but was embedded in ≈ 3.5 μm of moderately doped GaAs ($n \approx 4 \times 10^{16}$ cm^{-3}) on either side adjacent to which was the highly doped layer. GaAs rather than AlGaAs was used for the simple reason given above that the thickness of AlGaAs is limited but GaAs on the other hand can be grown much thicker. The structure is illustrated in figure 12.16, along with the refractive index profile. The highly doped GaAs layers

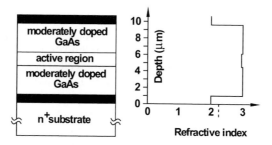

Figure 12.16. A waveguide structure based on carrier induced refractive index changes in the infra red. The heavily doped layers are shown in black.

(black) have a refractive index of 2 compared with 3 for the moderately doped GaAs, and this is sufficient to confine the optical field fully within this structure. The active region has a slightly lower refractive index because of the presence of AlGaAs barriers, but the effect is small and doesn't significantly alter the confinement properties.

Such high doping levels inevitably leads to absorption at long wavelengths, and in this structure the extinction coefficient, i.e. the imaginary part of the refractive index, is estimated to be $k = 0.1$ falling abruptly to 10^{-4} in the moderately doped regions. By way of comparison, the absorption depth at this wavelength for $k = 0.1$ is about 7 μm, but that strictly applies to a normally incident plane wave. Nonetheless, the cavity length will be several hundreds of micrometres long and the fact of such a high extinction coefficient will lead to considerable loss, calculated by the authors of this structure to be 1740 cm^{-1}, which is approximately 90% of the total waveguide loss despite an optical overlap in the heavily doped regions of 0.008. Losses in heavily doped cladding layers such as this therefore represent the most significant source of loss in the cavity, but are in fact an unavoidable consequence of this type of waveguiding structure. The loss itself is not the most important parameter; it has to be put into the context of the laser structure. That is to say, at threshold the modal gain matches the total loss,

$$g_m = \Gamma_{AR} g = \alpha_T = \alpha_M + \alpha_C \tag{32}$$

where g is the material gain, Γ_{AR} is the confinement factor for the active region, and the subscripts T, M and C refer respectively to the total loss, the mirror loss, and the cavity loss. The cavity loss includes the effects of the waveguide structure, so the threshold condition is given by

$$g = \frac{\alpha_M}{\Gamma_{AR}} + \frac{\alpha_C}{\Gamma_{AR}}. \tag{33}$$

The ratio of loss to optical confinement therefore represents a more important figure of merit than the loss itself, because if the optical confinement factor is increased by including a lossy layer the laser might still exhibit a lower threshold.

Waveguides for longer wavelengths utilise a particular property of thin metal films that at long wavelengths where the real part of the dielectric constant is large and negative such films, when bounded by a dielectric with a real and positive dielectric constant, will support an electromagnetic mode that will propagate over large distances [46,47]. In fact, the condition on the dielectric constants is not so strict and modes will propagate over a wide variety of wavelengths. Under the conditions described above the electromagnetic mode that propagates has a magnetic vector parallel to the interface and an electric vector perpendicular to the plane of the film, which is consistent with the polarisation properties of the emitted radiation. Therefore a metal contact placed over the active region serves two purposes; to act as a contact and to act as waveguide. This waveguide will act as such over a wide range of frequencies, but of course if the imaginary part of the dielectric constant of either the metal or the semiconductor is non-zero some attenuation of the propagating beam will occur.

The fact of a perpendicular electric vector requires a surface charge density on the metal. Induced by the electric feld and oscillating with it, this charge density corresponds to the collective oscillation known as a plasmon, hence the guide is often called a surface plasmon guide. The propagation depth into the semiconductor is quite long so the overlap from a mode supported by a single metal layer is quite small. Penetration of the optical field into the heavily doped substrate will lead to some attenuation but unless the substrate is doped so heavily that it becomes almost metallic it will not significantly alter the mode profile. Encapsulating the active region between two metal layers increases the overlap almost to unity, so whilst the attenuation of the mode

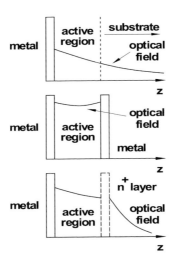

Figure 12.17. Surface plasmon wave guide structures using a single metal layer, two metal layers, or a metal layer in conjuction with a doped semiconductor.

might increase, the figure of merit α/Γ may well decrease. However, in order to achieve this sort of structure it is necessary to bond the wafer to a substrate after depositing the top metal contact in order to provide some mechanical support while the substrate is etched off and a metal layer deposited on to what was the bottom of the device. Such structures have been built and demonstrated at $\lambda \approx 100$ μm [48]. An alternative to a double metal wave guide is a heavy doped semiconducting or highly conducting intermetallic layer beneath the active region. Intermetallics such as NiAl compounds can be grown epitaxially on GaAs [49] and so have the advantage that the substrate does not need to be etched. Heavily doped semiconducting layers have the same advantage but suffer from the disadvantage that the mode field can leak out in the case of a thin layer and reduce the overlap. The optical fields from the different configurations are illustrated shematically in figure 12.17.

12.8 Summary

Quantum cascade lasers represent a radical departure from conventional diode lasers in which recombination of electrons and holes across the band gap results in photon emission. Necessarily the electron is removed from the active region by this process so that the maximum possible internal efficiency is one photon per electron-hole pair. In the cascade laser population inversion is achieved between electron subbands in a multiple quantum well structure, which means that the electron is available for further stimulated emission in another set of subbands. Quantum cascade lasers therefore contain more than one active region separated by injector regions designed for the sole purpose of transporting the electron from one active region to another.

The design of these systems is a complicated exercise in numerical computation involving solution of Schrödinger's equation, Fermi's Golden Rule for electron-phonon and electron-electron scattering, and in some instances Monte Carlo formulations of transport. For these reasons no simple design rules exist, but some principles of operation can be discerned. These are:

- At least three levels are needed in an active region.
- The lowest level serves the dual purpose of depopulating the lowest laser level and feeding the electron into the next injector region.
- The lifetime of the lower laser level has to be shorter than the lifetime of the upper laser level in order to achieve population inversion.
- Depopulation by optical mode phonon scattering is one of the most effective scattering mechanisms from the lower laser level, so the energy separation between the lowest levels should correspond to the phonon energy.
- Emission wavelengths can be extended out to the very far infra-red (≈ 100 μm) by careful design of the active region.

- Superlattices are usually used for the injector regions but they can also be used for the active regions in both bound-to-continuum and continuum-to-continuum designs.
- Using a superlattice as the lower laser level has the advantage that interminiband scattering times are extremely fast and population inversion is easier to achieve.
- Waveguide structures are more difficult to fabricate than in conventional edge-emitting devices but metal waveguides based on surface plasmons are common.

It should be borne in mind that whilst a very wide diversity of designs and operating wavelengths have been demonstrated, the quantum cascade laser is still a relatively recent invention and the technology is still very much under development. Many of the ideas described in this chapter may well find their way into commercial devices, but this is by no means clear at present.

12.9 References

[1] Gmachl C, Capasso F, Sivco D L and Cho A Y 2001 *Reports on Progress in Physics* **64** 1533–1601
[2] Harrison P and Kelsall R W 1998 *Solid State Electronics* **42** 1449–1451
[3] Harrison P 1999 *Appl. Phys. Lett.* **75** 2800–2830
[4] Hyldgaard P and Wilkins J W 1996 *Physical Review* **B53** 6889–6892
[5] Tripathi P and Ridley B K 2003 *J. Phys.: Condens. Matter* **15** 1057–1069
[6] Kinsler P, Harrison P and Kelsall R W 1998 *Phys. Rev.* **B58** 4771–4778
[7] Gorfinkel V B, Luryi S and Gelmont B 1996 *IEEE J. Quant. Electr.* **32** 1995–2003
[8] Yang Q K and Li A Z 2000 *J. Phys.: Condens. Matter* **12** 1907–1914
[9] Suchalkin S, Bruno J, Tober R, Westerfield D, Kisin M and Belenky G 2003 *Appl. Phys. Lett.* **83** 1500–1502
[10] Helm M 1995 *Semicond. Sci. Technol.* **10** 557–575
[11] Iotti R C and Rossi F 2001 *Phys. Rev. Lett.* **87** art. no 146603
[12] Ohtsuka T, Schrottke L, Key R, Kostial H and Grahn H T 2003 *J. Appl. Phys.* **94** 2192–2198
[13] Faist J, Capasso F, Sivco D L, Hutchinson A L, Chu S-N G and Cho A Y 1998 *Appl. Phys. Lett.* **72** 680–682
[14] Faist J, Capasso F, Sivco D L, Hutchinson A L and Cho A Y 1994 *Science* **264** 553–556
[15] Faist J, Capasso F, Sivco D L, Hutchinson A L, Sirtori C, Chu S-N G and Cho A Y 1994 *Appl. Phys. Lett.* **65** 2901–2903
[16] Faist J, Capasso F, Sivco D L, Sirtori C, Hutchinson A L and Cho A Y 1994 *Electronics Letters* **30** 865–868
[17] Faist J, Capasso F, Sirtori C, Sivco D L, Baillargeon J N, Hutchinson A L, Chu S-N G and Cho A Y 1996 *Appl. Phys. Lett* **68** 3680–3682
[18] Hofstetter D, Beck M, Allen T and Faist J 2001 *Appl. Phys. Lett.* **78** 396–398

[19] Yu J S, Slivken S, Evans A, David J and Razeghi M 2003 *Appl. Phys. Lett.* **82** 3397–3399
[20] Gmachl C, Capasso F, Faist J, Hutchinson A L, Tredicucci A, Sivco D L, Baillargeon J N, Chu S-N G and Cho A Y 1998 *Appl. Phys. Lett.* **72** 1430–1432
[21] Carder D A, Wilson L R, Green R P, Cockburn J W, Hopkinson M, Steer M J, Airey R and Hill G 2003 *Appl. Phys. Lett.* **82** 3409–3411
[22] Hofstetter D, Faist J, Beck M, Müller A and Oesterle U 2000 *Physica* **E7** 25–28
[23] Razhegi M and Slivken S 2003 *Physica Status Solidi (a)* **195** 144–150
[24] Kronig R de L and Penney W G 1931 *Proc. Roy. Soc. (London)* **A130** 499
[25] Eldib A M, Hassan H F and Mohamed M A 1987 *J. Phys. C: Solid State Phys.* **20** 3011–3019
[26] Steslicka M, Kucharczyk R, Akjouj A, Djafari-Rouhani B, Dobrzynski L and Davison S G 2002 *Surface Science Reports* **47** 93–196
[27] Kittel C 1976 *Introduction to Solid State Physics* 5th Edition (New York: John Wiley & Sons) p 192
[28] Manenti M, Compagnone F, Di Carlo A, Lugli P, Scamarcio G and Rizzi F 2003 *Appl. Phys. Lett.* **82** 4029–4031
[29] West L C and Eglash S J 1985 *Appl. Phys. Lett.* **46** 1156–1158
[30] Gelmont B, Gorfinkel V and Luryi S 1996 *Appl. Phys. Lett.* **68** 2171–2173
[31] Scamarcio G, Capasso F, Sirtori C, Faist J, Hutchinson A L, Sivco D L and Cho A Y 1997 *Science* **276** 773–776
[32] Tredicucci A, Capasso F, Gmachl C, Sivco D L, Hutchinson A L, Cho A Y, Faist J and Scamarcio G 1998 *Appl. Phys. Lett.* **72** 2388–2390
[33] Faist J, Beck M, Aellen T and Gini E 2001 *Appl. Phys. Lett.* **78** 147–149
[34] Rochat M, Hofstetter D, Beck M and Faist J 2001 *Appl. Phys. Lett.* **79** 4271–4273
[35] Kelsall R W, Kinsler P and Harrison P 2000 *Physica E* **7** 48–51
[36] Xin Z J and Rutt H N 1997 *Semicond. Sci. Technol.* **12** 1129–1134
[37] Lyubomirsky I and Hu Q 1998 *Appl. Phys. Lett.* **73** 300–302
[38] Köhler R, Iotti R C, Tredicucci A and Rossi F 2001 *Appl. Phys. Lett.* **79** 3920–3922
[39] Callebaut H, Kumar S, Williams B S, Hu Q and Reno J L 2003 *Appl. Phys. Lett.* **83** 207–209
[40] Xin Z J and Rutt H N 1997 *Semicond. Sci. Technol.* **12** 1129–1134
[41] Harrison P 1997 *Semicond. Sci. Technol.* **12** 1487–1490
[42] Scalari G, Ajili L, Faist J, Beere H, Linfield E H, Ritchie D A and Davis G 2003 *Appl. Phys. Lett.* **82** 3165–3167
[43] Köhler R, Tredicucci A, Beltram F, Beere H, Linfield E H, Davis G, Ritchie D A, Dhillon S H and Sirtori C 2003 *Appl. Phys. Lett.* **82** 1518–1520
[44] Rochat M, Ajili L, Willenberg H, Faist J, Beere H, Davies G, Linfield E and Ritchie D 2002 *Appl. Phys. Lett.* **81** 1381–1383
[45] Sirtori C, Kruck P, Barbieri S, Page H, Nagle J, Beck M, Faist J and Oesterle U 1999 *Appl. Phys. Lett.* **75** 3911–3913
[46] Burke J J, Stegeman G I and Tamir T 1986 *Phys. Rev. B.* **33** 5186–5201
[47] Yang F, Sambles J R and Bradberry G W 1991 *Phys. Rev. B.* **44** 5855–5872
[48] Williams B S, Kumar S, Callebaut H, Hu Q and Reno J L 2003 *Appl. Phys. Lett.* **83** 2124–2126

[49] Indjin D, Ikonić Z, Harrison P and Kelsall R W 2003 *J. Appl. Phys.* **94** 3249–3252

Problems

1. Assuming $m_e/m^* \approx 15$ for GaAs, show, by calculating the oscillator strength, that the strongest transitions between the subband of a quantum well are those occurring between states adjacent in energy, i.e. $1 \rightarrow 2$, $2 \rightarrow 3$, $3 \rightarrow 4$, etc. to confirm the requirement that the parity flips in inter-sub band transitions.
2. Using the idealised infinite square well quantum well estimate the thickness of two adjacent wells that from the active region of a laser emitting at $\lambda = 50$ μm. Assume a phonon energy of 36.0 meV similar to bulk GaAs. As an exercise, you might try calculating the equivalent quantities in a real quantum well formed from GaAs/AlAs just to illustrate how complicated is the procedure for designing cascade lasers.
3. Estimate the change in wave vector required for an electron to relax from the top of a miniband to the bottom in a GaAs superlattice with a period of 10 nm. Compare this with the kinetic energy of an electron in bulk GaAs at a similar wavevector.

Appendix I

Population inversion in semiconductors

The number of stimulated emissions per unit time per unit volume is

$$N_{st} = W_{cv} \cdot N_c \cdot f_c \cdot N_v(1 - f_v) \cdot \rho(\nu) \tag{1}$$

where W_{cv} is the microscopic probability for the transition from the conduction band to the valence band, $N_c f_c$ is the number of occupied states in the conduction band, N_c being the effective density of states at the bottom of the conduction band and f_c being the occupancy function, $N_v(1 - f_v)$ is the number of states in the valence band occupied by holes, by similar reasoning to the above, and $\rho(\nu)$ is the density of photons at frequency ν.

By similar reasoning, the number of absorptions per unit time per unit volume is

$$N_{ab} = W_{vc} \cdot N_v \cdot f_v \cdot N_c(1 - f_c) \cdot \rho(\nu) \tag{2}$$

where $N_v f_v$ is the number of occupied states in the valence band and $N_c(1 - f_c)$ is the number of empty states in the conduction band.

From Einstein, $W_{cv} = W_{vc}$ so for $N_{st} \geq N_{ab}$ we must have

$$f_c(1 - f_v) \geq f_v(1 - f_c) \tag{3}$$

hence

$$\exp\left(\frac{E_c - E_{fc}}{kT}\right) \leq \exp\left(\frac{E_v - E_{fv}}{kT}\right) \tag{4}$$

where $E_{fc,v}$ are the quasi-Fermi levels for the conduction and valence bands respectively. Therefore

$$E_{fc} - E_{fv} \geq E_c - E_v \tag{5}$$

which is known as the Bernard–Duraffourg condition.

Appendix II

The three-layer dielectric slab waveguide

Light incident on a transparent surface will be partially transmitted and partially reflected. The properties of these three beams of light, the incident, reflected, and transmitted, are summarised by Snell's well known laws of reflection and refraction:

1. the incident, reflected and refracted beams all lie in the same plane;
2. the angle of incidence θ_i is equal to the angle of reflection θ_r;
3. the ratio of the sines of the angle of incidence (measured with respect to the normal) and the angle of the transmitted beam are given by the ratios of the refractive indices in the two materials.

$$\frac{\sin \theta_i}{\sin \theta_t} = \frac{n_t}{n_i}. \tag{1}$$

For light travelling into a more optically dense medium $n_t > n_i$, and $\theta_i > \theta_t$. For light travelling from a dense medium to a less dense medium $\theta_t > \theta_i$ and θ_t can have a maximum value of $\pi/2$, i.e. the refracted beam travels along the interface. In this case,

$$\sin \theta_i = \frac{n_t}{n_i} \tag{2}$$

and θ_i is known as the critical angle. For angles of incidence larger than this the incident ray will be totally reflected and no energy will be transmitted across the boundary. If we now consider three dielectrics, as in figure 1, a light ray travelling from the core towards the outer layer will be reflected if the angle of

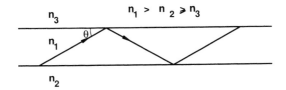

Figure 1. Total internal reflection in a waveguide structure.

incidence (relative to the plane of the interface) is below the critical angle. On traversing the guide again to the other interface a similar reflection will occur. It would seem that this ray can be transmitted down the middle dielectric. However, no account has been taken of phase changes.

Imagine two rays adjacent to each other. Points of similar phase can be identified such that a line joining them describes the wavefront. In traversing the guide at an angle the phase of the ray is advanced. There is also a phase change on reflection, the magnitude of which depends on the polarisation, the angle of incidence, and the refractive indices of the two layers. It is possible therefore to map out the point at which the phase has advanced by 2π and this then constitutes the new wavefront. If the new wavefront is facing the same direction as the old wavefront, but is simply displaced down the guide, then the wave will propagate. However, if the wavefront has changed direction the wave is effectively scattered. Only certain angles of incidence will maintain the wavefront down the guide, so even though a ray may be reflected it will be diverted over a number of reflections unless it propagates at one of the specific angles. Figure 2 illustrates this requirement.

As the first ray travels from C to D and undergoes two reflections in the process, the second ray must travel from A to B in order that the wavefront – the dotted lines joining regions of equal phase on the rays – is maintained down the guide. Hence

$$n_1(CD-AB)\frac{2\pi}{\lambda}+\Phi_1+\Phi_2=N2\pi \qquad (3)$$

where $CD - AB$ is the difference in geometrical path lengths, Φ_1 and Φ_2 are the phase changes on reflection at C and D, λ is the wavelength of light in free space, and N is an integer. It can be shown from elementary trigonometry that

$$CD-AB=\frac{d}{\sin\theta_1}-(\cos^2\theta_1-\sin^2\theta_1)\frac{d}{\sin\theta_1} \qquad (4)$$

so that equation (5) becomes

$$n\left[\frac{d}{\sin\theta_1}-(\cos^2\theta_1-\sin^2\theta_1)\frac{d}{\sin\theta_1}\right]\left(2\frac{\pi}{\lambda}\right)+\Phi_1+\Phi_2=N2\pi \qquad (5)$$

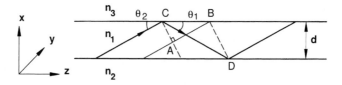

Figure 2. Maintenance of the wavefront during propagation down the guide.

This is known as the eigenvalue equation and leads to restrictions on the value of θ_1. Such values of θ_1 as are allowed are called modes of the waveguide. The phase angles Φ_1 and Φ_2 depend on the polarisation of the beam, and there are two distinct cases, known respectively as the transverse electric (TE), when the electric vector lies in the plane of the interface, and transverse magnetic (TM), when the electric vector is perpendicular to the interface. Analytical formulae exist for these phase shifts [1] but will not be given here.

The mathematics of the waveguide is normally taken further so the final form of the eigenvalue equation becomes

$$\tan(Kd) = \frac{K(\gamma + \delta)}{K^2 - \gamma\delta} \qquad (6)$$

for the TE mode, and

$$\tan(Kd) = \frac{n_1^2 K(n_3^2 \gamma + n_2^2 \delta)}{(n_2 n_3)^2 K^2 - n_1^4 \gamma\delta} \qquad (7)$$

for the TM modes, where we define for convenience

$$K = (n_1^2 k^2 - \beta^2)^{1/2}$$
$$\gamma = (\beta^2 - n_2^2 k^2)^{1/2} \qquad (8)$$
$$\delta = (\beta^2 - n_3^2 k^2)^{1/2}$$

where $k = 2\pi/\lambda$ is the free-space wavevector and $\beta = n_1 k \cdot \cos\theta_1$. For given values of n_1, n_2, n_3 and d, modes will only propagate at particular angles such that K, γ, and δ satisfy these equalities.

The eigenvalue equation can be appreciated by considering a light source inside the core emitting at all angles. Those incident on the interface at a large angle (i.e. near normal) will pass through to the outside of the guide. Those incident at angles below the critical angle will be reflected but will not propagate unless the angle of incidence coincides with the mode. For a thin guide, or a guide with a small refractive index difference, the mode angle will be relatively large and the ray will propagate some way into the cladding. If the width of the guide is increased then clearly the phase changed on traversing the guide would increase without any change in the angle of incidence so the mode is changed. The angle of incidence decreases. Similarly, if the width is kept constant but the refractive index difference is increased the phase change on reflection is altered so the mode changes. Again, the angle of incidence decreases, which means that the penetration into the cladding is reduced. Eventually in either process the conditions will be achieved where another ray at a larger angle matches the conditions for a mode to propagate, and as either the width of the guide or the refractive index difference increases the angle of incidence of this mode will decrease. Eventually a third mode will propagate, and so on. Looked at the other way around, in a wide guide several modes will propagate at distinctive angles. As the width of the guide decreases the angles will increase until total internal reflection no longer occurs and the mode is "cut off". In an asymmetric guide, $n_2 \neq n_3$ and all modes cut off if the guide is thin enough. In a symmetric guide $n_2 = n_3$ and at least one mode, the zero'th mode, propagates at all values of the core thickness.

Reference

[1] Marcuse D 1974 *Theory of Dielectric Optical Waveguides* (Academic Press)

Appendix III

Valence bands and effective masses

The following is intended to give some insight to the nature of the valence bands and hole effective masses. The physics of the quantum well presented in chapter 7 shows that the energy of quantum confined carriers depends on the effective mass, but it is also necessary to know the energy band dispersion in the plane of the well in order to calculate the density of states and the optical gain. The most convenient assumption is that a well-defined effective mass exists in the plane of the well, but often this is not true. In fact the behaviour is often very complicated, such that the concept of an effective mass is not always justified. Quantum confinement of three bands – the light and heavy holes, and the split-off band – causes the states to overlap and interfere with each other, leading to complicated non-parabolic behaviour. This is a complicated, and a very mathematical problem in semiconductor physics, but it is not necessary to understand the mathematics fully to appreciate the essential features. What follows is an outline, starting first with a bulk semiconductor.

The starting point is the recognition that the potential inside the semiconductor varies rapidly with the periodicity of the lattice. It is common in many texts in semiconductor physics to ignore this because it is an extra complication not usually needed. Many semiconductor devices can be understood by treating the semiconductor as a quasi-classical system with the electrons as quasi-classical particles. The conduction and valence bands are therefore portrayed as uniformly smooth throughout the volume of the crystal, the only variations being caused either by a change in material properties, for example at a heterojunction, or by the application of a voltage which causes a redistribution of charge within the semiconductor. The resulting electric fields cause a significant spatial variation in the band edges, but usually this variation occurs over large distance compared with the lattice spacing.

However, a solid is composed of atoms which are bonded together through a redistribution of charge between neighbours. In a fully ionic material there will be alternating positive and negative charges, so it is easy to appreciate that the potential "seen" by an electron (treating it as a classical charged particle) propagating through the material must depend on its position relative to these charges. In a fully covalent semiconductor, such as silicon, the valence electrons reside on average between the atoms so the potential is lowest there and a maximum at the centre of the atom. From a classical point of view, the electron would not occupy the same space as the atoms. III–V materials are not, of course, fully covalent and there is a displacement of charge towards the group V atom, but the principles are the same. In a quantum view, the atom is mostly empty space and the electrons propagate freely through this space but with an amplitude modulated at the periodicity of the lattice. Therefore, when the square of the amplitude is calculated there is a greater probability of finding the electron mid-way between the atoms than anywhere else. These quantum states are called Bloch functions. They are essentially plane wave representations with a periodic amplitude, i.e.

$$\psi_{nk}(r) = u_{nk}(r) \exp(jkr) \tag{1}$$

where n is the band index, r is the electron position and $u_{nk}(r)$ is a coefficient with the periodicity of the lattice, referred to variously as a unit cell wavefunction or an atomic wavefunction [1]. The unit cell is the basic building block of the lattice and therefore composed of the atoms of the material. The band index will be described in detail as we go.

This wavefunction can be substituted into the Schrödinger equation.

$$H\psi(r) = \left[\frac{-\hbar^2}{2m_0}\nabla^2 + V(r)\right]\psi(r) = E\psi(r) \tag{2}$$

and it will be shown below that this naturally leads to a term in $\mathbf{k}\cdot\mathbf{p}$ for the Hamiltonian. Operation once by the differential operator leads to

$$\nabla\psi_{nk}(r) = \nabla[u_{nk}(r) \exp(jkr)] = [\nabla u_{nk}(r) + jk u_{nk}(r)] \exp(jkr) \tag{3}$$

and operation a second time

$$\nabla^2[u_{nk}(r)\exp(j\mathbf{k}\cdot r)] = [\nabla^2 u_{nk}(r) + 2jk\nabla u_{nk}(r) - k^2 u_{nk}(r)] \exp(j\mathbf{k}\cdot r). \tag{4}$$

Concentrating on the middle term, multiplication by $-\hbar^2/2m_0$ required by the Schrödinger equation yields

$$\frac{\hbar}{m_0}\mathbf{k}(-j\hbar\nabla u_{nk}(r)) = \frac{\hbar}{m_0}\mathbf{k}\cdot\mathbf{p}u_{nk}(r) \tag{5}$$

where p, the momentum operator, is

$$p = -j\hbar\nabla. \tag{6}$$

Therefore

$$\left[H_0 + \frac{\hbar}{m_0}\mathbf{k}\cdot\mathbf{p}\right]u_{nk}(r) = \left[E_n(\mathbf{k}) - \frac{\hbar^2 k^2}{2m_0}\right]u_{nk}(r) \tag{7}$$

where

$$H_0 = \left[\frac{-\hbar^2}{2m_0}\nabla^2 + V(r)\right] = H(k=0) \tag{8}$$

and $E_n(k)$ is the energy eigenvalue for the state k in band n. Clearly when $k=0$ the solution to this eigen equation (8) are the functions $u_{n0}(r)$. The more general solutions $u_{nk}(r)$ can be constructed in terms of this basis set

$$u_{nk}(r) = \sum_m a_m u_{m0}(r) \tag{9}$$

where the band index has been changed to m to indicate the general nature of the solution and to show clearly that many different band states can contribute to the solutions for one particular band. This is called "band mixing".

So far this treatment is entirely general and logical, but the correspondence with the physical reality is not always easy to see. In particular, the band index n needs some further explanation. Recalling the discussion in chapter 1 on the formation of bands in semiconductors, a band arises from the atomic energy levels that broaden due to the proximity of neighbouring atoms. This is most easily imagined in an elemental semiconductor where all the atoms are the same and there is a one-to-one correspondence between the atomic levels and the bands, but in compound semiconductors such as the III–V materials, the constituent atoms obviously have different energy levels and the simple intuitive picture becomes a little more complicated. We can still appeal to physical intuition, though. The electrons involved in the bonding are one of the p electrons and the two s electrons of the group III element, and the p electrons of the group V element. There is an element of hybridisation, i.e. the mixing of these s and p bonds to give three similar covalent bonds, but there is also a difference in electronegativity that causes the electrons to transfer partially from the group III to the group V. In other words, the bonds are not fully covalent, and in so far as the electrons spend the majority of their time associated with the group V element their nature is now predominantly p-like.

If we further accept that electrons in the valence band are electrons residing in these covalent bonds it follows that the conduction band must correspond to the next atomic level up, which will of course be the higher s electronic orbital. Intuitively, then, we arrive at the picture described without explanation in chapter 1 that the valence band in III–V semiconducting materials is comprised of three p-like orbitals and the conduction band is comprised of s-like orbitals. There are therefore three valence bands and one conduction band. What is meant by s and p in this context means essentially the symmetry properties of the electrons; s electrons are spherically symmetric but p-electrons are orthogonal and anti-symmetric. The symmetry properties of these states can be very useful in the detailed analysis of valence and conduction band structure but are not the principal focus of the present discussion, which is centred on the band indexes. These correspond in most cases to the four bands described, but there exist more complicated treatments that also consider other bands, such as the *s*-orbitals lying below the valence electrons of the group V element. These are not immediately part of the band structure usually of interest but they can interact with, and therefore influence the properties of, the *p*-electrons above. For the most part, though, the four bands described above are the most commonly treated.

AIII.1 The Kane model: one-band, two-band, and four-band calculations

The simplest calculation that can be performed is a one-band calculation. Putting $k=0$ leads to the eigen equation (8) that has exact solutions $u_{10}(r)$. We don't really need to know what these functions are because they appear on both sides of the equation. All we really need to know is that the electrons are expressed as plane waves and that

$$E(k) = \frac{\hbar^2 k^2}{2m_0}. \tag{10}$$

The band is perfectly parabolic with a free-electron mass. This is obviously not a realistic model.

A two-band model is not realistic either, but serves to demonstrate some useful features. The wavefunctions are constructed according to equation (9), and equation (7) becomes

$$\left[H_0 + \frac{\hbar}{m_0} \mathbf{k} \cdot \mathbf{p} \right] \sum_m a_m u_{mk}(r) = \left[E_n(k) - \frac{\hbar^2 k^2}{2m_0} \right] \sum_m a_m u_{mk}(r). \tag{11}$$

At this stage use is made of the orthonormality property of wavefunctions, i.e.

$$\int u_{m0}^* u_{n0} dr = \delta_{mn} \tag{12}$$

where $\delta_{mn} = 1$ if $m=n$ and 0 if $m \neq n$, as required by orthonormality. Multiplication of equation (11) through by u_{n0}^* and integration over all space (in practise this is only the unit cell of the lattice) leads to

$$\sum_m \left(\left[E_n(0) - \frac{\hbar^2 k^2}{2m_0} \right] \delta_{nm} + \frac{\hbar}{m_0} \mathbf{k} \cdot \mathbf{p}_{mn} \right) a_m = E_n(k) a_n \tag{13}$$

where the term $E_n(k=0)$ appears as a result of H_0, and

$$p_{mn} = \int u_{n0}^*(r) \cdot p \cdot u_{m0}(r) d^3r \qquad (14)$$

is known as the momentum matrix element. The principles behind orthonormality can be applied to the momentum matrix element, but now, because the momentum operator is a differential any state with a definite parity, i.e. an odd or even function, will have that parity flipped, which is equivalent to shifting the phase through 90°. Therefore cross terms become non-zero and $p_{nn} = 0$. As we are considering only two bands, the summation resolves into two independent equations which, expressed in matrix form, become

$$\begin{bmatrix} E_1(0) + \dfrac{\hbar^2 k^2}{2m_0} & \dfrac{\hbar}{m_0} k \cdot p_{12} \\ \dfrac{\hbar}{m_0} k \cdot p_{21} & E_2(0) + \dfrac{\hbar^2 k^2}{2m_0} \end{bmatrix} \begin{bmatrix} a_1 \\ a_2 \end{bmatrix} = E(k) \begin{bmatrix} a_1 \\ a_2 \end{bmatrix}. \qquad (15)$$

The solution for the eigen energies requires that the determinant be set equal to zero, i.e.

$$\begin{bmatrix} E_1(0) + \dfrac{\hbar^2 k^2}{2m_0} - E(k) & \dfrac{\hbar}{m_0} k \cdot p_{12} \\ \dfrac{\hbar}{m_0} k \cdot p_{21} & E_2(0) + \dfrac{\hbar^2 k^2}{2m_0} - E(k) \end{bmatrix} = 0. \qquad (16)$$

Some simplifying assumptions can be made at this point. First, we assume that p_{12} is isotropic, which is true if the functions of the basis set are themselves isotropic. Thus we are restricted to s-like states, i.e. the conduction band. Second, suppose that the parabolic term is small, and that we are therefore restricted to small values of k.

$$\begin{bmatrix} E_1(0) - E(k) & \dfrac{\hbar}{m_0} k \cdot p_{12} \\ \dfrac{\hbar}{m_0} k \cdot p_{12} & E_2(0) - E(k) \end{bmatrix} \cong 0. \qquad (17)$$

Taking the cross-products and subtracting,

$$[E_1(0) - E(k)][E_2(0) - E(k)] - \left[\dfrac{\hbar}{m_0} k \cdot p_{12}\right]^2 = 0. \qquad (18)$$

The points $E_{1,2}(0)$ correspond to the band extrema, so we can define the band gap to be the difference between the two and the energy at the top of the valence band can be set arbitrarily to zero, i.e.

$$\begin{aligned} E_1(0) &= 0 \\ E_2(0) - E_1(0) &= E_g \end{aligned} \qquad (19)$$

yields an equation quadratic in $E(k)$

$$-E(k)[E_g - E(k)] - E_p \frac{\hbar^2 k^2}{2m_0} = 0 \qquad (20)$$

where

$$E_p = \frac{2}{m_0} |p_{12}|^2. \qquad (21)$$

At first sight equation (20) is not so easy to solve, but we have already made the assumption in its derivation that k is small. This immediately puts the energies of interest either at the top of the valence band, in which case $E(k) \sim 0$, or at the bottom of the conduction band, in which case $E(k) \sim E_g$. These substitutions therefore provide the solution. Dealing with the valence band first, the key substitution is the term in brackets, otherwise the solution is trivially zero, and leads to

$$E(k) = -\frac{E_p}{E_g} \frac{\hbar^2 k^2}{2m_0} \qquad (22)$$

or

$$E(k) = \frac{\hbar^2 k^2}{2\left(-m_0 \frac{E_g}{E_p}\right)} = \frac{\hbar^2 k^2}{2m^*} \qquad (23)$$

and m^* is now the effective mass. Similarly for the conduction band, the key substitution is the term outside the brackets, and

$$E(k) = E_g + \frac{E_p}{E_g} \frac{\hbar^2 k^2}{2m_0} \qquad (24)$$

where the effective mass has the same magnitude but opposite sign.

The effective mass is thus seen to arise from the addition of an extra band in the summation over states. It is this extra band that alters the parabolicity and leads ultimately to non-parabolicity, as this solution is strictly valid over a limited range of k close to the band extrema. Thus the effective mass is an approximation.

As for the eigen states themselves, the extent of band mixing becomes clear through a re-examination of equation (15).

$$\begin{bmatrix} 0 & \frac{\hbar}{m_0} k \cdot p_{12} \\ \frac{\hbar}{m_0} k \cdot p_{21} & E_g \end{bmatrix} \begin{bmatrix} a_1 \\ a_2 \end{bmatrix} = E(k) \begin{bmatrix} a_1 \\ a_2 \end{bmatrix}. \qquad (25)$$

The Kane model

Using the eigen energy solution for $E(k)$ (equation (17)) we then have

$$\begin{bmatrix} -E_g - \dfrac{E_P}{E_g}\dfrac{\hbar^2 k^2}{2m_0} & \dfrac{\hbar}{m_0} \mathbf{k} \cdot \mathbf{p}_{12} \\ \dfrac{\hbar}{m_0} \mathbf{k} \cdot \mathbf{p}_{21} & -\dfrac{E_P}{E_g}\dfrac{\hbar^2 k^2}{2m_0} \end{bmatrix} \begin{bmatrix} a_1 \\ a_2 \end{bmatrix} = 0. \qquad (26)$$

From equation (9)

$$u_{nk}(r) = \sum_{m=1}^{2} a_m u_{m0}(r). \qquad (27)$$

At $k=0$, equation (26) reduces to

$$\begin{bmatrix} -E_g & 0 \\ 0 & 0 \end{bmatrix} \begin{bmatrix} a_1 \\ a_2 \end{bmatrix} = 0 \qquad (28)$$

leaving

$$-E_g a_1 = 0 \qquad (29)$$

so that $a_1 = 0$. The states must be made up entirely of states from band 2 so $a_2 = 1$ and

$$u_{2k}(r) = a_2 u_{20}(r) = u_{20}(r). \qquad (30)$$

For $k \neq 0$, though, a_1 and a_2 are generally non-zero and formally equation (27) becomes

$$u_{2k}(r) = a_1 u_{10}(r) + a_2 u_{20}(r). \qquad (31)$$

This shows explicitly that away from $k=0$ the Bloch functions for the electrons in the conduction band contain an admixture of valence band states at $k=0$, and in fact the degree of mixing increases with increasing k.

The two-band model, whilst unrealistic, does at least demonstrate the essential technique. It is possible to construct a four-band model, but the correct choice of the basis states is important in order to ensure the correct bands are reproduced. Use is thus made of the symmetry properties of the s-like and p-like states. That is, s-like states are spherically symmetric and p-like states are anti-symmetric and orthogonal to each other. In addition, there are two states per basis state corresponding to spin up and spin down, and the 2×2 matrix for the Hamiltonian given in equation (15) therefore becomes an 8×8 matrix. There is also a contribution from spin-orbit interaction.

Spin-orbit interaction is most easily understood with reference again to the hydrogen atom in which the solution to the Schrödinger equation clearly represents a physical orbital. The potential arises in this case from a central charge which gives rise to a central force, and the electron orbits the charge. A moving charge constitutes, of course, an electric current so the orbiting electron effectively behaves like a current coil with a magnetic field being generated as a result. There is an angular momentum associated with the orbital motion. The electron also spins, and similarly, the spin motion gives rise to an angular momentum and a magnetic moment, which interact with the angular momentum and the magnetic moment due to the orbital motion. In a solid the interaction between spin and orbital angular momentum is less intuitive, but as the bands are constructed essentially from atomic orbitals it exists and must be included. A moving electron has angular momentum and gives rise to a magnetic moment, and its own motion within that moment is modelled by spin-orbit interaction.

Classically, angular momentum l is defined as

$$l = r \times p \qquad (32)$$

where r is the radius of the motion and p is the linear momentum, ($=mv$). If the spin angular momentum is defined as s, then the coupling is defined as

$$l \cdot s = (r \times p) \cdot s. \qquad (33)$$

The derivation of the spin-orbit Hamiltonian is very rarely given in undergraduate texts and instead is just quoted as

$$H_{SO} = \frac{\hbar}{4m_0^2 c^2} \nabla V \times p \cdot \sigma \qquad (34)$$

where V is the potential seen by the electron (essentially the conduction band profile which is periodic on the scale of the lattice) and σ is the Pauli spin matrix. You can begin to appreciate that this problem now becomes quite complicated as not only must H_{SO} be added to the left hand side of equation (7) but we must find eight basis states corresponding to the conduction band and three valence bands, each with two spins, and we must define the spatial symmetry of these states. The problem proceeds essentially as before, but will not be described in detail.

We find that one band decouples from the others and appears as a solution on its own with an energy $E=0$ at $k=0$ and an electron mass equal to the free electron mass. This is clearly a valence band, and in fact turns out to be the heavy hole band. This wrong result is one of the unfortunate features of this model and one of the reasons that a different approach is adopted, which will be described in due course. There are three other bands to describe;

1. a band at energy E_g, the conduction band;
2. a band at energy $E=0$, the light hole band;
3. a band at energy $E=-\Delta$, the split-off band.

In essence, then, each of the three p-orbitals forms its own band, two of them being degenerate, i.e. having the same energy, at $k=0$ and the third having a lower energy due to the spin-orbit interaction. The conduction band arises from the s-like states of the higher lying atomic orbitals.

The effective masses are given in the same way, but as there are more bands the terms are clearly more complicated. In the conduction band,

$$\frac{m_e}{m_0} \cong \frac{E_g(E_g + \Delta)}{E_p\left(E_g + \frac{2}{3}\Delta\right)} \quad (35)$$

where E_p is equivalent to equation (21) but with a slightly modified matrix element that reflects the additional complexity of the problem. The light hole effective mass is

$$\frac{m_{lh}}{m_0} \cong -\frac{3}{2}\frac{E_g}{E_p} \quad (36)$$

and for the split-off hole band,

$$\frac{m_{soh}}{m_0} \cong -3\frac{E_g + \Delta}{E_p}. \quad (37)$$

We shall not proceed to develop the eigen functions that lead to these effective masses. Suffice it to say they are quite complicated and contain admixtures of the different basis states. Again, effective mass arises from this type of admixture, which ultimately leads to a departure from parabolicity at a large enough energy away from the band extrema. To summarise, then, **k . p** theory is not concerned with the band structure throughout the Brillouin zone, i.e. for all electron wavevectors, but instead concentrates on particular points in the band structure. The most common points are the band extrema at Γ, i.e. the bottom of the conduction band and the top of the valence band corresponding to the direct transitions. Thus all the important properties of the electrons at the band edges are evaluated without reference to the band structure at all electron wavevectors. Most importantly the effective mass is defined in the theory.

The important features of the model thus revealed are:

- The effective mass of the conduction band electrons arises from the inclusion of other bands in the formulation of the problem.
- The effective mass of the light hole states scales with the band gap.
- The effective mass of the split-off valence band also scales with the band gap, in this case $E_g + \Delta$.

Luttinger–Kohn method

The above treatment, whilst giving the incorrect solution for the heavy hole effective mass, nonetheless is an exact solution within the approximations employed. An alternative approach is to adopt a perturbative method in which some bands are treated exactly and others are introduced as a perturbation. That is to say, an additional term is introduced into the Hamiltonian and the new wavefunctions developed recursively. Such a method is the Luttinger–Kohn method that uses a Hamiltonian expressed in terms of the so-called Luttinger parameters γ_1, γ_2, and γ_3. As such the method is less intuitive but mathematically more convenient. The three valence bands are treated exactly, and the states of the conduction band are introduced as the perturbation. The effective masses are also expressed in terms of the Luttinger parameters [2]. Hence, in the [100] direction of a zinc blende structure, which corresponds to most III–V materials of interest, the heavy hole effective mass corresponds to

$$m_{hh}^{*[100]} = \frac{m_0}{(\gamma_1 - 2\gamma_2)} \tag{38}$$

and the light hole effective

$$m_{lh}^{*[100]} = \frac{m_0}{(\gamma_1 + 2\gamma_2)}. \tag{39}$$

In other crystal directions, [110] and [111], respectively, the effective masses become

$$m_{hh}^{*[110]} = \frac{2m_0}{(2\gamma_1 - \gamma_2 - 3\gamma_3)} \tag{40}$$

and

$$m_{hh}^{*[111]} = \frac{m_0}{(\gamma_1 - 2\gamma_3)} \tag{41}$$

for the heavy holes and

$$m_{lh}^{*[110]} = \frac{2m_0}{(2\gamma_1 + \gamma_2 + 3\gamma_3)} \tag{42}$$

and

$$m_{lh}^{*[111]} = \frac{m_0}{(\gamma_1 + 2\gamma_3)} \tag{43}$$

for the light holes. These crystallographic directions are illustrated in figure 1. The crystal planes corresponding to the sides of the cubic cell, [100], [010], [001], are identical in an isotropic crystal.

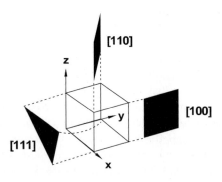

Figure 1. The crystallographic planes of a cubic (zinc blende) lattice.

The conduction band effective mass is identical to equation (35) if only the top three valence bands are considered. If deeper lying valence bands are included their effect is subsumed into a parameter, F, such that [2]

$$\frac{m_0}{m_e} = (1+2F) + \frac{E_p\left(E_g + \frac{2}{3}\Delta\right)}{E_g(E_g+\Delta)}. \quad (44)$$

Along with the Kane matrix element E_p (equation (21)), F is difficult to determine accurately as remote band effects can be calculated but not directly measured. Vurgaftman [2] has reviewed the band parameters of the III–V compounds of interest and has tabulated experimentally measured parameters, including the Luttinger parameters, the spin-orbit energy Δ, the band gaps, and effective masses, and determined E_p and F from the known experimental data on the band structure. Band calculations using Luttinger Hamiltonians are described by Heinamaki [1] who has also tabulated the Luttinger parameters for InP, GaAs, and InAs. The different valence band states correspond to well-defined spin states of; a total spin $J=3/2$ for the heavy hole and $J=\frac{1}{2}$ for the light hole, and spin $\frac{1}{2}$ for the SO band. The effective mass for the SO band is

$$\frac{m_0}{m_{SO}} = \gamma_1 + \frac{E_p\Delta}{3E_g(E_g+\Delta)}. \quad (45)$$

Quantum well valence band states

Confinement of electrons and holes in a quantum well leads to a modification of the energy levels, as discussed in chapter 7. Most importantly, the energy of the quantised state depends on the effective mass, which, from the preceding, can be seen to be a consequence of the detailed band structure. The quantum confined states are therefore related directly to the band states of the bulk material. There are various ways of deriving the confined states, some more complicated than others. The more complicated treatments are usually required when the simple treatments fail.

The first effect of confinement is to lift the degeneracy of the light and heavy hole states. The Luttinger representation can be used to show that an additional effect also occurs, and that is that the states normally associated with the heavy hole ($J=3/2$) can have their effective mass in the plane of the quantum well reduced, and that states normally associated with the light hole ($J=1/2$) can have their in-plane effective mass increased [3]. In fact the two normally cross over at some particular energy, but an exact treatment would show that an anti-crossing is the outcome. At the point of closest approach the states diverge away from each other, but in such a way that it seems as if the heavy hole state continues where the light hole state should and vice versa (figure 2).

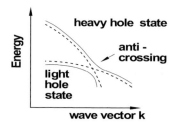

Figure 2. Anti-crossing of light and heavy hole states. The dotted line indicates the individual state and the solid line indicates the final state after anti-crossing.

Writing the wavefunction as

$$\psi(r) = \sum_m \chi_m(r) u_{m,k_0}(r) \quad (46)$$

$\chi_m(r)$ is the envelope function and $u_{m,k_0}(r)$ is the rapidly varying Bloch function appropriate to band m at the momentum point k_0, we recognise that in a lattice matched semiconductor the crystal structure of the well material is no different from that of the bulk material. This means that the Bloch functions of the well material are identical to the substrate material and can therefore be extracted from the wavefunctions leaving only the envelope functions. This is equivalent to ignoring the rapidly varying periodic potential within the semiconductor and treating both the conduction and valence bands as uniform in space unless the material properties change. In order to see how this develops, let the wavefunction within the barrier be

$$\psi(r) = \sum_m \chi_m^B(r) u_{m,k_0}^B(r) \quad (47)$$

and within the well

$$\psi(r) = \sum_m \chi_m^W(r) u_{m,k_0}^W(r) \quad (48)$$

where

$$u_{m,k_0}^B(r) = u_{m,k_0}^W(r) \quad (49)$$

and

$$\psi(r) = \sum_m \chi_m^{W,B}(r) u_{m,k_0}(r). \quad (50)$$

It is possible to separate out the directional components of the envelope.

$$\chi_m^{W,B}(r_{plane}, z) = \frac{1}{\sqrt{S}} \exp(j \cdot k_{plane} r_{plane}) \chi_m^{W,B}(z) \quad (51)$$

where S is the area of the quantum well and the pre-factor containing this term normalises the wavefunction. k_{plane} is a two-dimensional wavevector in the plane of the well, and r_{plane} is the position in the plane of the well. Hence the electron moving in the plane of the well is regarded as a plane-wave with an amplitude that varies in the growth direction z, i.e. is proportional to $\chi_m^{W,B}(z)$. These $\chi_m^{W,B}(z)$ are the only functions needed to describe the quantised state of the electron. The envelope function therefore describes the "outline" of the summation of the Bloch functions and shows none of the rapid variations.

The wavefunction of equation (51) is substituted into the Schrödinger equation taking into account only the changing potential at the heterojunction rather than the periodic potential of the lattice to give a series of states at well defined energies with a parabolic dispersion in the plane of the well. This is the so-called Ben Daniel–Duke model, as described by Bastard [4] and illustrated in figure 3. There are circumstances when this simple approach, which is intuitive and can easily be understood even though the practical implementation may be a little more complicated than is implied above, breaks down. One of the principal requirements for the envelope function approximation is that states from different bands are not too close to each other in energy, otherwise it becomes necessary to mix them together. Just as confined states can be constructed from the direct band gap states at Γ, confined states can also be constructed for the indirect band gap states at X, where the electron has a heavier effective mass. As the bulk energies for the X-point states lie well above the Γ-point states, the first quantised state from the indirect band gap must also lie much higher in energy in the quantum well, but if the well is narrow, leading to a large confinement energy, the Γ-like states will have similar energies to the X-like states. The envelope function method calculates these states separately so they remain separate in this formalism, but in reality where the states lie close in energy, such that they may be occupied by the same particle, it is necessary to construct a quantum state that takes into account all

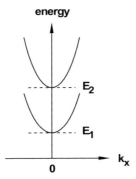

Figure 3. In-plane dispersion of quantised conduction band states according to the Ben Daniel–Duke model.

the possibilities of position and energy, so some mixing and anti-crossing occurs. Thus in narrow wells even the first quantised state of the conduction band may contain some X-like character, but in wide wells the first quantised state is always Γ-like and only the higher lying conduction band states may be mixed in nature. The heavy hole states are well served by the envelope function method at $k = 0$ in any heterojunction system [4].

The mixing of the states can be better appreciated by realising that equation (50) implies that the wavefunction has to be summed over several bands. This is similar to the Kane model already discussed, but there are important differences. First, the Bloch functions are of no interest, and second the summation is not just over the bands m but over the materials comprising the well and the barrier. The summation over m is the same for both barrier and well materials, so it is assumed that the quantum confined state is built from the bulk wavefunctions of both the barrier and well materials, as described above. However, the summation over the bands is limited to a small number of bands, and as with the Kane model, this predicts a parabolic band in the plane of the well which will only strictly apply to a small range of k close to the band edge. It has to be emphasised, therefore, that the envelope function method is an approximation. It gives an in-plane effective mass from which the total energy of the electron can be calculated, but the effective mass is not an accurate representation of the energy of either the electron or the hole across the Brillouin zone. If band mixing occurs there will be a strong nonparabolicity in the band so a more accurate representation of the wavevector across the whole Brillouin zone, i.e. from $k = 0$ to $k = 2\pi/a$, will be needed. Other, more complete methods must be used therefore to determine the wavefunctions and the energy dispersion. One such method is a tight-binding calculation which uses atomic wavefunctions of the constituent atoms to calculate the band structure. Such a treatment is beyond the scope of this book but the outcomes of such calculations can be described. Chang and Shulman [5] used such a method to calculate the dispersion properties of the valence band quantised states of a GaAs/Ga$_{0.75}$Al$_{0.25}$As superlattice in the [110] and [100] directions, both of which lie in the plane of the quantum wells. These are lattice-matched structures, but it is evident from figure 4, which shows the dispersion of the light and heavy holes for bulk GaAs for comparison, that the dispersion properties are far from simple. It is not even clear, for example, that an effective mass can be assigned to the quantised states.

In fact this is typical of many quantum well systems and shows up the dangers inherent in making simplified assumptions. The quantisation is apparent from the energy shift of the states relative to the bulk band edges, but the effective mass of the first quantised state of the heavy hole (HH1) is larger than in the bulk, though still negative, whereas the effective mass for the second quantised state of the heavy hole (HH2) is actually positive and becomes essentially an indirect band-gap. Moreover, the first light hole state (LH1) appears below HH2 and LH2 appears below HH4. The light hole states have

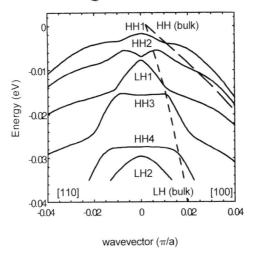

Figure 4. Valence band states of an AlGaAs – GaAs superlattice. (After Chang and Shulman [5].) The bulk bands are shown as dashed lines.

the most well-defined effective masses, but again they differ between the two states, and HH3 and HH4 do not behave at all well in respect of the effective mass. Finally, there is some anti-crossing evident in the HH1-HH2 and LH1-HH3 states. The full range of possible effects are illustrated in figure 5, where the degeneracy of the bulk light and heavy hole bands is lifted by quantisation, inversion of the effective masses in the plane of the well occurs, and because of the crossing of the band states, an anti-crossing behaviour occurs.

In summary, the valence band states in both bulk and quantum well materials are complicated. To a great extent it is not necessary to know the detail behind the bulk effective masses, but it is not sufficient simply to assume that what applies in the bulk will apply to a quantum well. The situation is much more complicated than the simple particle in a box problem described in chapter 7, and the presence of more than one valence band causes mixing and

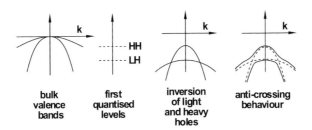

Figure 5. Possible quantisation effects in the valence band.

considerable deviations from parabolicity in the plane of the well. Of course, it is always possible to assume an effective mass. This will allow the quantised energies to be calculated, from which the wavelength of the laser transitions can be derived, but this is only one aspect of a laser. The dispersion properties of the states in the plane of the well affects the densities of states and the gain calculations, and any attempt to model a device in all but the most rudimentary manner requires a knowledge of the dispersion properties. Again, an in-plane effective mass can be assumed along with a parabolic band, but this will have limited validity. Moreover, the in-plane effective mass will be different from the effective mass in the growth direction, and the relationship between the two is not simple. If the full dispersion properties of the valence band are required, there is no alternative but to undertake a full quantum mechanical treatment.

References

[1] Heinamaki A and Tulkki J 1997 *J. Appl. Phys.* **81** 3268–3275
[2] Vurgaftman I, Meyer J R and Ram-Mohan L R 2001 *J. Appl. Phys.* **80** 5815–5875
[3] Weisbuch C and Vinter B 1991 *Quantum Semiconductor Structures* (Boston: Academic Press Inc.)
[4] Bastard G 1988 *Wave Mechanics Applied to Semiconductor Heterostructures* (France: Les Editions de Physique)
[5] Chang Y C and Shulman J N 1985 *Phys. Rev. B* **31** 2069

Appendix IV

Valence band engineering via strain

The band structure of semiconductors is modified considerably by strain. In a quantum well diode laser strain is introduced by the growth of lattice-mismatched layers, provided the layer thickness is below the critical thickness for strain relief. Most III–V materials of interest are cubic structures, and the growth along the side of the cubic lattice in the [001] direction (figure 1) is the most technologically important [1,2]. This simplifies matters enormously.

In one dimension, the ratio of stress to strain is a constant and is called Young's modulus. In three dimensions the equivalent elasticity modulus is a tensor but the symmetry of a cubic structure means that only three components are required; C_{11}, C_{12}, and C_{44}. These are experimentally determined parameters, tabulated values of which may be found in, for example, [3] for a limited number of III–V materials, or in Vurgaftman's review of the band parameters [4]. Slight differences may exist between different sources, but that is only to be expected with an experimental parameter.

The strain is defined as the relative change in the dimension of the material, and again in three dimensions this is a tensor. Symmetry reduces the important components to ε_{xx}, ε_{yy}, and ε_{zz} acting in the x, y, and z directions respectively. For growth in the [001] direction the strain will be biaxial (see

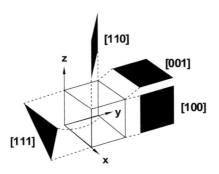

Figure 1. A cubic lattice with key planes indicated in black.

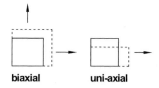

Figure 2. A plan view of the unit cell distortion caused by biaxial and uni-axial strains in the direction of the arrows.

figure 2) as opposed to uni-axial because the lattice parameter in two-dimensions is extended to match the substrate. The in-plane strain is therefore

$$\varepsilon_{xx} = \varepsilon_{yy} = \frac{a_s - a_l}{a_l} = \varepsilon_\parallel \tag{1}$$

where a_s is the lattice parameter of the substrate material and a_l is the lattice parameter of the growth layer. Strain in any direction will also cause a corresponding perpendicular strain, as illustrated in figure 2 for the uni-axial strain. For a biaxial strain in the plane of the layer there will be a corresponding deformation in the z-direction. The magnitude of this perpendicular strain is determined by Poisson's ratio, which is simply the ratio of the transverse change to the longitudinal change, and is

$$\varepsilon_{zz} = \varepsilon_\perp = \frac{2\sigma}{1-\sigma}\varepsilon_\parallel. \tag{2}$$

In terms of the elastic moduli,

$$\sigma = \frac{C_{12}}{C_{11} + C_{12}} \tag{3}$$

and hence

$$\varepsilon_{zz} = -\frac{2C_{12}}{C_{11}}\varepsilon_\parallel. \tag{4}$$

Equations (1) to (4) are all that is needed to characterise the strains in a cubic structure where the growth is on the (001) plane. However, the deposited layer will not remain strained for an indefinite thickness because the change in volume requires work to be done. This work can be expressed as an energy per unit area which increases linearly with the thickness of the film [2,5]. When the energy stored within the film exceeds the energy for the formation of dislocations, the atomic planes slip and the strain is relieved. The thickness at which this occurs is called the critical thickness, and several models exist within the literature, all of which give a similar expression but which differ in the detail. O'Reilly discusses the models and Köpf reviews the critical thickness in InGaAs on GaAs in relation to one of the most widely used models, that of Matthews and Blakeslee [6]. There is fairly widespread agreement between the MB model and experiment for InGaAs.

InGaAs is one of the most important strained layer materials. Historically the AlGaAs system was one of the first quantum well systems but it is lattice matched throughout the entire composition range and strain is not an issue. InGaAs is important because it allows wavelengths in the near infrared, especially for the 1.3 μm telecommunications window. Moreover is aluminium-free, which is an important consideration for the lifetime and stability of high power devices. For any In composition, InGaAs will be compressively strained with respect to GaAs. At low In composition the mismatch is low but so also is the carrier confinement. Critical thicknesses range from about 80 nm at ~ 8% In down to ~ 10 nm at about 35% In [7] and it is important to remain below this threshold. The maximum permitted quantum well width therefore varies with composition. Devices grown close to the critical thickness show a rapid degradation in performance, with the threshold current rising steeply as a function of the operating time. However, devices grown below the critical thickness are stable and exhibit very small increases in threshold current.

The critical thickness is not really such an important issue in a single quantum well as the devices often have an optimum thickness which is usually below the critical thickness. The useful range is typically about 5–10 nm. Thinner wells can be grown, of course, but then the effects of interfacial fluctuations assume ever greater importance. Thicker wells can be grown provided the composition allows it, but the energy separation between confined states decreases and the density of states increases. However, in multi-quantum well systems each well is strained and the strain energy increases with the number of wells. If the total thickness of all the wells exceeds the critical thickness relaxation will occur. The exception to this occurs in strain compensated materials where alternately compressive and tensile strained layers are grown. The strain energy cancels out to an extent depending on the thickness of each of the layers and if the system is such that each layer exactly compensates for its neighbours, infinitely thick pairs of layers could be grown.

Strain also has an effect on the band structure. Recalling that energy bands arise in solids from the interaction between neighbouring electronic wavefunctions, it is easy to imagine that these interactions are modified in the presence of strain. The effect of the strain depends on whether it is hydrostatic, i.e. related to a change in volume, or axial, i.e. a shear strain. These are, respectively [2]

$$\varepsilon_{vol} = \varepsilon_{xx} + \varepsilon_{yy} + \varepsilon_{zz} = \frac{\Delta V}{V} \tag{5}$$

and

$$\varepsilon_{ax} = \varepsilon_\perp - \varepsilon_\parallel. \tag{6}$$

The bandgap changes by an amount

$$\Delta E_g = a\varepsilon_{vol} \quad (7)$$

where

$$a = a_c + a_v \quad (8)$$

comprises a contribution from the conduction band and the valence band and is known as the band gap deformation potential. Therefore

$$\Delta E_g = \Delta E_c + \Delta E_v = a_c\varepsilon_{vol} + a_v\varepsilon_{vol}. \quad (9)$$

The review by Vurgaftman *et al.* [4] contains a discussion of the experimental determination of these parameters in III–V materials, and Köpf *et al.* [1] also contains an extensive discussion relevant to GaAs and InP. There are also two shear deformation potentials, b and d, but for growth along the [001] direction only b is important. The shear energy is

$$S = -b\varepsilon_{ax} \quad (10)$$

which has the effect of splitting the degeneracy of the light and heavy hole states. The combined effect of the hydrostatic (H) and shear (S) components on the band edges in the z-direction is illustrated in figure 3, neglecting the effects of spin-orbit interaction on the light and heavy hole states. Spin-orbit effects shift the energy of the light hole state slightly upwards, but not greatly [8,9]. Spin-orbit coupling causes the energy of the split-off band to be shifted down by the same amount. Note that under tensile strain the uppermost valence band is a light hole state, which can change the polarisation properties of the quantum well laser [9]. However, the situation is more complicated than a simple lifting of the degeneracy, because these effects apply only to the band edges. If the dispersion properties of the hole states are required throughout the Brillouin zone it is necessary to calculate the states from first principles using a strain Hamiltonian constructed from the hydrostatic and shear energies H and S.

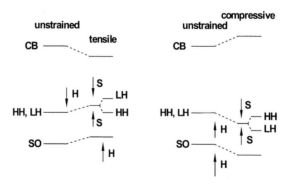

Figure 3. Schematic illustration of the effects of strain on the band edges.

References [2,8,9] explain the method. However, for small values of the wavevector in the plane of the well the light and heavy hole states are inverted. That is, the light hole state in the growth direction behaves like a heavy hole state in the plane and *vice versa*. In terms of the Luttinger parameters the light and heavy hole states are swapped.

These are bulk effects but they can be used to advantage within a quantum well laser. To recap, non-parabolicity in the valence band structure in the plane of the well arises from the presence of nearby heavy and light hole states that anti-cross. In unstrained material the confinement energies of these two valence bands are different purely because of the difference in effective mass, but when the degeneracy is lifted by strain the energy separation of the confined levels must take into account this additional separation. Thus in compressively strained material, the heavy hole is shifted up relative to the light hole band, so energy separation is enhanced and the mixing and anti-crossing occurs at a larger value of the wave-vector. The effective mass approximation in the plane of the well (figure 3) applies over a larger range of hole energies through the reduced interaction with neighbouring states. Moreover, because of the bulk anisotropy in the effective mass in the plane of the well is equivalent to the light hole mass, which has the effect of reducing the density of states. In figure 4 the effective mass and the band gap have been left unchanged by the presence of strain for the sake of simplicity and in order to emphasise the effect of degeneracy lifting on the density of states, but it is important to emphasise that strain can, and often does, change the effective masses so that additional difference between the strained and unstrained cases will be apparent.

By way of example, Kano *et al.* [10] mapped out the valence bands as a function of wave vector in both the [100] and [110] crystallographic planes for strained multiple quantum wells of InGaAsP/InP. Both of these directions lie in the plane of the quantum well. The well width was 50 Angstrom sandwiched in barrier layers corresponding to a band gap of 1.1 μm. The result is shown in figure 5 for three of the bands. The asymmetry is a result of the different directions being plotted on the same axes. In the lattice matched quantum well (a) the degeneracy of the heavy hole and the light hole ground states is lifted

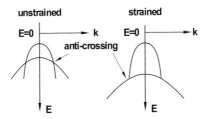

Figure 4. Schematic illustration of the dependence of energy on the in-plane wavevector for a compressively strained quantum well. Note the light hole behaviour in the plane of the well.

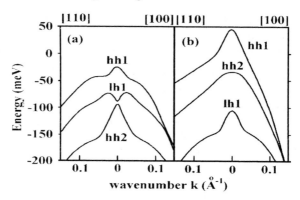

Figure 5. Modifications of the valence band due to strain in InGaAsP/InP quantum wells. (After Kano *et al.* [10].) The lattice matched valence bands are shown in (a) and the strained valence bands in (b).

by confinement and the heavy hole forms the band gap. Similar features as observed by Chang and Shulman [appendix III, figure 4] are evident. First, the heavy hole state has quite a small zone centre effective mass, but there is a strong non-parabolicity caused by anti-crosssing of the first light and heavy hole states. Second, the light hole has a positive effective mass. Third, the second heavy hole state shows a strong non-parabolicity but a relatively small effective mass at the zone centre.

In the strained quantum well (b) however, and the strain is only 2%, the heavy hole ground state is shifted up by 79 meV and the light hole state is shifted down as described above, but to the extent that the second heavy hole band lies above the first light hole band. It is quite obvious from visual inspection that at $k=0$ the heavy hole mass is much reduced and similar to the light hole mass as a consequence of the strain induced splitting and the reduced mixing.

For tensile strained material the light hole in the growth direction becomes the uppermost valence band in the bulk. Within the quantum well this state will tend to have a lower confinement energy than the equivalent heavy hole state and may well be pushed below it in energy so that the heavy hole state forms the band gap. Therefore effect of degeneracy lifting in the bulk is to reduce the energy separation in the quantum well but if the strain is large enough and the energy separation great enough, the light hole state will remain the uppermost state and will form the band gap. This will change the polarisation selection rules and allow operation of an edge-emitting laser in the TM mode. [9].

References

[1] Köpf C, Kosina H and Selberherr S 1997 *Solid State Electronics* **41** 1139–1152
[2] O'Reilly E P 1989 *Semicond. Sci. & Technol.* **4** 121,137

[3] Frederikse H P R 2002–2003 *The Handbook of Chemistry and Physics* 83rd Edition (CRC Press)
[4] Vurgaftman I, Meyer J R and Ram-Mohan L R 2001 *J. Appl. Phys.* **80** 5815–5875
[5] Grovenor C R M 1989 *Microelectronic Materials* (Bristol: Adam Hilger) p 140
[6] Matthews J and Blakeslee A 1974 *J. Crystal Growth* **27** 118
[7] Coleman J J 1992 *Thin Solid Films* **216** 68–71
[8] Coldren L A and Corzine S W 1995 *Diode Lasers and Photonic Integrated Circuits* (New York: John Wiley) p 532
[9] Chang C S and Chuang S L 1995 *IEEE J. Sel. Topics. in Quant. Electr.* **1** 218–229
[10] Kano F, Yamanaka T, Yamamoto N, Mawatari H, Tohmori Y and Yoshikuni Y 1994 *IEEE J. Quant. Elect.* **30** 533–537

Appendix V

The scalar potential and electromagnetic interaction Hamiltonian

The scalar potential A is given by the time derivative of the electric vector (chapter 4), and must therefore be oscillatory in nature. If

$$A = A_0 \, \varepsilon \cdot \exp\left[j\left(\frac{\omega}{c} n \cdot z - \omega t \right) \right] \qquad (1)$$

where ε is a unit vector describing the direction of polarisation and n is a unit vector describing the direction of propagation z, then for small distances r the exponential can be expanded

$$\exp\left(j \frac{\omega}{c} n \cdot z \right) \approx 1 + j \frac{\omega}{c} n \cdot z \approx 1. \qquad (2)$$

This is essentially the dipole approximation. It is possible to show from this that the electromagnetic interaction is of the form of an electric dipole $q \cdot r$. The justification for this is as follows.

Recalling from chapter 4 that the electromagnetic interaction Hamiltonian is

$$H' = \left(-\frac{q}{m} \right) A \cdot p \qquad (3)$$

the transition from an initial state i to a final state f is described by the probability integral

$$-\frac{q}{m} \int \psi_f(z) A \cdot p \psi_i(z) dr. \qquad (4)$$

Concentrating on the oscillatory terms of the vector field and neglecting the coefficients for the moment, then

$$\int \psi_f(r) \left[\exp\left(j\frac{\omega}{c} n \cdot z\right) \varepsilon \cdot p \right] \psi_i(z) dz \approx \varepsilon \cdot \int \psi_f(z) p \psi_i(z) dz. \tag{5}$$

That is to say, under the dipole approximation only the momentum is important. Consider now the action of the kinetic and potential energy operators on the state $r \cdot \psi$, where r is the position operator. It is left as an exercise to show that

$$H_0 z \psi = \left(-\frac{\hbar^2}{2m} \frac{\partial^2}{\partial z^2} + V \right) z \cdot \psi = z H_0 \psi - \frac{j\hbar}{m} p \psi. \tag{6}$$

Therefore

$$\int \psi_f(z) p \psi_i(z) dz = j \frac{m}{\hbar} \int \psi_f(z)(z H_0 - H_0 z) \psi_i(z) dz. \tag{7}$$

Expansion of the term on the right hand side requires a clear understanding of the properties of Hermition operators, but it is fairly straightforward to show [1] that this integral becomes

$$\int \psi_f(z) p \psi_i(z) dz = -j \frac{m}{\hbar} (E_f - E_i) \int \psi_f(z) z \psi_i(z) dz = -jm\omega_{if} \int \psi_f(z) z \psi_i(z) dz. \tag{8}$$

In other words, the momentum operator has been replaced by the position operator and the interaction Hamiltonian becomes

$$H' = \left(-\frac{q}{m} \right) A \cdot p = -j\omega q A \cdot z. \tag{9}$$

Given that

$$E = -\frac{\partial A}{\partial t} \tag{10}$$

then

$$H' = -qE \cdot z. \tag{11}$$

Reference

[1] Yariv A 1987 *Quantum Electronics* 3rd Edition (New York: John Wiley & Sons) p 663

Solutions

Ch. 2

1. i. $n = 4.3 \times 10^{12}$ cm^{-3} $\quad\quad\quad\quad p = 9.1 \times 10^{-1}$ cm^{-3}
    ii. $n = 2.35 \times 10^{17}$ cm^{-3} $\quad\quad\quad p = 8.3 \times 10^{-8}$ cm^{-3}
    iii. $n = 5.5 \times 10^{-7}$ cm^{-3} $\quad\quad\quad p = 3.5 \times 10^{18}$ cm^{-3}
2. 10.36 nm; $n = 1.1 \times 10^{17}$ cm^{-3}
3. $p = 6.10 \times 10^{-25}$ kg m s^{-1} (phonon); $\quad p = 5.98 \times 10^{-28}$ kg m s^{-1} (photon)
4. $p = 1.21 \times 10^{-24}$ kg m s^{-1}

Ch. 3

1. $m = 4096 \quad\quad \lambda = 830.08$ nm
2. $\Delta\lambda = 0.2$ nm
3. 24.1 cm^{-1}; $\tau = 4.72$ ps; $\Delta\nu = 1.71$ MHz
4. 12.2 cm^{-1}; $\tau = 9.26$ ps; $\Delta\nu = 0.44$ MHz
5. 9.08×10^3 cm^{-1}; 2.725×10^3 cm^{-1}; 2.749×10^3 cm^{-1}
6. 9.6×10^3 A cm^{-2}
7. 7.04×10^4 A cm^{-2}
8. 1.28×10^3 A cm^{-2}; 1.8×10^3 A cm^{-2}

Ch. 4

1. (a) $n^2 = 1 - j2.79$; $\quad\quad n = 1.41 + j0.99$
    (b) $n^2 = 2.56 - j1.48$; $\quad n = 1.66 + j0.44$
2. $n^2 = 2.46 - j0.99$; $\quad\quad n = 1.60 + j0.31$
3. $\Delta n = -2.83 \times 10^{-3}$

Ch. 5

1. $E_g = 1.972$ eV; (a) 0.053 eV; (b) 0.085 eV; (c) 0.156 eV; (d) 0.248 eV
2. $n = 8.94 \times 10^{17}$ cm^{-3}

3. (a) 7.55×10^{14} cm^{-3}; (b) 2.61×10^{15} cm^{-3}; (c) 4.09×10^{16} cm^{-3};
   (d) 1.44×10^{18} cm^{-3}
   (Note: in (d) the barrier is negative and the Boltzmann approximation is probably inaccurate.)
4. (a) 2.89×10^{12} cm^{-3}; (b) 5.79×10^{12} cm^{-3}; (c) 1.44×10^{13} cm^{-3};
   (d) 2.89×10^{13} cm^{-3}.
5. (a) 109 A cm^{-2}; (b) 376 A cm^{-2}; (c) 5.9×10^3 A cm^{-2};
   (d) 2.1×10^5 A cm^{-2}.
6. Differentiating gives

$$d_a^2 = \frac{2b}{(a+b)} \cdot \frac{1}{k^2(n_1^2 - n_2^2)}$$

so that $d = 188$ nm or 159 nm for $n_1 = 3.44$ or $n_1 = 3.54$, $D = 1.19$ in both cases, and $\Gamma = 0.414$ or 0.417.

Ch. 6

1. $n = 1$; $|\psi|^2 = 0.4$ (max) at $z = 0$
   $n = 2$; $|\psi|^2 = 0.4$ (max) at $z = \pm 1.27$; $|\psi|^2 = 0$ (min) at $z = 0$
   $n = 3$; $|\psi|^2 = 0.4$ (max) at $z = \pm 1.67$; $|\psi|^2 = 0$ (min) at $z = \pm 0.84$
2. $n = 1$; 14.9 meV; 219 meV
   $n = 2$; 59.7 meV; 877.9 meV
   $n = 3$; 134.3 meV; 1.97 eV
3. $\Delta E_c = 0.219$; $k_0^2 = 3.94 \times 10^{17}$ m^{-2}; $(k_0 L)^2/4 = 2.46$; transforming the eigenvalue equation as in figure 6.11 gives the intercept at $(kL)^2/4 = 0.933$ so that $k^2 = 1.49 \times 10^{17}$ m^{-2} and $\kappa^2 = 2.45 \times 10^{17}$ m^{-2}; this results in $E_1 = -0.137$ eV, i.e. 81 meV above the band edge.
4. $\Delta E_v = 0.329$; $k_0^2 = 3.91 \times 10^{18}$ m^{-2}; $(k_0 L)^2/4 = 2.44$; the intercept lies at $(kL)^2/4 = 0.932$ so that $k^2 = 1.49 \times 10^{17}$ m^{-2} and $\kappa^2 = 3.766 \times 10^{18}$ m^{-2} and $E_1 = -0.316$ eV, i.e. 12.5 meV below the valence band edge. The total transition energy is 1.424 eV + 12.5 meV + 81.8 meV = 1.518 eV or 816 nm. The band gap of GaAs corresponds to 870 nm at RT.
5. $k_e = 3.83 \times 10^8$ m^{-1} and $k_h = 3.86 \times 10^8$ m^{-1}; the penetration depth is 2.6 nm for both carriers.

Ch. 7

1. Without material dispersion

$$\Delta \lambda = \frac{\lambda}{2 \cdot n \cdot m_\lambda}$$

444   Solutions

with material dispersion (λ in μm)

$$\Delta\lambda = \frac{\lambda}{2 \cdot m_\lambda \left(\dfrac{dn}{d\lambda} - n\right)}.$$

Hence, for a one-wavelength cavity the mode spacing is either 123 nm or 141 nm.

2. $\lambda = 0.85\ \mu$m: $x = 0:0.9$; $r = 0.0846$;   27 pairs;   3.49 μm
                             $x = 0:0.8$; $r = 0.0760$;   30 pairs;   3.84 μm
                             $x = 0.1:0.8$; $r = 0.0651$;   35 pairs;   4.52 μm
    $\lambda = 1.3\ \mu$m: $x = 0:0.9$; $r = 0.0748$;   31 pairs;   6.38 μm
                             $x = 0:0.8$; $r = 0.0669$;   34 pairs;   6.94 μm
                             $x = 0.1:0.8$; $r = 0.0587$;   39 pairs;   8.02 μm
    $\lambda = 1.5\ \mu$m: $x = 0:0.9$; $r = 0.0739$;   31 pairs;   7.40 μm
                             $x = 0:0.8$; $r = 0.0662$;   35 pairs;   8.28 μm
                             $x = 0.1:0.8$; $r = 0.0574$;   40 pairs;   9.55 μm
3. $\lambda = 0.85\ \mu$m: $r = 0.381$;   6 pairs;   1.15 μm
    $\lambda = 1.3\ \mu$m: $r = 0.359$;   6 pairs;   1.78 μm
    $\lambda = 1.5\ \mu$m: $r = 0.356$;   6 pairs;   2.06 μm

4. The mobility of the electrons within the stack is unknown, but it will be much smaller than the mobility in either of the materials because of the activated nature of conduction across the barriers. Suppose as a generous estimate, $\mu = 50$ cm^2 s^{-1} then the scattering time $\tau = 1.93 \times 10^{-15}$ s and $\gamma = 5.18 \times 10^{14}$ s^{-1}, and $\omega_p^2 = 2.34 \times 10^{28}$ s^{-2}. If $\sqrt{\varepsilon_\infty} \approx 3$ then $k \approx 1 \times 10^{-3}$ and 85 cm^{-1}.

Ch. 8

1. Calcuating the optical confinement from chapter 5, $D = 1.28$ and $\Gamma = 0.45$. From equation (#8,30) $S = 1 \times 10^{15}$ cm^{-3}. Strictly it is necessary to solve for the waveguide mode and calculate the effective penetration of the light into the cladding but it is simpler to assume that the emitting area is the same as the stripe cross-section. Therefore $R_{stim} = 2.12 \times 10^{26}$ cm^{-3} s^{-1} and $I = 7.6$ mA or 507 A cm^{-2}.
2. The stimulated recombination rate becomes

$$R_{stim} = \Gamma v_g a(N - N_t)S \equiv \frac{\Delta N}{\tau_{stim}}.$$

Coldren and Corzine (chapter 8, [3]) give the differential gain of bulk GaAs as $\approx 4 \times 10^{-16}$ cm^2 so the effective lifetime for the stimulated recombination is 2.9×10^{-10} s and $\Delta N = 1.29 \times 10^{17}$ cm^{-3}. From question (#5,2) $N_t = 8.94 \times 10^{17}$ cm^{-3} and $N = 1.02 \times 10^{18}$ cm^{-3}. This formulation of

Solutions   445

an effective lifetime above demonstrates the idea that the carrier density remains close to transparency during steady state operation of the laser.

3. $\tau = 4.2$ ns. This sort of lifetime is not typical of radiative lifetimes, and would probably represent the combined effect of a number of non-radiative processes. However, assuming that all the electrons generate photons then $S_{spon} \approx 10^{21}$ cm^{-3}, five orders of magnitude below the stimulated photon density. Assuming that the quality of material has been improved so that non-radiative processes are insignificant the τ will be of the order of 1 μs, giving $S_{spon} \approx 4 \times 10^{18}$ cm^{-3}

4. It is a question of judgment as to what transit time is acceptable, but for $\omega\tau \geq 0.2$ the high frequency performance begins to be affected. A response of 90% occurs at $\omega\tau \approx 0.34$ corresponding to a confining layer thickness of ≈ 116 nm.

5. This function varies dramatically with both time constant and hole density. Under the conditions described the 80% current limits lie at 49 nm and 430 nm respectively for $\tau = 1$ ns and 10 ns. At 100 nm thickness the currents are 232 and 383 A cm^{-2}.

Ch. 9

1. About 45–50 mW
2. 30.8 GHz; 27.2 GHz
3. 214 nm
4. $\kappa = 41.9$ cm^{-1}; $L_{eff} = 111$ μm; $R = 87\%$; $\Delta\lambda = 0.6$ nm, $\Delta\beta = 6 \times 10^{-6}$ and the real part of the phase = 0.23. Hence $r = 0.217$ and $R = 0.05$

Ch. 10

1. This problem proceeds as (2) in chapter 8. A large optical cavity means that $\Gamma \approx 1$ so $S = 9.45 \times 10^{16}$ cm^{-3} and $\tau_p = 3.40 \times 10^{-11}$ s, giving $R_{stim} = 2.78 \times 10^{27}$ cm^{-3} s^{-1}. Hence $J = 4.5 \times 10^4$ A cm^{-2}, or a current of 11.2 A. However, this is not the total current. Calculating $\tau_{stim} = 3.1 \times 10^{-12}$ s gives $\Delta N = 8.6 \times 10^{15}$ cm^{-3}, i.e. just above transparency, so $N = 9.02 \times 10^{17}$ cm^{-3}. Therefore $F_{n2} = 50$ meV and for $\Delta E_c = 219$ meV (chapter 6) $F_{n3} = 169$ meV, and $n_3 = 6.1 \times 10^{14}$ cm^{-3}. Therefore $J_{leak} = 97$ A cm^{-2}, which is trivial compared with the stimulated emission current.
2. Now $F_{n3} = 111$ meV, giving $n_3 = 6.82 \times 10^{15}$ cm^{-3} and $J_{leak} = 1.09 \times 10^3$ A cm^{-2}. This is still less than 3% of the stimulated emission current.
3. 34.5 W

4. $\tau = 1$ ns: $\Delta N = 3 \times 10^{18}$ cm^{-3}, giving a total carrier density of 4.73×10^{18} cm^{-3} and a leakage current of 5.7×10^{3} A cm^{-2} for $x = 0.4$, and 5.2×10^{4} A cm^{-2} for $x = 0.3$.
$\tau = 0.1$ ns: $\Delta N = 3 \times 10^{17}$ cm^{-3}, giving a total carrier density of 1.17×10^{18} cm^{-3} and a leakage current of 1.7×10^{2} A cm^{-2} for $x = 0.4$, and 1.6×10^{3} A cm^{-2} for $x = 0.3$.

5. The change in band gap follows that of GaAs to a first approximation and is estimated at 23.5 meV giving a wavelength of 820 nm. The ban offsets do not change appreciably, so the main effect on the leakage is through the increased injection over the barrier, i.e. the barrier electron density is 3.5×10^{17} cm^{-3} giving a leakage current of 5.5×10^{4} A cm^{-2}.

Ch. 11

1. 0.035 nm; 91 μm and 140 μm
2. 3.579; 2.57
3. Assuming $n \approx 2.6$, $R \approx 19.5$, $\alpha_m \approx 33$ cm^{-1}. The equivalent cavity length ≈ 700 μm.

Ch. 12

1.

n	m	f
1	2	14.406
1	3	1.7094
1	4	0.461
2	3	28.006
2	4	3.6016
3	4	40.825

2. The transition energy is 24.8 meV so the two states of the first quantum well must be 60.8 meV apart, taking into account the phonon energy. This leads to a quantum well width of 2.14 nm. The zero-point energy of the adjacent well must lie at the zero-point energy of this well plus the phonon energy, i.e. 56 meV, leading to a thickness of 1.29 nm.

3. The momentum change is $\pi/d = 3.142 \times 10^{-8}$ kg . m s^{-1}. For an electron in bulk GaAs this corresponds to a kinetic energy of ~ 55 meV. This is greater than the width of the miniband of the superlattice, but can you think of a good reason why this must be the case?

Index

Absorption, 46, 54, 60, 62, 65, 67, 112, 312–317, 366, 374, 389, 400, 401, 406
Acceptors, 15
Active region, 39, 40–43, 52, 61, 66, 187, 222, 381, 388
AlGaAs, 62, 108, 168, 171, 207–213, 214, 221, 259, 268, 300, 304, 325, 383, 386, 394
AlGaInP, 116, 216, 327
AlN, 356–374
Aluminium free lasers, 324, 331, 343
Aluminum oxide, 207, 335, 357
α-parameter, *see* Phase–amplitude coupling coefficient
Ambipolar transport (*see also* Diffusion), 236, 240, 248
Amplitude modulation (AM), 257, 264, 281
Antimonide-based lasers, 332–339
Arrays, 293, 295
Auger recombination, 120, 271, 224, 332, 333, 338

Band bending, 37–39, 99
Band gap, 8, 27
   Renormalisation, 163
Band filling, 61, 175, 363
Band mixing, 22, 161, 164, 419
Band offset, 98–99, 147, 175, 336, 364, 369, 387
Band tails, 18, 366
Ben Daniel–Duke model, 429
Bernard–Duraffourg condition, 34, 412
Bipolar transistor, 244
Bloch function, 140, 154, 392, 418, 428

Bowing parameter, 366, 369
Bragg mirrors, 189–191, 209, 211, 369, 374, 381, 387, 389
Broad area lasers, 115, 294, 297, 338, 375
Buried heterojunction, 271–272

Capture time, *see* Quantum well
Carrier confinement, *see* Electrical confinement
Carrier density, 11–18, 41, 45–49, 226, 236, 298, 300–303, 310, 317, 326, 330, 347
Carrier scattering, 161, 385
Carrier transport, 10, 236, 382, 385, 387
Carrier
   capture time, 172–177
   escape time, 168, 172–177
Cascade laser, *see* Quantum cascade
Catastrophic optical damage (COD), 178, 295, 311–322, 331, 375
Cavity lifetime, 223
Characteristic temperature, 106, 117, 120, 123, 171, 233, 331, 335–338
Charge neutrality, in active region, 102, 75, 241, 386,
Chirp, in high-speed lasers, 268, 277, 279, 290
Chirp width, 270
Conduction band, 8, 10, 17, 34–39, 43, 309, 319
Conservation of momentum, 73, 375
Contacts, electrical, 36, 169, 178, 189, 191, 204, 212, 217, 326, 328, 360, 375, 389
Coupled mode theory, 194, 282–285
Covalent bonding, 7, 352

447

Covalency, 352
Critical layer thickness, 83, 163, 168, 358, 433–434

Dark line defects (*see also* Catastrophic optical damage), 313, 331
Density of states (DOS), 151, 191, 195, 366, 417
   Effective, 12
   Joint, 227
   Reduced, 73
Dielectric mirrors, *see* Bragg mirror
Differential gain, *see* Gain
Diffusion
   Ambipolar, 239, 240, 242
   Coefficient, 25, 43, 171, 172
   Doping, 36
   Electron, 38, 39
   Length, 25
   Minority carrier, 24
Diode pumped solid state lasers (DPSSL), 339–342, 344
Dislocations, 83–84, 96, 118, 163, 169, 313, 332, 336, 357, 358, 367, 434
Distributed Bragg reflector (DBR) (*see also* Bragg mirror), 191, 211, 217, 281
Distributed feedback (DFB), 281, 307
Donors, 14
Doping, 14
Double heterostructure (DH) lasers, 79–128, 188, 268, 332, 334, 343, 373
   Electrical confinement in, 102–106
   Optical confinement in, 108

Effective index, 176, 201, 277, 286
Effective mass, 16, 20–23, 75, 140–147, 364, 374, 398, 401, 417–432
   Reduced, 158
Elastic constants, 353, 433
Electrical confinement, 41, 45, 167–177, 179, 216, 235, 332, 336, 373, 387
Envelope function, 145–147, 156, 392, 428
Epitaxy, 81–96, 381, 387
   Of nitrides, 357–362
Epitaxial lateral overgrowth (ELOG), 358

Equivalent circuit of laser diode, 253–258
Erbium doped fibre amplifier, 177, 263, 342
Etching, 208, 373, 389, 390

Fabry–Perot cavity, 30, 281
   Modes, 31, 47, 195, 277, 287, 370
   semiconductor lasers, 269
Facet
   Cleaved, 40, 373, 387, 389
   Reflection at, 40, 42, 272
Fermi energy, 12, 13, 26, 76
   Quasi-, 26, 34, 48, 75, 165, 177, 195, 309, 412
Fermi function, 11, 319, 372
Fermi's Golden Rule, 71, 154, 174, 177, 385, 404, 408
Filamentation, 127, 269, 297, 320–311, 343
Free-carrier absorption, 64, 65, 218, 221, 283, 389, 390, 406
Free electrons, 6, 10
Fusion bonding, 214

GaAs, 1, 4, 7, 36, 43, 62, 82, 90, 108, 116, 146, 158, 169, 170, 185, 188, 240, 246, 300, 314, 320, 325 , 351, 383, 386, 394, 405, 430
GaAsP, 93, 166, 331
GaInAs, 4, 82, 85, 92, 116, 146, 166, 172, 177, 188, 191, 233, 235, 248, 252, 288, 304, 312, 314, 320, 325 , 332, 343, 389, 399, 402, 434
Gain, 41, 65–66, 76, 100, 338, 373, 375, 387, 399
   Differential, 165, 166, 175, 230, 235, 238, 251, 273, 276, 303, 309, 400
   Modal, 111, 112, 163, 272, 368, 375
   In quantum wells, 157, 195
   threshold, 42, 118
   threshold modal gain, 112
Gain compression, 224–227, 249, 274
Gain guiding, 191, 228
Gain spectrum, 30, 40, 47, 51, 399
Gain switching, 275
GaInP, 108, 178, 342, 343
GaN, 186, 351–374

GaSb, 332, 338
Gateway states, 245
Grating, 281

Hamiltonian, 8, 70, 440
Heavy doping, 15, 366, 405
　And band gap shrinkage, 17, 37, 39, 350
Heavy holes, (*see* Holes)
Heisenberg uncertainty principle, 15
High power lasers, 293–344, 375
High-speed modulation, 185, 268
Holes, 11, 21, 147, 154–157, 309, 338, 424–425, 430

InAs, 4, 82, 125, 174, 375
Index guiding, 109, 271
InGaAs, *see* GaInAs
InGaAsP, 83, 87, 92, 112, 166, 170, 174, 213, 232, 261, 271, 281, 309, 314, 323, 325, 331, 437
InGaN, 349, 363–374
InGaP, 166, 217, 315
InN, 355–374
InP, 83, 87, 112
InSb, 4, 124
Interband transitions, *see* Gain
Intervalence band absorption (*see also* Auger recombination), 214
Intrinsic region
　reduction in dimensions of
Intrinsic voltage, 196–197
Inversion carrier density
IR lasers (*see also* Quantum cascade), 331–338, 383

Kramers–Kronig transformation, 65
Kronig–Penney, 390
k-selection, 40

Laser oscillator, 29–30
Lattice-matching, 84, 192, 349
Lattice parameter, 82, 117, 163
Linewidth, 161, 267, 277–281, 339, 386, 398–400
　enhancement factor, 51, 302–305, 309, 311, 343
　function, 161

Lead salt lasers, 333–334
Liquid phase epitaxy, 85–87, 332
Localised state, 15
Low frequency roll-off, 238, 253
Luttinger–Kohn (LK) parameters, 426, 437

Matrix element, 72, 75, 154–158, 174, 385, 398, 404, 421, 425, 427
Maxwell equations, 202
Metalorganic
　chemical vapour deposition (MOCVD), 85, 92, 95, 328, 334, 360
　molecular beam epitaxy (MOMBE)
　vapour phase epitaxy (MOVPE), *see* MOCVD *above*
Mirror loss, 50–52, 112, 228, 273, 406
Mode field diameter, 44, 208
Mode filtering, 210, 272, 305, 327
Mode spacing, 41
Modulation, 173, 273
　see also High-speed modulation;
　large-signal modulation
　small-signal amplitude
Molecular beam epitaxy (MBE), 88, 327, 337, 349, 360, 362
MOPA (master oscillator power amplifier), 326
Multiple quantum well (MQW) lasers, 162, 168, 233, 334, 371, 372
　carrier transport in, 168, 246

Non-absorbing mirrors, 189, 322, 325, 332

Optical confinement Γ, 45, 112, 114, 163, 167, 168, 179, 187, 223, 230, 278, 406
Optical confining layers (OCL), 169–176, 239, 240, 369, 373
　Carrier density in, 236
　quantum well barrier height, 170
　state filling in, 170, 176
　superlattice, 176
Optical fibre, 260–268
Optical loss, 45, 113, 187, 188, 194, 214, 273, 276, 373

Index　　449

Optical transitions, 54, 58, 66–76
Organometallic chemical, *see* metalorganic
Orthonormality, 136
Oscillator strength, 75, 366, 375, 397, 398, 400
Oxide aperture, 208–213, 215

Parabolic band approximation, 20, 74, 140, 152, 164
Pauli exclusion principle, 8
p-doping, 15, 217, 359, 360
Pendeo-epitaxy, 359
Phase-amplitude coupling coefficient, 278, 303, 309, 343
Phonon, 19, 33, 48
  Scattering, 383, 388, 403
  Cascade, 35
Photon density, 222,
  rate equations, 223, 226
Plasma frequency, 64, 405
p–n junction, 2, 18, 20, 35, 350
Polarisation selection rules, 156, 195, 398
Population inversion, 33, 34, 368, 383, 386, 401, 412
Potential well, 132, 138, 139, 147, 391
  electronic states in, 147–150
Proton implantation, 203

Q-switching, 329, 341
Quantum cascade laser, 381–409
Quantum confinement, 23
Quantum confined Stark effect, 363
Quantum dots, 356, 364, 374–376
Quantum operator
  Energy, 9, 26, 391, 441
  Momentum, 68, 154–156, 396
Quantum well (*see also* Potential well), 381
  barrier height, 170, 175
  capture and emission time, 171–177, 236, 238, 246, 248, 253
Quantum well (QW) semiconductor lasers, 132, 188, 195, 373
  density of states for, 151, 153, 164, 165
  optical confinement in, 167
  gain coefficients, 160, 163
  transition matrix elements, 154–158
Quarter-wavelength pairs, *see* Bragg Mirrors

Rate equations (*see also* Photon density), 298, 330
  DH lasers, 222
  for quantum well laser structures, 172
  small-signal, 228
Recombination, 19, 25
  At surfaces and interfaces, 106, 120, 317, 332
  Lifetime, 23, 27, 222, 236, 368
  Non-radiative, 43, 124, 223, 275, 313
  Radiative, 23, 330
  Stimulated, 223
Reflection high energy electron diffraction (RHEED), 88
Refractive index, 31, 43, 46, 54–66, 170, 217, 277, 332, 333, 369, 386
Relaxation oscillations, 49, 268, 249–250, 290, 301
Relaxation resonance frequency, 232, 251, 269

Sampled grating, 290
Scattering
  Electron–electron, 240, 384–386, 404, 408
  Electron–phonon, 383, 404, 408
Schawlow–Townes linewidth, 50
Schrödinger equation, 67, 137, 385, 390, 408, 418
Semiconductor Bloch equations, 197
Separate confinement heterostructure (SCH) lasers, 168, 235, 247, 338
  carrier transport, 171
  graded-index (GRIN), 169, 315
SiC, 357, 359
Side-mode suppression ratio, 272, 287
Single heterostructure laser, 81
Single mode laser, 281
Spectral hole burning, 326
Split-off band, 21, 75, 147, 271, 333
Spontaneous emission, 32–33, 43, 47, 223, 247
  Coupling factor, 223, 233

State filling, 178
Stimulated emission, 31, 222
  Lifetime, 42
Strain, 82, 336, 343
  In quantum well structures, 163, 176, 177, 233, 235, 309,
  And band structure, 163, 309, 343, 433
Stranski-Krastanow self-organised growth, 375
Stripe Laser, 269, 271, 298–305, 315
Sum rule for optical transitions, 400
Superlattice, 168, 176, 336, 373, 381, 389, 390–396, 401–403
Surface emitting lasers, *see* Vertical cavity

Terahertz lasers, 403–405
Thermal rollover, 322, 343
Threshold current, 40, 45, 79, 115, 122, 168, 170, 171, 178, 195, 217, 232, 233, 332, 335, 337, 368, 373, 375, 402
Threshold modal gain, 41, 287, 406
Transparency current, 41, 188, 194
Tunable laser diodes, 289

Unipolar lasers, 386, 387

Valence band, 7, 16–22, 309, 417–432
Vapour phase epitaxy, *see* MOCVD
Vector potential
  and interaction Hamiltonian, 70, 440–441
Vertical cavity surface emitting lasers (VCSEL), 185–219
Visible lasers, 216–218, 327
Vegard's law, 121

Wavefunction, 9, 21, 135, 374, 390
Waveguide, 44, 387, 413
  Metal-clad, 405–408
  Ridge, 272, 315, 325, 343, 390
  Tapered, 326
  In VCSEL, 200–202
Wavelength detuning, 311
Wavelength division multiplexing, 263, 268, 288

ZnS, 349
ZnSe, 349